CASE FILES OF THE ROCKY MOUNTAIN PARANORMAL RESEARCH SOCIETY

VOLUME 3

CASE FILES OF THE ROCKY MOUNTAIN PARANORMAL RESEARCH SOCIETY

VOLUME 3

Robert Lewis and Bryan Bonner

POLYMATH
— PRESS —

Aurora, CO

Case Files of the Rocky Mountain Paranormal Research Society Volume 3
by Robert Lewis and Bryan Bonner

This book contains the true case files of the Rocky Mountain Paranormal Research Society and is intended for educational purposes, to stimulate interest and conversation concerning the paranormal, history, and science. The views and opinions expressed herein are solely those of the authors and do not necessarily reflect those of any of Rocky Mountain Paranormal's clients. Every effort has been made to ensure accuracy of the information presented, but neither the Rocky Mountain Paranormal Research Society nor Polymath Press offer any guarantees and shall not be held responsible for the use or misuse of any information contained herein.

Cover art: Bryan Bonner
Interior photographs: Bryan Bonner & Robert Lewis
Interior artwork: Elderlemon Design
Design and layout: Robert Lewis

First edition
September, 2025

ISBN (trade paperback): 978-1-961827-12-7
ISBN (eBook): 978-1-961827-13-4

To all who seek the truth,
wherever the evidence may lead.

Other Polymath Press titles by Robert Lewis & Bryan Bonner

Case Files of the Rocky Mountain Paranormal Research Society Volume 1
September, 2023

Case Files of the Rocky Mountain Paranormal Research Society Volume 2
September, 2024

Other Polymath Press titles edited by Robert Lewis

In the Woods: A Fiction Foundry Anthology
November, 2023

Arithmophobia: An Anthology of Mathematical Horror
March, 2024

Visit www.polymathpress.com or wherever books are sold to get your copies of these and other fine books.

Praise for *Case Files of the Rocky Mountain Paranormal Research Society Volumes 1 & 2*

"Forget everything you've seen on T.V. or movies about paranormal investigating – this book is the real deal, and it's smarter and scarier (and sometimes funnier). The Rocky Mountain Paranormal Research Society's case files are detailed but also exciting, written not just with a scientist's quest for truth but also a novelist's attention to pacing and background. This is a book that I suspect many of us will return to over and over in the years to come." -**Lisa Morton**, author of *Ghosts: A Haunted History* and *Calling the Spirits: A History of Seances*

"The *Case Files of the Rocky Mountain Paranormal Research Society* shows us what real paranormal research looks like, a testament to the search for the paranormal and the extreme passion and dedication behind it. Overall, these case files make a telling record of hauntings, history, investigation, frustration, and tantalizing, teasing mysteries." -**Craig DiLouie**, author of *Episode Thirteen*

"No flashy gimmicks here! What we have are the first-hand raw and real experiences from Colorado's finest paranormal investigators. Hands down, the most comprehensive compilation of Colorado haunts!" -**Jimmy Lee Combs**, director of *Terror Tales*

"Gives [the subject] the respect it deserves. I am looking forward to reading future volumes as they come out!" -**Jennifer Griffin**, *Horror Tree*

Table of Contents

List of Photos and Illustrations

Introduction

"They're back." In addition to being a tagline for the classic horror film *Poltergeist II*, it is also an apt description of the current state of affairs. Or perhaps we should instead quote another classic sequel (in this case, *Ghostbusters II*) and use the first-person plural: "yes, we're back." Many of you, we assume, picked this book up because you've read and enjoyed our first two volumes in this series, but if this happens to be your first introduction to our work, we welcome you with equal eagerness. And never to fear: you needn't read these volumes in any particular order. Each stands alone. In fact, by and large, each *chapter* stands alone. Though we have detected and often comment upon certain patterns or commonalities across our various investigations, the reports as presented in these books are designed to be read independently of each other. Each of these investigations was its own unique adventure.

Adventure seems like the right word. Sometimes people ask what keeps us going in this often frustrating line of work, and though our usual tongue in cheek answer is something like "because we're too stupid to know when to quit" or "because we lost a bet," the real truth of the matter is that this work has introduced us to people and places we wouldn't otherwise get to see or know. Sure, sometimes it's frustrating and sometimes it's uncomfortable, but at the end of it all, we invariably walk away from every case with some kind of memory and some kind of story to tell. That's the reason for these books. The stories deserve to be told.

When we set out to start writing these books (initially they were going to be a single book, but that ship sailed the very minute we realized just how much information we've collected over the years), we were just coming out of the Covid 19 pandemic and so our operations had all but entirely shut down. Yes, we still gave our educational lectures during those years, thanks to Internet video conferencing technologies and video sharing platforms. Yes, we still read everything about the paranormal we could get our hands on, from believers and skeptics alike. But the real adventure we always crave is in the investigation itself, and that process had come to a screeching halt.

We're pleased to report that this is no longer the case. Our investigative apparatus is back in full swing. While the first two books consisted almost (but not quite) entirely of older cases from our more than a quarter of a century of investigations, this book presents a mix of some of our very oldest as well as several of our very newest cases. Perhaps the scales still tilt a little bit toward the older, but we're proud to be able to show off some of what we've been working on as recently as the one-year period between the publication of Volumes 2 and 3 of this series. Of course, we haven't given everything new away in this one, either. We have to save some things for Volume 4. And yes, there will be a Volume 4 (though whether or not we maintain our one-book-per-year publication schedule for Volume 4 is, as of this writing, an open question).

Though the bulk of this book, like the other volumes in the series, is dedicated to looking back at our past investigations and experiences, we wanted to take a moment here at the start to reflect for a moment on the state of the paranormal more generally in the present. Are people more or less prone to believe in the paranormal these days? Have the specific details changed? Perhaps people in the past believed more in ghosts than UFOs or vice versa and now those trends have reversed.

It's well known that belief in the paranormal can be affected by external factors. Over the years, we've observed (admittedly anecdotally) that paranormal beliefs tend to associate positively with periods of significant disaster, social unrest, or widespread fear and anxiety. It makes sense when one thinks about it. When national or international political events out of one's immediate control seem particularly threatening, people may spend more time thinking about things like the afterlife, the meaning of life, and all those big questions with which human beings have grappled for millennia. Such thoughts can lead directly to religious faith, belief in paranormal phenomena, and even other related ideas like conspiracy theories. It must be noted, of course, that regardless of the reasons, good or bad, people might come to these beliefs, such sociological musings say little to nothing about the truth or falsehood of the beliefs themselves, so each claim would still need to be analyzed independently.

But as we were preparing this manuscript, the levels of social and political unrest and uncertainty did not escape our notice. Ukraine and Russia remain at war, as do Israel and Hamas, and that's just to name two prominent international crises. Social unrest at home currently isn't at the level it was in 2020, but many people (of all political orientations) seem to feel like the world is nearing some kind of tipping point if trends continue. The point of this book isn't to make any political pronouncements on any of these issues. Our point is only to observe that there seems to be a high degree of political tension in the current environment. At the same time, our inboxes have started filling up faster than they have in recent years with new claims of paranormal activity and requests for investigations or assistance. That's good for our business, of course, but taken together these two observations made us think about our hypothesis that paranormal belief is associated with social unrest (or uncertainty or fear) and we wanted to see what we could figure out about the state of the paranormal world in recent years.

Fortunately, Rocky Mountain Paranormal is not alone in our interest in such questions. A research group at Chapman University conducts an annual Survey of American Fears, which assesses the level and nature of American anxieties about a wide variety of fears ranging from small scale personal phobias to incidents of international or even apocalyptic importance.[1] Much to our delight, we learned that the survey sometimes includes questions specifically measuring individuals' religious faith and belief in a variety of paranormal claims.

The first thing we noticed is that our initial observation was correct: Americans are indeed more frightened now than at any time in the recent past. Indeed, for the first time since the group began conducting their annual survey a decade ago, all of the top ten fears (plus several more that didn't even make the top ten list) were experienced by more than 50% of the American population.[2] We're not sure whether this is because the world has actually become a more frightening place or whether it's because people have, perhaps due to social media and increased connectivity, become more sensitive to fears. Support for the former interpretation can be found in the Bulletin of Atomic Scientists' recent decision to move their famous "Doomsday Clock" to "89 seconds to midnight," indicating their belief that humanity is now closer than it has ever been to global apocalyptic disaster, citing concerns over nuclear war (and particularly the conflicts in Ukraine and the Middle East), climate change, technological disruptions, and more as areas of marked concern.[3] On the other hand, it is also well documented at this point that increased exposure to social media is associated with heightened sensitivity to fear under at least some circumstances,[4] though this relationship is not fully understood. For our purposes, it hardly matters. What's important for us to realize is simply that Americans are more afraid now than in recent memory. If our hypothesis is correct, interest and belief in a variety of paranormal claims should also be on the rise.

Unfortunately, the researchers do not include the survey items concerning paranormal belief on every year's questionnaire and the most recent administration was not one in which they included those questions. However, fear of ghosts, though not one of the top ten fears (for 2024, those were: corrupt government officials, people I love becoming seriously ill, cyberterrorism, people I love dy-

1   Chapman University (n.d.) The Division on the Study of American Fears. *Chapman University.* <https://www.chapman.edu/wilkinson/research-centers/babbie-center/survey-american-fears.aspx> (accessed June 3, 2025).

2   Mouchard, A. (2024). Chapman's annual survey finds Americans more afraid today than at any time in recent history. *The Orange County Register.* <https://www.ocregister.com/2024/10/25/chapmans-annual-fear-survey-finds-americans-more-afraid-today-than-at-any-time-in-recent-history/> (accessed June 3, 2025).

3   Mecklin, J. (2025). Closer than ever: It is now 89 seconds to midnight: 2025 Doomsday Clock Statement. *Bulletin of Atomic Scientists.* <https://thebulletin.org/doomsday-clock/2025-statement/> (accessed June 3, 2025).

4   For example:
Intravia, J., Wolff, K. T., Paez, R., & Gibbs, B. R. (2017). Investigating the relationship between social media consumption and fear of crime: A partial analysis of mostly young adults. *Computers in Human Behavior, 77*: 158-168.
Näsi, M., Tanskanen, M., Kivivuori, J., Haara, P., & Reunanen, E., (2020). Crime News Consumption and Fear of Violence: The Role of Traditional Media, Social Media, and Alternative Information Sources. *Crime & Delinquency, 67*(4): 574-600.
Haidt, J. (2024). *The Anxious Generation: How the Great Rewiring of Childhood Is Causing an Epidemic of Mental Illness.* New York: Penguin Press.

ing, Russia using nuclear weapons, not having enough money for the future, US becoming involved in another world war, North Korea using nuclear weapons, terrorist attack, and biological weapons[5]), did increase from 2023 to 2024, albeit only by 0.3 percentage points.[6] That was the only paranormal fear listed in the most recent survey, and it assessed Americans *fear* of ghosts only, not their *belief* in them. We needed to find other information.

Alas, the precise information we required wasn't readily available, so we've added this sort of systemic study of paranormal belief to our ever growing to-do list, but for now, we can try to piece together a little bit of the puzzle from limited information.

The most recent Survey of American Fears administration that included the paranormal belief questions was in 2021 so we can't determine what's been happening in the few years of greatest interest to us, but we did spend some time examining their data set.[7] To make analysis even more difficult, the precise questions have varied to some degree from one survey to the next, but we found relevant data ranging from 2014 to 2021 that demonstrated some interesting trends. Belief that extraterrestrial aliens visited in the ancient past increased every year they asked the question (landing at 42.3% who either slightly or strongly believed in 2021). Similarly, belief in modern alien visitation was on a general upward trend during the same period (with 34.8% believing in 2021). Belief in Bigfoot was also on an overall upward trend for much of that time, though it appears to have peaked at 20.7% in 2018 and declined to 17.2% in 2021. Indeed, 2018 seems to have been something of a peak year for many paranormal beliefs, including ancient advanced civilizations (such as Atlantis), protection by guardian angels, psychokinesis, and that the government is covering up what it knows about aliens (though for some of these questions, data are only available from a couple years). One of only a few of these questions they asked every year was whether houses or other locations can be haunted by spirits; this statistic also peaked in 2018 with 57.7% agreeing either slightly or strongly, and declined to 52.8% in 2021.

It's unfortunate that we've been unable to find a more systematic study of paranormal belief and its correlates because it would be fascinating to learn what drives these belief systems. For now, though, the most we can say with any measure of certainty is that though the individual beliefs ebb and flow over time, believe in paranormal phenomena in general remains quite strong.

As long as there are paranormal beliefs, bizarre claims, or weird phenomena to discuss, we at Rocky Mountain Paranormal will be more than happy to investigate. In the meantime, we are proud to present this third volume of such

5   Chapman University (2024) Chapman Survey of American Fears 2024: Key Findings. *Chapman University*. <https://www.chapman.edu/wilkinson/research-centers/babbie-center/_files/2024-csaf-key-findings-final.pdf> (accessed June 3, 2025).

6   Chapman University (2024) Chapman Survey of American Fears, Waves 9 and 10 Compared. *Chapman University*. < https://www.chapman.edu/wilkinson/research-centers/babbie-center/_files/2024-csaf-wave-9-10-compared.pdf> (accessed June 3, 2025).

7   ARDA (n.d.). Chapman Survey of American Fears. *The Association of Religion Data Archives*. <https://www.thearda.com/search-the-arda?searchterms=chapman+survey+of+american+fears> (accessed June 3, 2025).

investigative case files for your enlightenment and entertainment.

Following the same pattern established in prior volumes, we've divided these case files into three rough categories. First of those is our "Public Venues and Cemeteries" section in which you will find stories of places either publicly owned or open to the public (or at least with a fair amount of public fame or recognition). Following those will be our "Private Residences" section in which we discuss (anonymously) cases in which private individuals have invited us into their homes to look into a variety of paranormal experiences they've been having. Finally, "Media Analyses and Other Activities" represents a grab bag section into which we put everything that isn't one of the more common on-site investigations (whether public or private).

Just to give you a little tease of what's ahead, the twenty-seven chapters to follow include tales of ghosts and aliens and other bizarre phenomena (because unlike many paranormal investigators, we do not limit ourselves to the study of only a subset of potential paranormal phenomena), murders, encounters with the police, and even the cricket demon case we hinted about in a previous volume. One location we'll be discussing shortly was the site of *tens of thousands* of deaths prior to our investigation, leading directly to innumerable ghost stories. In another, we accidentally got involved as witnesses in a truly ghastly criminal case. And in another still, you'll read about one of the rare hoaxes we have perpetrated ourselves in order to prove an important point (albeit in a humorous way). As always, these represent a good mix of solved and unsolved cases and are presented in an order we think will make for a textured reading experience rather than in chronological order.

As we bring this introduction to a close, all that remains is for us to briefly explain our philosophy and methodology for any readers who are joining on this journey now and haven't yet read the first two volumes. If that describes you, by all means please keep reading. But if you *have* been with us for the previous volumes, the remainder of this introduction can be safely ignored as it has been adapted directly from Volumes 1 and 2. For those of you leaving the introduction here, we wish you happy haunting and we'll see you in the next section. For everyone else, read on.

The Rocky Mountain Paranormal Philosophy

There seem to be two major factions when it comes to paranormal claims. On the one hand, you have the believers who tend to seek confirmation of the existence of the ghosts/aliens/entities in which they believe. Counted on the other hand are the self-styled skeptics comprised primarily of non-believers who seek to debunk the paranormal claims. Rocky Mountain Paranormal proudly sits right between these two groups. When we happen to be in a contrarian sort of mood, we like to joke that we anger the believers and non-believers equally. In truth, though, we'd like to think we actually fit in rather well with both of these groups because our philosophy is compatible with what both sides at least claim to be doing.

One way to think about our overall philosophy is to distinguish between skepticism and what we can call cynicism. In this context, cynicism refers to rigid adherence to a foregone conclusion and not to Cynicism as in the ancient Greek philosophy pioneered by such thinkers as Antisthenes, Diogenes, and Crates of Thebes. In our sense of the word, "cynic" is often used as a derogatory term for those who disbelieve in whatever paranormal claim is being discussed. And sometimes the label applies. However, we maintain that there are cynics on both sides of the paranormal belief spectrum.

The cynical believer is one who absolutely believes in the paranormal claim in question and no amount of disconfirming evidence can change this individual's mind. Similarly, the cynical non-believer is one who absolutely does not believe in the paranormal claim and no amount of confirming evidence will ever be satisfactory. In the extremes, these individuals are rare. Most people, upon considering a certain weight of evidence, are at least in theory capable of changing their minds. However, slightly less to the extremes of the spectrum exist a lot of people who may be *capable* of changing their minds but certainly don't seem very interested in giving alternative explanations a fair and open hearing.

True skepticism requires us to treat each case from a completely neutral point of view. That's not to say we need to disregard everything we know from prior experience, but we do need to consider the possibility that any given claim may or may not turn out to be an example of a genuine paranormal phenomenon (however we choose to define that word).

An example seems to be in order, so let's consider a simple ghost story. Imagine Building X is supposed to be haunted by the spirit of someone who was murdered in the building some fifty years ago. The believer tends to want the story to be true, so even the slightest anomalous experience will be taken as proof positive that the entire story is true. The disbeliever tends to reject even the possibility of a genuine ghost or haunting and so considers even the slightest discontinuity in the story as proof positive that the whole thing is false. The true skeptic, on the other hand, takes a decidedly different approach and says: maybe the story is true, maybe the story is false, and we need to investigate (as impartially as humanly possible) to find out where the truth lies.

Rocky Mountain Paranormal insists upon the skeptical approach. We're open to the possibility of a ghost. We even hope that it's true. But we're going to look for any potential naturalistic explanations first. In our hypothetical example, one thing we'd certainly want to do is verify whether the alleged murder even took place. If we confirm it, that doesn't prove the ghost story; if we debunk it, it certainly means at the very least that part of the story was wrong. We'll also examine Building X and try to account for every phenomenon people have witnessed therein.

And it goes even further than that. Imagine that, during one of our investigations, we're able to explain every portion of a claim through purely natural and mundane mechanisms. Does that mean ghosts don't exist? Hardly! It's always possible the next investigation could be a real ghost. Indeed, it doesn't even disprove the existence of the ghost of a murder victim at Building X. All it means is that, of the various phenomena once thought to be evidence of the ghost, they all turned out to be caused by something else.

The late great magician and skeptic James Randi once gave another marvelous example. He imagined trying to investigate whether or not Santa Claus could use flying reindeer as a mode of transportation. As an experiment, he imagined taking a group of, say, one hundred reindeer to the top of a skyscraper and, one by one, pushing them off. If one manages to fly, well, that's case closed. But what happens if, as seems so much more likely, they all instead fall to their rather messy deaths on the street below? Does that prove that reindeer cannot fly? It does not. It merely proves that our particular group of one hundred reindeer couldn't fly *or chose not to*.[8]

Does that mean we're cursed to never know anything? If we're looking for absolute certainty, the answer is painfully close to "yes." Outside of pure mathematics, absolute certainty doesn't really exist. Scientists tend not to speak of "proof," but of evidence and the strength thereof. Even in courts of law, where we do use the word "proof," we only insist upon proof beyond a reasonable doubt, not beyond all doubt, because there's always room for *some* doubt. What, then, are we to make of something like the reindeer experiment? We can't claim absolute proof, particularly of a negative, but we can say that the outcome of the experiment seems to make it less likely that flying reindeer exist. We must remain open to the possibility that they exist out there, somewhere.[9] But if we keep experimenting often enough, eventually we keep moving that dial closer and closer to the "disbelief" side. Just as long as we never claim to move it all the way, we should be in good philosophical shape.

What about evidence in favor of something like a ghost? Imagine we watch an inanimate object move, seemingly of its own volition. Occurrences of exactly this type, in fact, will be discussed in the pages to follow. Imagine further that we do our due diligence and rule out all the natural explanations we can think of (which might include hoaxes, a draft in the room, optical illusions, hallucinations, videographic anomalies, and the list goes on). Does that prove that a ghost exists and has moved the object? It does not. The more natural explanations we rule out, the more we might move our dial toward the "belief" side of the spectrum. But without the kind of absolute proof that simply doesn't seem to exist, we can never move it all the way.

An important thing to remember is that just because we maintain that two distinct interpretations are both possible, we don't necessarily have to consider them equally likely. This isn't going to turn into a lecture on inferential statistics (we do plan to go into some greater detail on that subject, albeit in a manner still appropriate to the lay reader, in a forthcoming book giving a more complete and technical explanation of the methods of paranormal investigation), but that's essentially the whole point of statistical testing. We may not be able to remove all doubt, but we can at least use sophisticated mathematical techniques to quantify exactly how uncertain we are.

When it comes to paranormal research, there aren't a whole lot of statistical tests to be done. The kinds of evidence obtained during paranormal investigations are often qualitative rather than quantitative. We strive to obtain quantitative

8   Randi, J. (1992). *Pseudoscience and the Paranormal: Inaugural Skeptics Society Lecture* [DVD]. The Skeptics Society Distinguished Lecture Series.
9   Indeed, a believer in Santa Claus might object to our experiment on the grounds that of course *those* reindeer didn't fly because only those in the employ of Father Christmas possess that ability.

results as often as possible—measuring things is a key part of science—but the nature of paranormal claims frequently precludes the possibility. Nevertheless, even if it must be done qualitatively or informally, that gradual adjustment of the belief/disbelief ratio is how we gradually move toward truth. Such incremental adjustments of belief are the foundation of a branch of statistics known as Bayesian analysis, but a full explanation of that would take us beyond the scope of this book.

Along the same lines, it's necessary to think not in terms of certainties, but in degrees of certainty. For example, we can be 100% certain of a mathematical theorem that has been proved.[10] Similarly, we can each at least arguably be certain of our own individual existence (that's Descartes' "cogito ergo sum," or "I think therefore I am"). We can be a little less certain, but still as close to certainty as any sane person should ever require, of the existence of things like objective reality and the physical objects that surround us. We can also be reasonably assured that the sun will rise and set on schedule tomorrow. And then we get to other things that are substantially less certain, but we nevertheless seem to take for granted: that we won't die in our sleep tonight, that our car will start tomorrow morning. And so on, all the way down to things we can be pretty certain aren't the case: flying reindeer seem like a good example, as would the idea that the Earth is flat (which we would consider a paranormal claim, to be sure, but which we have thus far felt is beneath our dignity to investigate as one of our case files).

Paranormal claims, too, exist on a spectrum of plausibility. Extraterrestrial aliens make a great example. Given the immensity of the universe, many people are comfortable with the idea that it seems likely for there to be some extraterrestrial life out there somewhere. It's much less likely that they've visited us on Earth. And is even less likely that they're shapeshifters that have secretly infiltrated our governments. That's not just to pick on those particular claims for being less plausible than others. It's a mathematical necessity that each successive claim in this chain is less likely than its predecessor. For aliens to have infiltrated our government, they must also have visited Earth. To have visited Earth, they must exist. By mathematical necessity, the probability of two things both being true is less than the probability of either one of them being true. This is especially easy to see when one claim is a subset of the other. Mathematical thinking like this, however, does not come naturally to the human mind. One requires substantial training to think in terms of mathematical logic.

Even when we don't have mathematical reason to say that one paranormal claim is more or less likely than another, there still seems to be that spectrum of plausibility. In our estimation, a cryptid like Bigfoot seems more likely to exist than a ghost (that's just our estimation—you're free to estimate otherwise). It's entirely possible that ghosts could exist and Bigfoot not, but given the spotty data, we estimate the best we can.[11]

10   At least within the constraints of our axiomatic system of mathematics— though if you really want to tumble down a mathematical rabbit hole, we suggest you should read up on Gödel's incompleteness theorems, which exposed the limitations of that axiomatic system. Be forewarned: such reading will, as the Oracle told Neo in *The Matrix*, really bake your noodle.

11   We base this estimate on the idea that Bigfoot, while certainly paranormal

Regardless of those estimates, though, you will never find any of us claiming knowledge we do not possess. Even if we might think a particular claim seems unlikely to be true, we will do our best to approach our investigation from a position of neutrality. And to the extent that true neutrality might be impossible for the human mind to achieve, we're very careful to build safeguards against bias directly into our methods, so we can at least maintain methodological neutrality even if our attempts at philosophical neutrality fail.

At some point, though, we need to make decisions about what to believe or what not to believe. Though we will always keep investigating as much as possible, when the time comes to reach a tentative conclusion, the best approach is to apply Ockham's Razor (also sometimes spelled Occam's Razor and also known as the principle of parsimony). This is a philosophical precept which tells us that, all else being equal, the simplest explanation is usually the correct one. This is often misunderstood. "Simplest," in this context, doesn't mean "easiest to understand." It means, instead, that the explanation we should consider more likely is the one that relies on the fewest unproven assumptions. Another way to think about it is that we should always try to choose the path from Idea A to Idea B with the fewest possible intermediate steps.

Because paranormal claims, by most definitions anyway, include the existence of unproved entities, Ockham's Razor tells us we should be reluctant to accept them. When mundane or naturalistic explanations exist, they are more likely to be the correct ones.

All of that having been said, and while keeping all of that skeptical philosophy firmly in place, there's another piece to the Rocky Mountain Paranormal philosophy that's a bit friendlier to the believers' side of things: we really want at least some of these claims to be true. A big part of the reason we look into paranormal claims is in the hope that someday we might be able to confirm one of them. We may not think it's particularly likely (again, depending on how "out there" the claim in question might be), but that doesn't mean we aren't hoping.

We're horror fans through and through. There are few things in the world that bring us more joy than a good ghost story or tale of alien abduction. Even if we're somehow able to conclusively disprove one of the claims (as we have done in some but not all of the case files you're about to read), we still love the stories themselves. They're a part of our collective mythology—perhaps even part of what Jung referred to as the collective unconscious. And they're just plain fun.

The world is a big place, and undoubtedly there are still phenomena our science has yet to discover. Combine that with the number of people reporting paranormal experiences, and there's a strong temptation to think there must be something to all of this. When you add that to our predisposition to want these claims to be true, it would be very easy for us to abandon our skepticism and take even a whisper of evidence as proof of the claim. So, we're constantly walking that tightrope between open-mindedness and skepticism.

(as in, not of the normal), is not usually thought to be supernatural. The idea of an undiscovered animal—even one as remarkable as a Sasquatch—seems to us more facially plausible than the idea of the spirit of a dead person returning to Earth. On the other hand, belief in ghosts seems to be more popular in the United States than belief in Bigfoot, so we remain open to either possibility.

It's probably true, as the Bard of Avon reminded us, that there are more things in Heaven and Earth than are dreamt of in your, my, or anyone's philosophy. And with as many people as there are in the world, even if some paranormal event is exceedingly rare, there's a possibility that someone (perhaps even a lot of different "someones") may have experienced it. On the other hand, widespread belief in or experience of a claim does not necessarily mean it's true. Human senses are flawed and easy to deceive, and the human brain is remarkably bad at evaluating claims objectively. Paranormal investigation, properly done, is a balancing act between these competing ideas.

Beyond the question of the legitimacy of paranormal claims, though, is another piece of our philosophy: the storytelling piece. We take the position that, even if all the paranormal claims turn out to be false, they're still stories worth telling because our stories tell us a lot about who we are. Not only are they fun, but they're often psychologically deep at a metaphorical or symbolic level.

And then there's the history. As the saying goes, those who don't learn from history are doomed to repeat it (unfortunately, someone once pointed out to us that those who do learn from history are often doomed to watch while everyone else repeats it, as history itself will often attest). Alas, a lot of our history is being lost. Wonderful historic buildings are constantly being torn down, either due to disrepair or to make room for some new development. Some of the buildings described in this series are no longer standing, making these case files a kind of ghost story in and of themselves.

Perhaps even worse, people don't take the time to read the stories of the past. Consider all those "man on the street" videos floating around the Internet in which people fail to answer basic questions about their history. To many people, the study of history seems quaint or irrelevant. We maintain—no, we insist—this is not the case. Not only does our history provide the context for why things are the way they are today, not only does our history provide lessons for how we should proceed into the future, not only does history tell us about our own heritage, both individually and collectively, but the stories from history are the very voices of our ancestors. If you want to talk about ghosts or speaking with the dead, you really need to talk about history. Ghosts as literal entities may or may not exist—we continue to take our neutral stance—but the ghosts of the past can and do speak directly to us through our historic stories, documents, and buildings, if only we would take the time to listen to them.

Particularly when it comes to ghost stories, paranormal investigation offers a window not just into the fun and spooky tales of the ghosts themselves, but into our rich historical heritage. Part of our mission at Rocky Mountain Paranormal is to document and preserve that history to the greatest extent we can. As you can read about in Volume 1, Chapter 5 of this series, when a piece of history was almost lost, it was Rocky Mountain Paranormal, working in conjunction with a team of scientific experts, who managed to give a voice back to the dead in some small way. That's something of which we're incredibly proud, and it's something we try to do whenever we can. Both in the context of historical research and in the context of reminding people of history through our lectures (and now these books), we're using the paranormal stories as a framework to help preserve our history. Chapter 1 of the present volume continues this pattern of attempting to document and preserve almost-lost history.

Given that paranormal stories are fun on their own and that they have such value in providing an excuse to talk about history and science, it's little wonder that some of us have found such passion in this field. And that passion is directly related to our guiding philosophy. To sum it up as succinctly as possible, our philosophy is that there's great value in paranormal stories, but they must be evaluated skeptically and scientifically to determine what is (or what is likely to be) true.

Or to put it another way: our guiding philosophy rests on equally important pillars of history, horror, storytelling, science, and skepticism.

A Brief Note on Methodology

There's no way we can provide a complete course in paranormal investigation here in this book.[12] However, it's worth taking a moment to briefly explain some of our methods so you'll understand our approach in the case files to follow. As we discuss this method, you should be aware that the precise order of events is variable. Everything always depends on the specific paranormal claim under study. But just to provide a sketch of a "typical" investigation, we're going to walk you through the steps we would ordinarily take.

In the case of a private residence, the process usually starts with a client reaching out to us, typically via email. Hopefully we're the first team they reach out to, because sometimes other teams with insufficient experience or ethics have confused or frightened the clients even more by the time the case reaches our desk. Investigations of public venues follow similar procedures, though in some cases we might have approached them with a proposal. For this brief description, we'll describe a private case but note a few key differences along the way.

Regardless of whether we're the first contact or not, we open a case and request further information from the client. Typically the claims made in these initial letters are vague enough that we don't know whether we'll be able to help or not, so we reach out to the client and request additional details. As things progress, if the case sounds like something with which we may be able to assist, we'll schedule an interview.

During the initial contact and interview phases, we try to collect detailed information regarding what is happening and when it's happening. We almost always request the client should keep a diary of occurrences, so we can check for any patterns. If the phenomenon always occurs on Wednesday evenings at 10:00 p.m., for instance, it wouldn't do for us to try to investigate at noon on a Saturday.

We also try to collect as much information about the clients themselves as they're willing to share. This includes any potential substance abuse problems, psychological issues, or major life changes. Additionally, we often counsel people to speak with a licensed psychological professional. This is emphatically not because we're calling them crazy. Though the possibility exists that people's paranormal issues could stem from deep psychological troubles, the more common reason we need to ask those questions and provide that advice is that paranormal events (whether we assume them to be genuine or psychological in nature) tend to occur during times of profound stress. Furthermore, the experience of

12   Stay tuned. One is in progress and publication is planned for the near future.

paranormal phenomena itself can prove stressful or strain family relations, and it's important people should seek the help they need. It's also essential that we should understand the clients' own religious or spiritual background so we can better interpret their claims.

Once we have a detailed account of the claimed phenomena, including a diary or timeline, and any photos or videos the clients may send us (if they have them), we schedule a time for an on-site investigation if we deem it beneficial. The scheduling can be difficult because we need to coordinate times that work for our team as well as for the clients, but most importantly which correspond with the timing of the phenomenon in question. Though it can be challenging, we absolutely will rearrange our schedules for these investigations when necessary.[13]

Before we arrive at the site (and often again after leaving the site), we go through an extensive process of background research. This includes digging through newspaper archives, public records including real estate transactions, and genealogical records if they seem relevant. We spend a lot of time on the Internet, at the state archives, or at the libraries[14] finding any information that might be relevant to the case. If possible, we try to obtain blueprints of the site. If these are unavailable, we'll make our own approximation of a site map during our investigation. Background information concerning public venues is almost invariably easier to find and more abundant.

When we arrive at the site, we're not doing what you often see on television. First of all, we don't bring a television crew (much to the surprise of some of our clients). We also ask them not to invite guests (see Chapter 22 for a humorous story about that). Though we realize our work is interesting to people, we need the site to be as controlled as possible, and that means not having a bunch of people running around contaminating data. At public venues, we try to conduct investigations after hours when only ourselves and employees are present (at least whenever possible; there are exceptions to every rule, and some venues do stay open 24/7, so we adjust as necessary). We'll then tour the building, determine locations for cameras and other monitoring equipment, set up our base camp, and prepare for the investigation. We ask the clients to pretend we're not there to the greatest extent possible and continue with their normal routine—it is, after all, during their normal routine that the phenomenon is claimed to occur.

Most importantly, once everything is set up, we shut up for the bulk of the investigation. We don't try to carry on conversations with the ghosts or run around talking to each other. Noise contaminates data. Furthermore, if we start talking to the alleged ghosts, we'd be psychologically priming ourselves not only that there is a ghost (which is what we're trying to determine in the first place) but

13   A recent case, tentatively slated for publication in Volume 4 of this series, involved a three-hour drive to a residence (and a three-hour drive back) in order to be present at the client's home when the paranormal phenomenon was thought to be active between the hours of midnight and 3:00 a.m. Our dedication is such that we are happy to investigate on whatever schedule is necessary.

14   To anyone interested in any kind of research (paranormal or otherwise), take note: librarians are your best friends. They don't just spend their time rearranging the bookshelves; they're almost like wizards in their ability to find whatever information you need.

also to expect specific answers to our questions. We might announce something like "if there are any spirits, feel free to make yourselves known to us" (with heavy emphasis on the word "if,") but that would be the extent of it. If the particular claim necessitates a more direct interaction with the alleged entity, we are careful to always do so hypothetically.[15]

Equipment we might use on site includes still cameras, video cameras, 360-degree or spherical cameras, electromagnetic field (EMF) detectors, seismometers, thermometers, audio recorders, and any other tools that might seem relevant to a specific case. We sometimes also include air quality or other environmental monitors to alert us if the house may have something like a gas leak (which can, under some circumstances, cause hallucinations). Though we seldom need to use them, we often bring sterile specimen jars in case we need to collect a sample of some potential environmental hazard for later laboratory analysis.

Because the paranormal is an ill-defined field of study, we don't know how to measure a ghost, alien, or demon. What we can do is establish baseline readings on our various instruments and then look for anomalous readings. If an anomaly should occur, we don't immediately assume it's supernatural. Instead, we note it and follow up to see if we can find a way to explain it.

There are plenty of devices and gizmos claimed to be part of the ghost hunter's arsenal that we tend not to use most of the time. These are the devices that were specifically designed for paranormal investigation. Such devices are either hoaxes or intended for entertainment purposes rather than forensic investigation. There's no way to calibrate a device to detect a ghost since we've never even conclusively proved that ghosts exist or figured out their properties. We do own these devices and will experiment on or with them, but they're not part the gear we'll use on site unless the claim in question is specifically related to one of them.

At the end, we'll present the client with a report of our findings. This will include all of our background research as well as all of our measurements. The most important part will be our lists of findings, and they come in three categories. First, we list explainable events, or apparently paranormal phenomena we were able to explain naturally. Next, we list any phenomena or events we witnessed but have not yet been able to explain. Finally, we list any phenomena the clients reported but we were unable to witness or recreate. Along with this final list, we'll include any possible explanations we can think of, with the caveat that until we actually witness or test it, those explanations are speculative.

After we present our report, the clients often still have difficulty dealing with their issues. We're happy to help them to the extent we're able, but we always encourage them to speak to a psychological professional or work with their own clergy. Should it be necessary, we're always happy to return to conduct follow-up investigations if the anomalous phenomena continue.

15   Whether or not to engage in these kinds of interactive activities on an investigation has been a source of some controversy among other investigators and much discussion within Rocky Mountain Paranormal. We maintain that there *is* a way to do this sort of thing legitimately (though we tend not to do so for the reasons indicated) but that it's difficult and requires sophistication in the study's design. A more complete treatment of the subject will appear in our forthcoming "how-to" book.

You'll note that our methods differ from what you tend to see on television in which the so-called investigators run around in the dark scaring themselves and each other. We mark a distinction between the activities of "ghost hunting" and "paranormal investigation." Ghost hunting is what you see on TV. It's a lot of fun, but it doesn't yield quality data. Paranormal investigation is methodical, scientific, and often quite boring or tedious to do.

Ghost hunting does have its place. It actually has two purposes. The first is purely for entertainment. As horror fans, we can appreciate running around in the dark and scaring ourselves silly. We just don't claim that to be part of a scientific investigation, no matter how many beeping gizmos we carry with us when we do so. The second is that ghost hunting can precede a proper scientific investigation and can be valuable in that context. It gives the phenomenon a chance to manifest itself under less stringent controls. The important thing to realize is that, when those controls are loosened, everything observed must be taken with a grain of salt.

In mathematics, there are two kinds of data analysis: exploratory and confirmatory. The former is essentially mining data to see what we can find. The latter is to confirm the validity or truthfulness of whatever we found in the former. Ghost hunting, in its proper context, can be thought of as a kind of exploratory analysis. Go to a haunted place and see what happens. But, as is the case in mathematics, you must follow it up with more rigorous confirmatory analysis.

Essay
Investigation With Dignity and Respect

In the previous volume of this series, we included an essay on ethical considerations in paranormal research and developed a list of "Ten Commandments" for ethical paranormal investigation. Omitting the lengthy explanations of each, we reproduce those Commandments here:

1. Thou shalt report thy findings honestly,
2. Thou shalt comport thyself as a scientist,
3. Thou shalt comport thyself as a professional,
4. Thou shalt not claim false expertise,
5. Thou shalt not investigate without an invitation,
6. Thou shalt not damage thy neighbor's property,
7. Thou shalt respect the religious beliefs of thy clients,
8. Thou shalt not harm thy client,
9. Thou shalt respect thy client's property, and
10. Thou shalt not charge for thy services.

This essay can be thought of as an addendum or follow-up to that one, because we have unfortunately seen too many other investigative groups fail to adhere to basic ethical guidelines when doing their research (see Chapter 27 for a detailed example). For the purposes of this paper, though, instead of thinking of ethics more generally, we want to zoom in on the issue of dignity and respect for the clients and any other individuals who might be involved in a case.

Respect is important in any line of work, but paranormal investigation can be a minefield of potential (even unintentional) abuses if one isn't careful. Consider that paranormal claims often intersect with religious beliefs, and different faiths may interpret a matter in idiosyncratic ways. Ghost stories often involve claims that a spirit is that of a particular deceased person, whose family might still be alive and in the area (if they are not the clients themselves). Stigma surrounding certain paranormal experiences might make individuals reluctant to let their stories be known. All of these issues and more should be at the forefront of the ethical paranormal researcher's mind.

An ounce of prevention is worth a pound of cure in paranormal investigation just as it often is in health. It's genuinely easy in most cases to treat individuals with dignity and respect as long as one keeps in mind that this line of work can be fraught with certain issues and takes the time to think about them before speaking or acting.

The first potential minefield we need to discuss has to do with religion. Faith is a deeply personal matter to many people so while a researcher may have his or her own faith (or none at all), it's important to tread carefully around the religious beliefs of others involved in an investigation, whether they're the clients or otherwise peripherally involved. That's easy enough to say and should be self-evident, but the problem occurs when the paranormal claims in question directly or indirectly involve religious ideas. This occurs with great frequency.

Recently, the co-authors of this book made a humorous observation. One of us had been watching a true crime documentary or listening to a true crime podcast (the precise details fade from memory and are unimportant). At some point during the investigation of the crime, police were shocked to discover the suspect had occult items in his house, and this was taken as informal evidence that they were looking into a deeply disturbed person. In the case at hand, that may very well have been the case, but we looked at each other while discussing this case and one of us said something along the lines of "God forbid either of us should ever be suspected of a crime; can you imagine what they'd think if they saw all the weird items in *our* collections?"

That got us thinking. Police and first responders have a stereotype of being somewhat superstitious people (as incidents in multiple chapters in this volume will illustrate), so it makes a certain degree of sense that they might be a little freaked out by a collection of occult items. Depending on one's own religious faith, some of those items might even be thought of as demonic in origin. For example, Catholics consider all forms of magic, sorcery, or divination (including Ouija boards, astrology, fortune tellers, wearing charms, consulting mediums, etc.) to be gravely sinful.[1] Others, however, consider various practices of divination not only harmless but a deep and important part of their own religious practice. These theological issues can be debated when discussing comparative religions, but misunderstandings can lead to serious problems in a paranormal investigation.

Consider a hypothetical situation. Imagine a paranormal researcher is a Catholic and visits a client who claims to be hearing strange sounds and feeling unwell at home. During the investigation, the researcher discovers the client has been using a Ouija board. How does the researcher proceed?

There are different options depending on the philosophy and demeanor of the client. But even if the researcher's own faith might suggest that the Ouija board is the source of the client's trouble and has invited some demonic force into the house, immediately telling the client that their problem is diabolic would exceed the rightful place of a researcher. At that point you're no longer acting in an investigative manner but as an evangelist for your own religious views.

That doesn't mean you need to completely check your own views at the door every time you investigate. But it does mean you need to tread carefully and treat

---

1   Paragraphs 2115-2117, in: Catholic Church (2000) *Catechism of the Catholic Church: revised in accordance with the official Latin text promulgated by Pope John Paul II.* Washington, DC: United States Catholic Conference.

the client's own beliefs with dignity and respect.

Such a researcher as the one in our hypothetical might gently ask the client about their use of the Ouija board to first determine how and why they use it. If they simply own a board because they bought it at a garage sale and never used it, that's a different situation from one in which it's a fundamental part of their own religious observance. The Catholic researcher can explain the Catholic perspective on the matter but should refrain from pushing his own religious views on the client. If the client turns out to share the researcher's religion, that might be a productive conversation, but if the client turns out to be an agnostic or a member of another faith, jumping to the conclusion that the Ouija board is the source of their disturbance on no basis but the researcher's own religion is going to either anger or offend the client.

Conversely, imagine the paranormal investigator is a Spiritualist or pagan of some description.[2] Though it would not ordinarily be a part of our own methodology (as outlined briefly in the introduction to this volume and as will be outlined at length in a future work, though exceptions can be made as in the case to be described in Chapter 14), researchers of such a religious persuasion might very well want to use a Ouija board as part of their investigative methodology. Those who would belong more to the category of "ghost hunters" than "paranormal investigators" are often seen making use of such tools. But what if, in this hypothetical, it is the *client* who is a devout Catholic and considers the use of a Ouija board at least ill-advised if not gravely sinful? In such a case, the researcher should seek alternative means to complete the investigation without insulting the client's religious sensibilities by bringing into his or her home an object of which he or she deeply disapproves.

We ran into a similar problem in a real case which we described in Volume 2 (Chapter 21). Though we'll not rehash the entire case here, the relevant portion in brief was that a prior investigative team had visited these clients before we were called in, and the client discovered that the investigators had been discussing the religious symbolism in the home's décor while the clients were in another room.

2   Spiritualism, formally speaking, is a religious movement originating in the nineteenth century which held that deceased individuals could be contacted through mediums. Though numerous publications and stunts exposed many of these mediums as fraudulent, certain variations of the movement still exist today. For more on the history of Spiritualism, see: Morton, L. (2020). *Calling the Spirits: A History of Seances*. London, UK: Reaktion Books.

Informally, spiritualism (with a lower-case "s") is often defined in modern times as a system of individualized religious beliefs outside the practice of any particular organized faith or church structure. Many people call themselves "spiritual but not religious."

Paganism, on the other hand, is a word used predominantly in modern times to denote membership in any religion that does not align with one of the major world religions (Judaism, Christianity, Islam, Hinduism, etc.) and is not formally defined as a specific faith with specific dogmas. In our usage here, many beliefs or practices associated with the paranormal, occult, or supernatural could be thought to belong to this "catch all" category.

We don't believe anything they said was intended to disparage the clients or their religion but they were nevertheless offended by the whole episode.

On one of our own cases, we entered a home and found little bowls of rice and milk left out all over the place. But instead of immediately jumping to any conclusions, we remembered our own advice about treating the clients with dignity and respecting their culture and simply asked in a polite but straightforward manner about them. They told us they were leaving those out as offerings for the ghost(s) they believed they had in their home. Such practices are uncommon in the United States but are a part of some cultures that engage in various practices of ancestor veneration.

Sometimes the clash of cultures rears its head even without the researcher doing anything wrong. One of our past cases involved a family who had (at the time) recently immigrated to the United States and had subsequently begun experiencing strange phenomena in the house. They explained to us that in their culture, that wouldn't have been anything to be afraid of because their ghosts are just ancestors. Sometimes they might be mischievous but they're fundamentally harmless or even welcome visitors. But, they explained, they'd been watching American horror movies and they were terrified because "American ghosts will kill you!"

Obviously, we still investigated that case as usual, but in a situation like that, the best thing to do was to contextualize the phenomenon they'd been experiencing within their own cultural background. American horror movies may be one thing, but as long as the phenomena in their home never otherwise gave them any cause for alarm, we saw no reason they shouldn't interpret any unusual phenomena in the United States exactly as they would have at home.

Matters might be different if we had sufficient evidence to *prove* one interpretation or another. Even if the evidence disproves a religious claim, we do believe that part of investigating with respect is to respect the client enough to deliver the truth unabridged. Absent that kind of overwhelming proof, though, respect for the client does mean a willingness to interpret their phenomena through their own lens. When their background differs from our own, if we must discuss those differences, it should always be done politely, calmly, and hypothetically.

Another major issue that comes up in a number of cases is that ghost stories might be attached not only to a particular location but to a particular person. Someone might not only think that *a* ghost is in their home but that it is *the* ghost of so-and-so. That's fine. It's part of the lore surrounding that case and it can be investigated using the usual methods. Where it becomes tricky, though, is when the ghost is thought to be behaving in a way the deceased individual might not have, when the deceased individual's family might hear of the case, or when the deceased individual's own (or his/her family's own) religious views disagree with the idea that the individual might be a ghost.

Exactly that situation occurred in the story of the supposedly-haunted Macky Auditorium in Boulder, Colorado (Volume 2, Chapter 5). The trouble in that case was that the alleged ghost was supposed to have been the victim of a ghastly murder which really did take place there in the 1960s and the victim's family maintained that her religious beliefs were such that even if it were possible to come back as a ghost, she never would have. What does one do in a situation like that? We still need to investigate the paranormal claims, but they involve someone whose family are still around and want nothing to do with the ghost stories.

Our solution there was indeed to proceed with the investigation, but to do so quietly and respectfully. When it came time to report our findings, we were able to dispel some of the myths that might have annoyed the family, and then we gave them the last word on the subject.

If an individual believes that a ghost *is* someone from his or her own life, that can be a similarly sticky situation but for the opposite reason. In that case, presumably they really want the ghost to be that of their spouse or grandmother or whoever they think it is, and it's difficult to have to disappoint them if our findings run to the contrary. Death visitations, wherein an individual witnesses an alleged ghost of someone who recently passed, are among the most common paranormal sightings reported, so this is a constant concern for us, and how we recommend dealing with these cases falls into two categories.

Category one is the case in which a person tells us the story of a death visitation that happened either once or just a few times in the past and asks our opinion of the subject. Under such circumstances, we can honestly tell them that there's no way we can investigate a singular event that happened so long in the past, so we can't say whether we think it's real or not. On the one hand, if one believes in ghosts, those stories are common enough to make it plausible within the believers' perspective. On the other hand, the psychological trauma of losing a person can easily conjure such a vision out of nothing. So our advice to those people is to interpret the experience in whatever way seems best to them.

Category two is the case in which the phenomenon is recurring and so there's a chance we can investigate. Here, we conduct our investigation as usual, but the difficulty can occur when it comes time to deliver our results. If we did find something unusual, we can report this and let the person interpret it as they may. But if they're hopeful the phenomenon is the ghost of some loved person and we instead manage to find a natural explanation for everything they've been experiencing, there's a risk that our findings could be at best disappointing and at worst traumatizing.

The best thing to do there is not to lie to the client or avoid the subject. That also violates our rules about dignity and respect. But the report can be delivered in a thoughtful manner. Just because the phenomenon we studied turned out to be natural doesn't mean anything about what the client is still free to believe about the existence or nonexistence of ghosts or the status of their deceased loved ones.

Finally, the last issue we want to talk about in this essay has to do with stigmatized beliefs. Often this comes up when someone witnesses something like a UFO. Popular culture often views those who have had ghostly or alien experiences as "crazy." For this reason, many witnesses may be reluctant to come forward with their claims or stories for fear of being ostracized.

Our approach here is quite simple and two-fold. First, we make it quite clear that not everyone who has such experiences is crazy. Maybe some of them have mental health issues, but the majority of paranormal claimants, even if they might sometimes be mistaken about the nature of what they saw, are perfectly sane and well-adjusted people. There needn't be any shame in it at all as far as we're concerned. Second, and perhaps even more important, is that we maintain the clients' privacy. Except when discussing historic or public figures, we employ a practice of anonymizing reports. Employees at public venues, unless they choose to be identified, are simply described as employees instead of by name. Clients in the private residence cases have their names and any other potentially identifying features altered or removed from our write-ups.

Fringe fields like the paranormal can present unique ethical challenges, but we will continue to remind anyone interested in this kind of work that it's essential to treat clients with dignity and respect. If you're a would-be investigator, we encourage you to follow our advice. On the other hand, if you're a potential client looking for an investigator, don't be afraid to insist upon working with someone who knows how to treat you properly.

PART ONE:
Public Venues and Cemeteries

While most paranormal claims are experienced and dealt with privately and seldom make the newspapers (and even if they do, tend to be quickly forgotten), there are some places that have become famous for their alleged paranormal phenomena. Many hotels and former hospitals are thought to be haunted. Same goes for plenty of cemeteries, event halls, restaurants, and schools. In this section's fourteen chapters, we're going to look at some of Colorado's most famous haunts as well as some lesser-known but still fascinating settings for alleged paranormal phenomena.

Essay
Paranormal Investigators as Historians

Readers who've been following our adventures in the first two volumes of this series (or individuals who have attended any of our public events or lectures) already know that we fancy ourselves as some sort of lay historians. Ghost stories without any history are great, but we love to tell the real history behind the ghost stories whenever we can. Sometimes we've been accused of going perhaps a step too far (Volume 1, Chapter 6 contains a paragraph which opens: "Our tale begins some 500 million years ago…") but we think preserving the history of these locations is as important a component of our work as telling the ghost stories or investigating the paranormal claims themselves.

Some might wonder why anyone ought to care about history. Unless one is a historian, that might be a fair question. The standard answer, so common now as to be a cliché, is that we study history lest we repeat it. Fair enough. That's true, but it often rings false to those of us who are not in positions of political power. Maybe we should all know some history so we can vote more intelligently, but part of the reason we elect representatives rather than deciding every matter by direct democracy is so that we can all specialize in our own careers and not have to worry about the minor details of, for example, what happened during the Second Peloponnesian War.[1]

Protection from the risk of repeating the grand mistakes of history, though, isn't what we consider the most important reason for studying history. Our own primary reason, probably, is just that we find history interesting and think it's worth studying for its own sake, though we admit that's not the most persuasive of arguments. Instead, we think the real reason we should want most people to be interested in and literate concerning history is three-fold: we have a duty to remember our ancestors as we would hope our progeny would remember us, we have a duty to understand our past so we can shape our future, and studying the characters of history helps build our own character.

The Jewish author Dennis Prager, in his commentary on the book of Exodus, made a similar point, outlining nine reasons "remembering" (his word for what we're describing as historical literacy) is important, and ultimately making

---

1   Though you ought to read about it anyway. It's fascinating.

the argument that remembering the past is an important if not a necessary element of becoming a good person and an upstanding citizen. Elsewhere in his commentary, he credits systematized memory, at least in part, with the longevity of the Jewish people.[2] History, then, is important both for the formation of quality individuals and for the long-term survival of societies. We agree.

We mentioned in the introduction to this book (as well as in our prior volumes) that we're often dismayed to learn of the limited knowledge of history possessed by far too many people. The introduction alluded to those depressing (if sometimes humorous) videos we've all seen online in which some social media personality or journalist interviews people on the street to ask questions of basic knowledge. We're often shocked and horrified by how few people know even the most basic facts of their own history (the same can be said of literature, science, mathematics, and other areas of general knowledge, but, while we're interested in all of those things, history is our concern for the purpose of this essay). Some might suggest those social media videos are selectively edited to present an unduly pessimistic view of people, and we have little doubt that social media creators do indeed edit their productions to maximize interaction with their videos. Unfortunately, formal scholarly research has also borne out the same devastating trends.

As recently as 2024, the American Council of Trustees and Alumni conducted a study called "Losing America's Memory 2.0" in which they surveyed more than 3,000 American college students from all fifty states. Some frightening highlights from their findings:

- Only 27% could correctly identify the then-President of the Senate,
- Only 40% could correctly identify the length of terms for Senators and Representatives,
- Only slightly more than a third could identify the sitting Chief Justice of the Supreme Court,
- Fewer than a third could identify James Madison's role as "father" of the Constitution,
- Only 28% knew that the 13th Amendment abolished slavery in the United States,
- Only 23% could correctly identify the Gettysburg Address as the origin of the phrase "government of the people, by the people, for the people," and 30% incorrectly thought the phrase originated in the Declaration of Independence, and
- Only 32% knew when the Constitution was written.[3]

These results are not anomalous. A 2019 report by the Woodrow Wilson National Fellowship Foundation found that two-thirds of Americans were incapable of passing the U.S. citizenship test, which requires a passing grade of only 60% of questions correctly answered from a quiz of ten questions (pulled from a pool of 100 including such elementary items as naming the first President, list-

---

2   Prager, D. (2018). *The Rational Bible: Exodus: God, Slavery, and Freedom.* Washington, DC: Regnery Faith.

3   ACTA (2024). Losing America's Memory 2.0: A Civic Literacy Assessment of College Students. *American Council of Trustees and Alumni.* <https://www.goacta.org/wp-content/uploads/2024/08/pdf-acta_civiceducation_07_01_2024_collegepulse.pdf> (accessed June 5, 2025).

ing one thing Benjamin Franklin was famous for, or identifying who the United States fought in World War II).[4]

Worse, things seem to be moving in the wrong direction. According to the "Nation's Report Card" from the National Center for Education Statistics in 2022 (the most recent administration of the report), students' scores in history have declined since the previous assessment in 2018 in all but the 90th percentile group and across all history themes or subjects.[5]

Low proficiency statistics in educational surveys are bad enough on the surface, but they become outright terrifying when one realizes that the level for proficiency is often set ridiculously low itself. Still, many students (and, indeed, many adults and even college graduates) often fail to meet the standards. In his biting critique of the American educational system, economist Bryan Caplan collected statistics from a variety of educational achievement studies and reported (limiting ourselves here to the subject of history though he presents similar statistics from other subjects), among other disheartening facts:

- On a four-choice question, only 21% knew that the right to vote was not in the Bill of Rights,
- Also choosing from four options, only 24% knew that the Constitution establishes the United States as a republic,
- Only 26% knew that the foundation of Jamestown, VA came chronologically before the Declaration of Independence (the other options were the Civil War, the Emancipation Proclamation, and the War of 1812),
- Only 55% knew the *decade* in which the American revolution began, and
- Only 8% could identify which of five items was prohibited by the Bill of Rights.[6]

Basic historical literacy is clearly in trouble. Lest anyone get the wrong idea, by the way, though we have focused on American statistics because we are Americans writing for a primarily American audience, similar problems do exist in countries around the world. This is a human problem, not a uniquely American one.

Perhaps some readers who pick these books up only for the ghost stories (and there is nothing at all wrong with that; we love the ghost stories) might wonder why we spend so much time talking about the state of historical literary (or, in other places, the state of scientific literacy, mathematical literacy, or just plain old literacy). The answer is that, though we're the kind of intellectual people who've always cared about these kinds of issues on our own private time, we accidentally got thrust into the world of history as a direct result of paranormal

4   Woodrow Wilson National Fellowship Foundation (2019). Reimagining American History Education. *Institute for Citizens and Scholars.* <https:// citizensandscholars.org/wp-content/uploads/2022/07/WW-American-History-Report. pdf> (accessed June 5, 2025).

5   NAEP (2022). Results from 2022 NAEP U.S. history assessment at grade 8. *National Assessment of Educational Progress.* <https://www.nationsreportcard.gov/ ushistory/supporting_files/ushistory_2022_infographic.pdf> (accessed June 5, 2025).

6   Caplan, B. (2018). *The Case Against Education: Why the Education System is a Waste of Time and Money.* Princeton, NJ: Princeton University Press.

research, because we have become the *de facto* historians of many of the locations we've researched.

Along with the declining rates of historical literacy we already mentioned, another ghastly thing we recently noticed is that we seem to be losing our public historians faster than the younger generations are willing or able to replace them. Once upon a time, there were any number of local, state, national, or international historians possessed of encyclopedic knowledge of their specialty within history and a passion for sharing that knowledge with the world. Alas, many of them have passed away or retired. Younger people haven't given up the fight entirely, and perhaps one of the benefits of social media is that it gives some of these people access to an audience, but there do seem to be too few historians around to preserve the stories of the past before it gets lost.

On the preservation side of things, consider that some of the places we've written about in this series are no longer standing. Remarkable buildings with rich and storied histories are now remembered only because some people like us happen to be keeping those stories alive. Indeed, were it not for the ghost stories, we question whether anyone would have bothered to document some of these places at all.

A good example of such a location is the so-called "Changeling House." We wrote about that location in Volume 1, Chapter 5 as a sort of addendum to our research on Cheesman Park (which is across the street from where the Changeling House once stood). The house is now long gone, having been demolished to make way for, of all things, the parking lot of an apartment complex. It was a delightful old mansion which we think ought to have been preserved even just for architectural reasons, but it was also the home of a once-prominent family whose names most locals have by now forgotten. It's tragic that any person or any house should pass so quickly from memory. But through the purest good fortune, the house happened to have inspired a ghost story which itself went on to inspire the classic horror film *The Changeling*.[7] These days, this house and its people are remembered not so much in the formal history books but in ghostly works such as ours and a few others. Though we set out to learn about a famous ghost story, we accidentally became the honored custodians of a small piece of local history. It's a duty we don't take lightly.

A similar story will be featured in the very first case in this volume (which is the reason for including this essay here). A particularly famous and important building was set to be demolished. Though this location is unlikely to ever have been forgotten entirely, we realized that no one else was going to do the work of documenting its final days and preserving those records for posterity. Once again, we came for the ghost stories and became accidental historians.

Similar experiences are common in the educational side of our work. Whether in these books or in our public events, we often hear from people who came to our work because they wanted to hear a spooky story but who go on to thank us for telling them about some of the history they might never otherwise have learned. Especially when we speak to children, we find that we're able to "trick" our audiences into learning history by couching the history lesson in a ghost story.

7   If you haven't seen it, do so. It's one of the best ghost stories ever put on film and stars George C. Scott in, we think, one of his finest roles.

Teachers often enjoy bringing us out to speak to their classes because they know we have found a way to bridge the gap between exciting topics (ghosts) and "boring school subjects" like history. Not that we'd ever agree that history is boring except perhaps in the hands of a poor teacher. Even at that, though, sometimes you need to add some sugar to the medicine before children are willing to take it.

The same is true even for adults. One acquaintance of ours said she never could wrap her head around what the Teapot Dome Scandal was all about until she read our description of it as part of our explanation of the background to one of our ghost stories (Volume 2, Chapter 9).

Obviously not all of our cases are wrapped in deep and important lessons from history. Private cases, for instance, often have very little history that we're able to dig up. Nevertheless, it's a constant source of amazement for us how often we *do* find some rich historical treasure hidden beneath the layers of a paranormal investigation. Often, sometimes even in cases where we don't expect the location to have much history, an investigation crosses paths with some noteworthy event in history. One of our cases in Volume 2 (Chapter 12), much to our surprise given the location's own relatively mild history, tangentially intersected both a plane crash that may very well have been Osama bin Laden's inspiration for the terrorist attacks of September 11, 2001 *and* one of the earliest documented cases of a "Nigerian" advance fee email scam. This is strange work we've chosen for ourselves, but as we said before, we consider it an honor to be the unofficial custodians of these stories from our history.

1
Documenting History: Children's Hospital

Among the most common questions we get during the Q&A time at our public lectures is "what's the most frightening experience you've ever had on an investigation?" It's difficult to answer because different cases have been frightening in different ways and for varied reasons. But one contender is an experience we had in the case at hand. When telling the story, we can make it even more frightening by starting our tale with a simple but true statement that we once investigated a place at which as many as twenty *thousand* people died. As horrible as it sounds, and as much as that feeds into the ghost stories, it's less troubling when we explain that we're talking about a well-respected trauma hospital.

The Children's Hospital, now Children's Colorado, is a top-ranked system of pediatric hospitals. Its flagship hospital is now located on the Anschutz Medical Campus in Aurora, but at the time of our investigation was located at the intersection of 19th Avenue and Downing Street in Denver, and has since been (mostly) demolished. After the Hospital's move to their new location, but prior to its demolition, we were invited to investigate the myriad of ghost stories that were passed down by patients and staff throughout the Hospital's history.

Figure 1.1. Children's Hospital exterior (photo: Bryan Bonner).

The History

The Children's Hospital, now Children's Colorado, is a remarkable institution with a long history. As such, to do it justice would require an entire book of its own. Nevertheless, we'll do our best, as is our way, to present as detailed as possible of an abridged history. Part of the trouble in that endeavor, though, is determining exactly where to begin the story. Perhaps the best place to start is before anyone even had a concept to start such a pediatric hospital in Colorado.

The need for what was eventually to grow into such a massive institution started to become clear in the late 1800s. At that time, it was common practice for those suffering a variety of illnesses to come to Colorado in hopes that the often cooler, drier, and thinner air might ease their ailments. Indeed, longtime readers may recall from our prior books that F. O. Stanley, founder of the famous Stanley Hotel, came to the state with a similar goal of recovering from his tuberculosis (Volume 1, Chapter 14). There's a certain irony to this reasoning in that people *from* Colorado often fled the state for lower elevations and more humid climates for the same reason and both groups of people experienced similar levels of success. Likely the thickness of the air had little to do with anyone's recovery, but simply getting away to rest and to stop passing contagions around their communities worked wonders for them.

Regardless of the medical validity of their reasoning, a lot of ill people were arriving in Colorado at around this time. Recognizing the increasing need to care for these people, particularly the ill children who were arriving from out of state, a group of volunteers established what they called the "Babies Summer Hospital" in 1897, consisting of a series of tents stationed in City Park and staffed by six medical professionals and volunteer nurses.[1]

That's the story as it's told by the Hospital itself, and it's true as far as it goes, but it also reflects a deliberate omission of the substantial role played by one of the Hospital's founders.[2] Understandable, we suppose, as the individual involved is quite a complicated and controversial figure in Colorado history, but our belief at Rocky Mountain Paranormal is that history should be told as-is, for us to admire or condemn as our own morality requires but on the basis of a complete presentation of the facts.

The individual in question was Minehaha Cecelia Francesca Tucker Love, better known as Minnie C. T. Love (1855 – 1942). Born in La Crosse, Wisconsin on November 9, 1855, she married Charles Guerley Love on August 16, 1876 in Washington, D. C., then relocated to Denver in the late 1870s, where she would become involved in a number of social, political, and medical causes.[3]

Her work in medicine is quite well documented. After studying medicine at Howard University, she served for twelve years as the Chief Physician of the Florence Crittenton Home in Denver, of which she also served on the board of

1   Children's Hospital Colorado (n.d.). Our History. *Children's Hospital Colorado.* <https://www.childrenscolorado.org/about/history/> (accessed June 8, 2025).

2   Beaton, G. M. (2012). *Colorado Women: A History.* Denver, CO: University Press of Colorado.

3   State Historical and Natural History Society of Colorado (1927). *History of Colorado.* Denver, CO: Linderman Co.

directors. She would also serve on the Colorado State Board of Health and would eventually be elected to the Colorado legislature in 1921 (and, after an electoral defeat in 1922, again in 1924) in which capacity she served as the chair of the Committee on Medicinal Affairs and Public Health. In addition to medical issues, she was a noted member of the women's suffrage movement and a member of the Equal Suffrage Association, the Women's Christian Temperance Union, and the Denver Women's Club.

It was through her combined association with medical and women's organizations that she was one of the key figures (if not *the* key figure) to play a role in the creation of the Babies Summer Hospital which would ultimately become the Children's Hospital.[4]

Writing in the *Denver Medical Times* in 1906, Love herself explained the need to expand beyond a temporary tent hospital and the steps she and others were taking to make a permanent pediatric hospital a reality:

> An association for the purpose of maintaining a hospital for sick and crippled children was organized in Denver last April. A board of thirty-two directors was chosen and Mrs. Thos. S. Hayden elected president of the association, which was incorporated under the name of the Blanche Roosevelt Hospital for Children.[5] The movement, which is a direct outgrowth of the philanthropic work of the Woman's Club of Denver, and is endorsed by it, will, it is hoped, reach a degree of success that will fully repay the members of the club for the effort put into the preliminary work.
>
> The hospital association is composed of many well-known Denver ladies noted for their broad and efficient work among the poor and needy of the city. While they realize the magnitude of the undertaking, they are enthusiastic and determined because of the great need for such an institution in our city. The membership is rapidly growing, ladies from all sections of the city, anxious to assist in so laudable a work, becoming members and inviting their friends to do likewise. The women's federated clubs of the city have, many of them, signified their willingness to help, financially and otherwise. Donations of both money and furnishings come in almost daily, thereby holding the enthusiasm of the workers to a high pitch, and enabling them to see the goal of their ambition down the vista of the weeks and months of the very near future.
>
> The cause of sick and crippled children needs no advocate, and the proposition to properly care for them meets with ready and universal approval. As soon as sufficient money and furnishings are pledged to warrant a beginning on a modest scale, a suitable house

---

4   Beaton, G. M. (2012). *Colorado Women: A History*. Denver, CO: University Press of Colorado.

5   Blanche Roosevelt (1853 – 1898), born Blanche Isabella Pauline Tucker, was a noted opera singer and journalist best remembered as the first to perform the role of Mabel in *The Pirates of Penzance* by Gilbert and Sullivan. She took the name "Roosevelt" after her mother's maiden name, and it's unclear whether these Roosevelts were related to any other prominent Roosevelt families. She was Minnie C. T. Love's sister.

will be procured and fitted out with all modern appliances for the successful treatment of the acute and chronic diseases of children, other than contagious, of both sexes, from infancy to sixteen years of age. The management desire to begin with at least twenty beds, half of which shall be free. The charge for the pay beds will be at as reasonable rates as obtain in similar institutions, and will be at the disposal of the profession at large.

The minimum membership fee is one dollar, which was placed at this small figure in order to have a large body of workers. Life memberships are one hundred dollars each, of which five have already been subscribed, and one free bed, besides the furnishings for about ten memorial rooms. To what better use can one put money than to relieving the sufferings of little helpless children?

Any one familiar with hospital work can appreciate the necessity of an institution such as that proposed. To place a sick child in the room with older people who are ill is a detriment to both; yet a child cannot be placed in a room by itself unless a nurse be there at all times to care for it. In a hospital devoted exclusively to children the little patients are free from subjection to sights and sounds not best for them to see and hear, but which are unavoidable among the adult sick, and the older patients are relieved from the nervous strain of having a sick child about them.

This hospital is to be absolutely non-sectarian, and not in any sense a private enterprise. The staff will be selected from among the regular profession at large, men and women alike, solely with regard to their desires and training in this most interesting branch of practice—pediatrics. M. C. T. L.[6]

The excitement and growth Love anticipated were soon to follow. She was quickly able to raise substantial support, including donations in the amount of $5,000 and $1,000, respectively, from Lawrence Phipps and Thomas Patterson (about whom see more in Volume 2, Chapter 13 about the Croke-Patterson Mansion).

Despite these good works, Love's own legacy is more complicated and her name has been stripped from many of the histories written of The Children's Hospital because they don't wish to be associated with the more controversial elements of her life.[7] What controversy could cause such a benefactor to be removed from the history books? Well, it comes down to her support of two causes upon which history looks unkindly. First, she was a member of the Colorado chapter of the Women of the Ku Klux Klan (the KKK played a prominent role in Denver politics during Love's life, particularly in the 1920s, though the full details of that organization's involvement in local business and government is beyond the scope of this book).[8] Second, she was interested in the eugenics movement,

6   Love, M. C. T. (1906). A Children's Hospital. *Denver Medical Times*, *XXVI*(1): 27-28.

7   Beaton, G. M. (2012). *Colorado Women: A History*. Denver, CO: University Press of Colorado.

8   *Ibid.* See also:

and while that particular issue didn't yet carry the stigma during Love's time as it does in the post-World War II era, it is well known that Love herself advocated for the forced sterilization of the mentally ill and hosted meetings of the Denver Women's Club at her home to discuss the ideas of eugenics and "better breeding."[9] History is a messy business, and few people are purely good or purely evil, as Love's mixed legacy demonstrates so vividly. But the history of The Children's Hospital largely diverges from Love's own biography at this point.

Not long after Love's announcement, the name changed from the Blanche Roosevelt Hospital for Children (for reasons unknown) and was officially incorporated as The Children's Hospital on May 9, 1908.[10] Demand quickly outgrew the size of available facilities. In 1909, The Children's Hospital purchased a former residence at 2221 Downing Street (which had once been home to the Denver Maternity and Women's Hospital). This new facility admitted its first patients to some of its thirty beds on February 17, 1910.[11]

The new institution honored Love's promise to offer free beds to those poor children who could not afford medical care under other circumstances. One of the earliest such patients was a young girl named Constantia, a polio patient whose treatment quickly became a focal point for the media due to her quality of care and unexpectedly speedy improvement; she arrived at the Hospital bedridden but was able to walk on her own with the aid of a brace in as little as four months.[12]

Demand continued to outgrow the facility. At its peak, its initial thirty-bed capacity was overrun by as many as forty-five children at a time. Yet another new facility was needed. The Hospital raised a fund of $200,000 to purchase a new facility at the intersection of 19th Avenue and Downing Street in 1917. Initially consisting of one hundred beds and fully equipped with modern medical instruments, this new facility allowed The Children's Hospital sufficient space to care for current patients but also to expand into the massive and well-respected institution it would soon become. Despite Love's initial thought that the institution she helped to create would not treat contagious diseases (presumably out of fear of infecting other patients), the new facility would include isolation areas for patients with infectious diseases as well as all the most modern medical equipment, indoor plumbing, a modern kitchen, and a system of dumbwaiters. As a hospital, this new institution was dedicated both to treating patients and to medical research and continues its research mission today in affiliation with the University

Goldberg, R. A. (1981). *Hooded Empire: The Ku Klux Klan in Colorado.* Urbana, IL: University of Illinois Press.

Goodstein, P. (2006). *Denver from the Bottom Up Vol. 3: In the Shadow of the Klan: When the KKK Ruled Denver, 1920-1926.* Denver, CO: New Social Publications.

9   Bruinius, H. (2007). *Better for All the World: The Secret History of Forced Sterilization and America's Quest for Racial Purity.* New York: Knopf Doubleday.

Beaton, G. M. (2012). Colorado Women: A History. University Press of Colorado.

10   Children's Hospital Colorado (n.d.). Our History. *Children's Hospital Colorado.* <https://www.childrenscolorado.org/about/history/> (accessed June 12, 2025).

11   *ibid.*

12   Packard, C. S. (1934*). A Little Story of the Children's Hospital of Denver.* Denver, CO: Children's Hospital Association.

of Colorado School of Medicine.

The thirty current patients from the 2221 Downing location were hand-carried by a team of nurses to the new facility which celebrated its opening day on February 12, 1917. It was a good thing the initial rules regarding the care of infectious patients were changed with the opening of a newer and larger facility because the new Hospital's opening happened to coincide with imminent outbreaks of influenza (particularly the "Spanish flu" outbreak of 1918), tuberculosis, and smallpox.

Figure 1.2. This kidney dialysis room was once a tuberculosis ward (photo: Bryan Bonner).

Enter another pair of benefactors at a time when the Hospital was once again in need of expansion. Fortunately, the current property included enough land to facilitate this expansion without moving to yet another new facility, but funds were required to add a new wing to the Hospital. Such funding was provided in the form of a gift of $100,000 from Harry Heye Tammen (then the publisher of the *Denver Post*) and his wife Agnes Reid Tammen in 1920. The story of that donation is quite touching. Harry had given his wife Agnes a check for $100,000 dedicated to the purchase of a string of pearls for Christmas. At about the same time, The Children's Hospital had asked Mrs. Tammen for a donation of $1,000 as part of their attempt to raise a total of $50,000 for their new wing. She told her husband it would be sinful to spend twice as much on such a personal luxury as was needed for the hospital wing and asked if she could donate $50,000. Mr. Tammen's response has been reported as: "You never cease to amaze me. We will give the entire $100,000."[13]

A groundbreaking ceremony was held on June 6, 1921 and the new wing opened on February 16, 1924. When Mr. Tammen died in 1924 only a few months after the new wing opened, he bequeathed his entire estate to the hospital and

13   Children's Hospital Colorado (n.d.). Our History. *Children's Hospital Colorado.* <https://www.childrenscolorado.org/about/history/> (accessed June 12, 2025).

Mrs. Tammen continued to support the institution until her own death in 1942, after which the Tammen Trust has continued the same mission.[14]

The next great development also occurred in the 1920s. Hospital staff and patients' families realized that many of the patients were spending a lot of time in their hospital beds which necessarily meant they were also spending a lot of time away from school and thus missing out on their education. Therefore, volunteers from the local Junior League of Volunteers were brought in as teaching staff to provide lessons and tutoring for the patients so they wouldn't fall too far behind in their schooling. This model of education for the patients continued until the Great Depression brought it to an end. To fill that gap, the Hospital eventually contracted with the Boettcher School (for more about the Boettcher family, see Chapter 6) to teach the patients and a schoolhouse was opened in 1940 across the street from the Hospital, with a tunnel to allow patients to make the trek to their classes during the colder months.

An outpatient facility also opened in 1924 but similarly closed due to the Great Depression in 1930. But the 1924 expansion did include several houses on Ogden Street which were used for a variety of functions including an isolation hospital for potentially contagious patients and some were rented out to raise funds for the hospital's continued maintenance and operation.

To thank the Tammens for their beneficence (and using funds made available thanks to the Tammens' generosity), the Hospital planned a new Tammen Hall with a groundbreaking in late 1930 and which opened in 1932. Designed by noted Denver architect Merrill Hoyt (about whom we've written in several other of our case files; the man built a disproportionate share of Denver's noteworthy structures), this new addition included private rooms for the nursing staff, a library, classrooms, an auditorium, gym facilities, laboratories, and a delightful rooftop garden. It was primarily used as a nursing school, but was additionally expanded in 1936 to include a swimming pool as well as hydrotherapy facilities and salt pools for polio and tuberculosis patients.

A new X-ray and radiology department was added in 1932, once again thanks to the purchase of X-ray equipment by the Tammens (specifically Agnes, and Harry had passed by then). Agnes Tammen also provided the Hospital's first iron lung in 1937. Then, 1942 saw the addition of a blood bank which allowed for the first bone marrow aspiration in 1944 and the first exchange transfusion in 1947. The following year, the Hospital opened its infant surgery ward under the guidance of Dr. George B. Packard, Jr. Later, in 1965, this ward would become the Children's Hospital Colorado Newborn Center, known for treatment of birth defects and which would treat 1,000 infants per year by 1975. Unfortunately, the period of the 1940s saw severe staffing shortages due to World War II, but field surgeons were often able to bring techniques they'd learned in the War back home and apply them in the care of trauma patients at Children's Hospital.[15]

By this time, the Hospital was on sure footing and was becoming internationally recognized for its excellence. Much of the rest of its history is tied intimately with the history of medicine (particularly pediatric medicine) itself, and is beyond the scope of this book, but a few highlights are nevertheless worth

14   *Ibid.*
15   *Ibid.*

mentioning.

Dr. John Grow made history in 1953 when he performed the Hospital's first open heart surgery, after which time the Hospital would become known for such specialties as pediatric cardiac surgery. Another new wing opened in 1958. It was named the Oca Cushman Wing after the Hospital's first superintendent and added an additional 72,000 square feet of operating and recovery rooms to serve the Hospital's growing patient population. The same year saw the introduction of one of America's first child abuse identification and prevention programs under the direction of Dr. C. Henry Kempe and in conjunction with the Child Protection Team. Additional expansions in the 1960s and 1970s included a new building to house departments for speech pathology and audiology (1967), Colorado's first amputee ski school at Arapahoe Basin (1968), a separate adolescent department, Colorado's first burn program (1974), and more. Indeed, Dr. James Todd, working at Children's, discovered and named toxic shock syndrome at the Hospital in 1978.[16]

Despite those medical breakthroughs, the 1960s were apparently a period of financial loss for the Hospital and a consulting firm was brought in to help keep the institution afloat. Plans were developed (but ultimately abandoned) to move Children's Hospital to a new location and merge with the University of Colorado Medical Center. However, though Children's remained where it was, the partnership between the institutions was somewhat productive anyway and the University relocated its own pediatric department to Children's. The cooperation between Children's and the University of Colorado continues to this day—in 1991 several departments were merged and by the time we get to the end of this history section you'll see how integrated these two institutions have now become.

The Children's Hospital Colorado Foundation was established in 1978 with the sole mission of fundraising to advance the mission of The Children's Hospital.[17] Longtime Colorado residents probably remember the Children's Hospital telethons. These were a flagship program of the Foundation and undoubtedly helped the Hospital to continue and expand its operations throughout the years.

During those years, numerous celebrities, media personalities, and other prominent people have visited the Hospital. Various entertainers have taken time out of their schedules to visit or perform for the sick children. Once, an entire circus came just so the patients would have a chance to "attend" the circus. Movie and TV stars, costumed superheroes, and even Santa Claus himself often make such visits.

In 2005, Tammen Hall was designated a Denver Individual Landmark.[18] Then, though no longer a part of the Hospital (see below), it was added to the

---

16   *Ibid.*

17   Children's Hospital Colorado Foundation (n.d.). About Us. *Children's Hospital Colorado Foundation.* <https://www.supportchildrenscolorado.org/about/> (accessed June 12, 2025).

18   City and County of Denver (2023). Denver Individual Landmarks. *City and County of Denver.* <https://www.denvergov.org/files/assets/public/v/4/community-planning-and-development/documents/landmark-preservation/individual_landmarks_list.pdf> (accessed June 12, 2025)

National Register of Historic Places in 2019.[19]

However, despite being able to expand so many times and to achieve so many remarkable things over the course of the prior century, a decision was made in 2007 that the Hospital needed to be relocated to a new facility. The new flagship location is now part of the Anschutz Medical Campus in Aurora (with satellite facilities throughout the region). The old building was vacated on September 29, 2007.

It was at this time that Rocky Mountain Paranormal got involved in the case because after the Hospital had been cleared of patients and staff, we were called in to investigate the various paranormal claims the Hospital attracted over the years. We would have done it anyway, but as we worked, we realized that in addition to being paranormal investigators, we were needed as de facto historians because we were to be among the last people on site to document the Hospital's final chapter.

We'll get to our investigation in a moment, but one of the important things we realized as we worked was that many of the patients, nurses, and other staff had written their final goodbye messages to the Hospital on the walls on their way out. As such, we spent much of our time photographing and documenting those heartfelt messages because we knew no one else was going to preserve them for posterity.[20] While a complete list of those messages would be far too long to fit in this book, we wanted to take a moment to highlight a few (the remainder have been photographed and preserved in our own archives):

- On a wall in a patient named Maria's room: "15 days old to 15 years old you've been here for us" followed by (in different handwriting), "The best is yet to come!"
- In a speech bubble accompanying hand-drawn portrait: "Goodbye from Joshua! Thanks for making me better."
- Written by a staff member: "A part of me will always remain. Love to all."
- Above a nurses' station: "How many rounds have we had?"
- On a fifth floor wall (written in several distinct hands): "Things we won't miss: Ghosts, the dark cave, broke tube station, teeny rooms 62, 63, whooshing sound when the doors close, heavy doors that try to eat you, people sneaking around to the door that's locked and scaring you while you're working."
- In giant print to fill an entire wall: "CANCER SUCKS!"
- A mother marked her and her two children's (aged fifteen months and four years, respectively) heights on a wall.

19   Tomasso, D. W. (2019). Tammen Hall: National Register of Historic Places Inventory – Nomination Form. [Register No. 100004612]. *National Register of Historic Places*. < https://npgallery.nps.gov/GetAsset/8e8414ff-eb50-46ad-b91b-d645e6daa0af > (accessed June 12, 2025).

20   While we ordinarily disapprove of graffiti, these heartfelt messages written on the walls of a building already slated for destruction and with the full knowledge and cooperation of patients and staff alike are an entirely different matter and we thought they needed to be documented as an important piece not only of the Hospital's history but of the lives of so many people.

- "May God bless all who enters this room."
- "How many loves have we cared for in here?!?!"
- Above a time clock: "How many hours have we all clocked here?"
- "In this room…Nathanyel learned to sit up, walk, sing to his nurses, & say 'bye-bye'…what a love bug."
- "This is where the magic happens!"
- With an arrow pointing to a built-in bench: "Mom's spot."
- "Nurse Nancy fell down here."
- A list of "Our Children" with many names. One named Amanda has a halo drawn above her name along with "God Bless You" written below. She was one of the children who sadly never left the Hospital.
- "Goodbye Children's Hospital. I love you all!"
- And finally: "Last patient out @ 1715 9/29! Yah, we made it!"

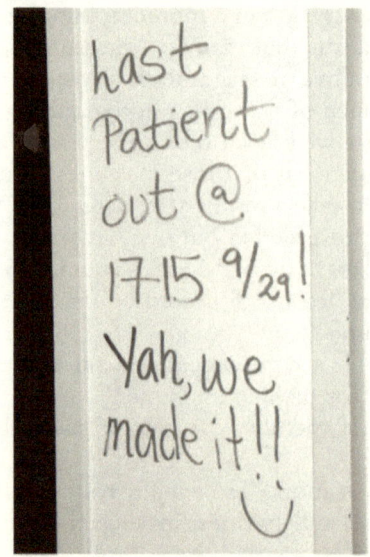

Figure 1.3. "Last patient out" message at Children's Hospital (photo: Bryan Bonner).

Once the final patients and staff had been moved to the new facility, the old building (but not all of the ancillary buildings like Tammen Hall and the administrative building) was slated for destruction. Everything was stripped out either to be brought to the new facility or to be sold, donated, or thrown away depending on the value, condition, and nature of the item in question.

The new facility started treating patients before the old facility closed to allow for as seamless a transition as possible. Thus, the history of Children's Hospital continues even though we bring our present history to a close because our case file extends only to the old building. One thing you may have noticed as you read those messages left on the walls, though, is that one of the things the staff said they wouldn't miss about the building was "ghosts." Indeed, there are plenty of ghost stories connected to the Hospital and it is to those that we now turn our attention.

Paranormal Claims

Ghost stories at Children's Hospital might not be the first thing on a patient or parent's mind when they checked in for medical care, but if you think about it for a while, it should probably come as no surprise that such a place would be filled with tales of spirits. Indeed, we think most hospitals around the world, if they've been in business for more than a few years, probably have some ghost stories. They may not be well known or popularized, but any place that's the site of so much illness, sadness, and death is bound to attract such tales.

Believers in the paranormal will say this is either just probabilistically true or that the character of such places becomes such that paranormal events are more likely. The former interpretation goes something like: even if only a small minority of people who die come back as ghosts, locations where so many people have died because of the nature of the location's work must have a disproportionate share of ghost stories. And the latter interpretation might suggest that the emotional "energy" (not that we like that word in this context, but it is the word that's typically used) in the place can attract spirits or cause spirits to linger.

Skeptics also won't have any problem understanding why places like hospitals should have ghost stories. Their own interpretation would suggest that even if there's no such thing as a ghost, the number of deaths and the amount of tragedy in a location is sure to help it develop a haunted reputation for psychological reasons.

Whichever interpretation you like, Children's Hospital is going to be a great example. Ghost stories at Children's Hospital go all the way back to the very beginnings of the institution, and almost every room and hallway is the site of some paranormal claim or other. Many are just vague claims of weird feelings or someone who once saw something out of the corner of his or her eye and so are difficult to document in any sort of detail, but a few of the tales are a bit more fleshed out. In both the vague and the specific ones, though, it can be somewhat difficult to trace the exact origins of the stories because they've been passed down by staff members and patients alike in a sort of oral tradition for generations. Almost anyone who spent enough time at the Hospital could repeat at least one story of phantom nurses and children haunting any given portion of the building.

One legend tells of a nurse who lived in Tammen Hall when it was used as housing for the nursing staff. According to the tale, she was pushed down an elevator shaft by her fiancé. Another variation of the story says that rather than her fiancé, a construction worker attacked her while the building was under construction. He is supposed to have abducted, raped, and killed her and then buried her at the bottom of the elevator shaft. Despite their marked differences, both stories do insist she was a bride-to-be and that she now haunts Tammen Hall.[21] Witnesses fortunate (or unfortunate, depending on whether one is scared of ghosts) enough to encounter her spirit will then later come across a wedding

---

21   Though most of the old Children's Hospital was demolished following our investigation, Tammen Hall itself and still stands (as do a few other parts of the campus), so the present tense is appropriate in this case. We don't make any claims

dress lying in one of the rooms of Tammen Hall.

Another story reported to us, this time from the main Hospital building, came from a medical researcher who reported seeing a "strange woman" at the end of a third-floor hallway. As one might do, she walked down the hallway to investigate, perhaps to see if the person was lost or needed some assistance, but she insists that just as fast as she noticed the person and moved to approach, the woman "just vanished."

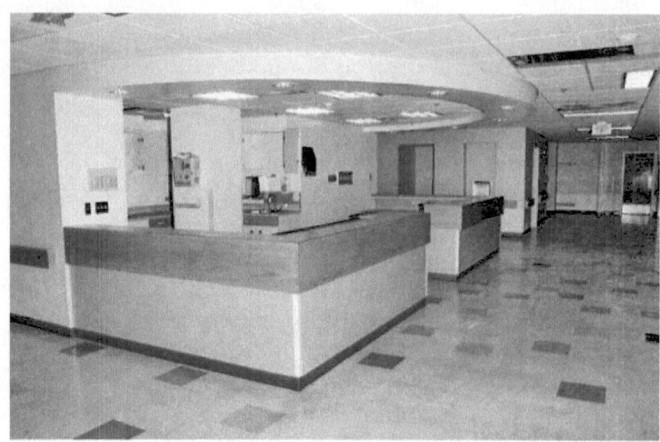

Figure 1.4. An abandoned nurses' station (photo: Bryan Bonner).

Even though most of the deaths at Children's Hospital were, unfortunately but expectedly, those of children, one thing we noticed is that many of the ghost stories we've heard over the years about the location centered not on children but on nurses. Perhaps children might be thought less likely to become ghosts, or perhaps the lore just associates hospitals more with ghostly nurses than children. Whatever the reason, we learned of many such stories of nurses haunting those hallways.

One worker installing wiring during a period of construction said that whenever he was working alone, lights would turn on and off and doors would open and close on their own. He became convinced this behavior was caused by the ghost of a nursing student who'd taken a liking to him. But even though, if that's true, this would have been a friendlier rather than threatening spirit, he nevertheless refused to work alone again for the duration of his job.

Another nurse is alleged to have committed suicide sometime in the 1950s in Tammen Hall, shortly after it opened. According to those rumors, her name was Anita—there's even a room named after her—and her spirit is said to still wander those halls.[22]

Something we've noticed over the years is that police and other first responders often skew toward the superstitious side. That observation is going to

regarding what might happen to any of the alleged spirits who haunted the now-demolished portions of the Hospital now that their "home" no longer stands.

22   Denver Terrors (2021). Denver Children's Hospital. *Denver Terrors.* <https://denverterrors.com/denver-childrens-hospital/> (accessed June 12, 2025).

be particularly relevant both later in this chapter and in Chapter 4. Perhaps the same is true also of security guards, as it's from a member of that profession we (indirectly) heard another story. While he was doing his rounds, he stepped out from an elevator (reports vary as to which floor the elevator was on at the time) and saw a "foggy entity" which looked at him and then passed through a wall. He immediately went to his boss and quit his job on the spot.

Speaking of elevators, we're not sure whether it's the same elevator as the security guard's story mentioned, but there's also a claim that the elevator which used to connect Children's Hospital proper to Tammen Hall often moved from floor to floor, seemingly of its own accord (or perhaps operated by an incorporeal person instead of a living one).

Figure 1.5. Abandoned waiting room play area (photo: Bryan Bonner).

Many additional stories tend toward the vague side of things. Plenty of people have reported cold spots throughout the Hospital. Others have described seeing, hearing, or feeling things they attributed to the ghosts of both children and nurses alike. Though stories were attached to just about every corner of the entire hospital, a disproportionate number seemed to attach themselves to the fifth floor.

Of those—and indeed, of all the ghost stories of Children's Hospital—by far the most prominent is the story of Shane. Reports claim that sometime back in the 1950s, a patient named Shane stayed in Room 562 at the Hospital, part of the Oncology department. Unfortunately, despite the best efforts of all the medical staff and the high quality of care for which the institution is known, Shane is said to have died in his room. Ever since—for the decades spanning whatever year that was in the 1950s right up until the Hospital's move in 2007—children staying in Room 562 would wake up in the middle of the night screaming or crying. They all told a similar story: when they woke up, they saw another child in the room with them. Often that other child was sitting atop their chest and trying to choke them.

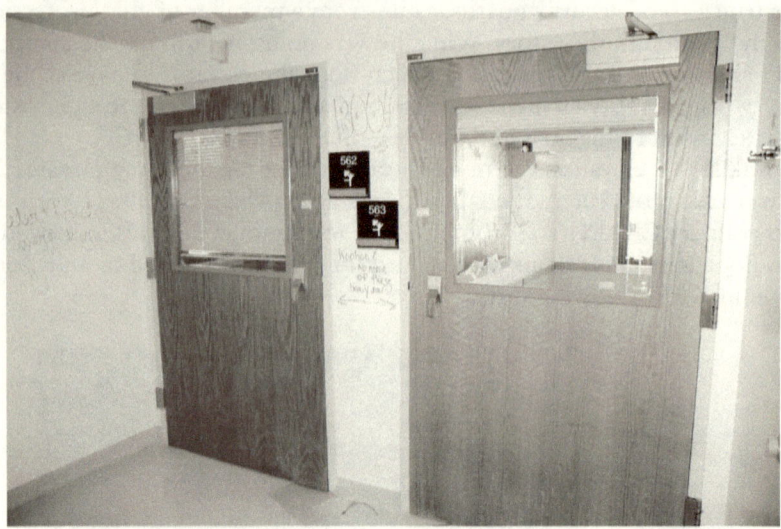

Figure 1.6. Shane's room. Note: someone has written "boo" on the wall (photo: Bryan Bonner).

If that's not scary enough—we think it is—there's even more. The stories told by these children, sometimes decades apart from each other, shared details. Children describing these experiences invariably described a young boy with blond hair and blue eyes as the ghost or culprit. Some might suspect that the children learned of the story somehow and then either made up their own experiences or had a nightmare. If that were the case, it would certainly explain the commonalities between their reports; after all, they'd just be repeating what they'd heard from others. But we have it on good authority that the nurses never told their patients the story of Shane before any of these experiences.

For what it's worth, though we have no way of knowing for certain, we believe the nurses that they never said anything to the children. Nurses caring for ill children are not ordinarily in the business of frightening their patients with ghost stories. So while "repeating a story heard elsewhere" is often a solid explanation for two ghost stories sharing some details in common, we don't think that's a good explanation in this case.

Spookier still, at one point after we began working on our investigation into the paranormal claims at Children's Hospital, a woman reached out to us who'd heard that we were looking into the potential ghost of a young boy named Shane. She told us she had a relative die at Children's Hospital in the mid-1950s. He was a young blond boy with blue eyes named Shane, and she thought—but couldn't absolutely confirm—that he'd been treated and ultimately died on the Hospital's fifth floor. Unfortunately, this is one of those stories we have to just present for whatever it's worth because we have no way to confirm or disconfirm it. Records, if they still exist, would be confidential and we'd have no means to access them.

But as we bring our ghost stories to a close and prepare to describe our own investigation, we'd like to give the Hospital nurses and staff the final word on the Shane story. As we mentioned in the previous section, nurses and patients had written their final messages on the walls of the Hospital before moving to the new location. On their way out, they decided the most appropriate thing to do would be to leave a final farewell for Shane himself. The message on the wall of Shane's room consists of a smiley face along with the words, "Good night, Shane!"

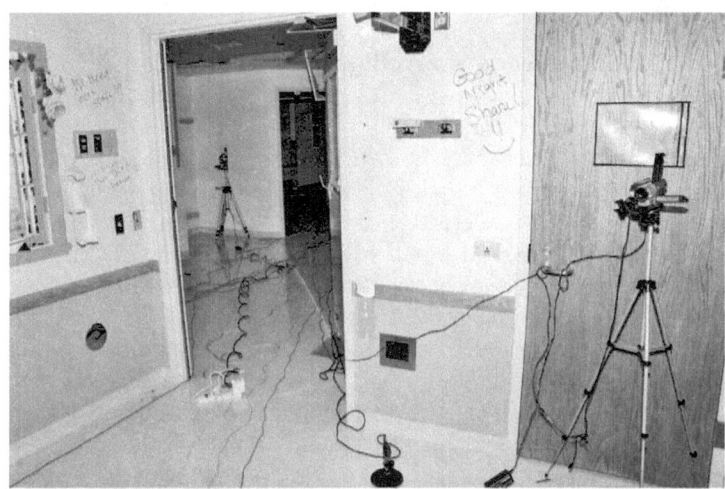

Figure 1.7. Good night, Shane! (photo: Bryan Bonner).

Our Investigation

Though we had long been aware of the haunted reputation of Children's Hospital, we never made much of an effort to schedule an investigation while the hospital was open for business. The prospect of trying to investigate paranormal claims in such a busy facility never really appealed to us.[23] Worse, bringing in a large crew to run around looking for ghosts might have either frightened the children or, worse still, served as a reminder of mortality to the desperately ill children and their families. Our senses of humor may be sick and twisted, but not *that* sick and twisted.

Patience is often generously rewarded, however. In 2007, after the staff and patients transferred to the new facility, we received a communication from the company the Hospital contracted to complete the demolition and asbestos abatement at the old facility. Like us, they were aware of the facility's haunted reputation and thought it would be a good idea for a paranormal investigation team to check it out before the building was torn down. They also knew of our interests

---

23   We've done cases almost that difficult in the past. For example, the Bullock Hotel in Deadwood, South Dakota (Volume 2, Chapter 4) remained open for business as a hotel and casino during our investigation. But unlike a hospital, where any given room might be urgently needed for patient care, we were able to carve out some places to conduct our work alone in that case.

in history and photography and thought it would be wise to bring someone in who could document the building's final days. Thank God they did because we completely agree with both of those assessments and, as we've discussed earlier in this book, we consider it an honor to be the custodians of history which might otherwise have been forgotten.

In the case of Children's Hospital, the "big picture" history is of course well-documented elsewhere, but no one had bothered to photograph all those messages the patients, nurses, and other staff wrote on the walls on their way out. Those heartfelt messages deserved to be remembered, and we were excited to take part in the project even just for that reason, whether or not we'd have any success on the paranormal front.

Our team met with several employees of the contracting firm a few weeks before our investigation was to take place. During this time, we accomplished several objectives. First, they were able to tell us about their own operations. The building was structurally sound but demolition had already begun in the sense that many large appliances were being removed and power had been shut off, so we needed to know about the conditions under which we'd be working. They also told us the building was under constant guard by off-duty police officers. More about them shortly. For now, the important point is that this advance meeting gave us the opportunity to plan our investigation. We discussed which areas we could monitor and inquired as to whether any were off limits to us. Because hospitals are such large facilities and we couldn't possibly cover every inch of the place, we needed this opportunity to plan. At the same meeting, they also told us what they knew of the paranormal claims (which we have incorporated into the previous section).

We scheduled our on-site investigation to begin in late February, 2008, shortly before the final demolition was to be complete. Having some extra time between our initial consultation and the on-site investigation proved to be a god-send for planning purposes. It gave us some opportunity to begin historic and public records research that might affect our approach to the project.

One thing we were particularly interested in was the nurse, perhaps named Anita, who supposedly committed suicide in Tammen Hall. Though Tammen Hall was not one of the properties to which we were going to have unfettered access during our investigation, we wanted to know to what extent we should gear our research toward those claims. Public records and news archive searches, however, yielded no specific information about a nurse suicide on the property—either in Tammen Hall or the Hospital proper. Neither were we able to find any records of a nurse named Anita who may have worked there and committed suicide elsewhere. None of that is to say the story is untrue, but we couldn't find any corroborating evidence, so we're confident at the very least that if there is some truth to the story, sufficient details have been changed over the years that the rumors now circulating among paranormal enthusiasts bear little similarity to reality.

Similarly, we were unable to find any records mentioning a nurse who may have been murdered by her fiancé on the property. Again, it's not impossible that there could be truth behind the legend, but such a murder ought to have been documented in newspaper articles and public records, so our inability to find anything suggests it is mere legend. That said, we don't consider this preliminary research to be dispositive when it comes to the paranormal claims. Perhaps the ghost stories are true but only details about the history have been misreported, so we still wanted to look into everything on-site with open minds.

The last of the "specifically named" or potentially identifiable ghosts we were aware of is, of course, Shane. But privacy laws would prevent us from being able to access any patient records so it's quite impossible for us to perform the kind of records search we'd typically like to do in a case like that. Unlike a murder, a child who died of natural causes might not make the newspapers and wouldn't be described in publicly-available police documents. All we really had to go on there was the one story we heard from a family member who thought the description might have matched someone in her family, but this cannot be confirmed.

After the investigation, if we can be forgiven for breaking the linearity of our narrative, we did learn that the Denver Public Library offers an index of obituaries that have been published in the *Denver Post* and/or the *Rocky Mountain News*. Given Shane was supposed to have died sometime in the 1950s, we searched this index for the name "Shane," focusing on the years 1945 to 1964 (to give ourselves a bit of extra wiggle room in case the reported decade wasn't quite right). We found several entries whose *surnames* were Shane, McShane, Shaner, or other comparable variants, but we could not find any individuals (of any age) whose first name was Shane.[24] That was somewhat surprising to us. Never mind whether or not we could find the specific Shane we were looking for; we were surprised to find that there were no Shanes at all! Our next stop, therefore, was to a searchable index of baby name popularity by year, where we found that Shane as a first name, while not unheard of, did not become popular until a couple decades later. Specifically, in the 1940s, there were five Shanes per million births, increasing to sixty-three per million in the 1950s, and only peaking at 1,526 per million in the 1970s.[25] Given that Shane, according to the ghost stories, would have been a pediatric patient in the 1950s, we assume he would have been born, most likely, in the 1940s, making him (by name) a five in a million rarity.[26]

Again, this doesn't disprove the story. The lack of obituary could simply mean no obituary was written—not every death is so reported. And the relative rarity of the name Shane doesn't disprove anything, either. Our Shane could have been one of the first to use that name in the United States. Or, given that Shane is an Anglicized derivative of the Irish name Seán, it's entirely possible the story could be true in all respects *except* the exact spelling of the name or the exact decade of his death—perhaps instead of the 1950s it was the 1960s or 1970s. That we had an individual tell us she thought the individual described in the story matched someone from her only family but couldn't remember all the details lends some degree of credibility (again, credibility is not the same as certainty

24   Denver Public Library (n.d.). Denver Obituary Project. *Denver Public Library Special Collections and Archives.* <https://history.denverlibrary.org/research-tool/denver-obituary-project> (accessed June 26, 2025).

25   Engaging Data (n.d.). US Baby Name Popularity Visualizer. *Engaging Data.* <https://engaging-data.com/baby-name-visualizer/> (accessed June 26, 2025).

26   It's likely the use of Shane as a given name was popularized by Jack Schaefer's 1949 western novel *Shane* and the 1953 film adaptation of the same title.

or even probability) to the latter interpretation. Similarly, that numerous patients reported similar experiences of "Shane" over the years without having the opportunity to coordinate their stories lends some degree of credibility to their testimony, regardless of whether we eventually look for a natural or paranormal explanation.[27]

Returning to our on-site investigation, we arrived in the evening on Friday, February 22, 2008 at about 6:00 to meet the work crews and begin setting up. Upon arrival, we had our first interaction with the police officers who were guarding the facility. They'd been hired as around-the-clock security to deter both curiosity seekers who might want to explore the hospital as well as would-be thieves who might want to steal large amounts of copper in the building. Furthermore, because the power had been disconnected, the building had no operational fire alarms and so on-site personnel were mandated at all times for safety reasons. Our own entry was delayed as we watched officers conduct an arrest right as we first arrived. After it was all over, we were informed an employee of the contracting firm had been caught loading his truck with copper from the facility. Never a dull moment in paranormal investigation!

Once things settled down, we made introductions between ourselves and the officers with whom we'd be sharing the building. Earlier, we mentioned that police and first responders stereotypically have a bit of a superstitious streak. It was on full display here. Several of the officers said they didn't like to patrol areas reputed to be haunted. Some said they even refused to visit certain areas. One of them walked right up to us and said (paraphrasing here): "Thank God you guys are here. There's something about that fifth floor. If I go up there and I see *anything*, lead's flying."

Given the fifth floor was where we planned to spend a good portion of our time—that's where Shane's ghost was supposed to be—we had some ambivalent feelings about this. On the one hand, vigilant security would be welcome. On the other hand, we didn't want to be accidentally mistaken for a ghost by a potentially trigger-happy peace officer.

The officers did provide us with some eerie information, though. One officer, during our initial interviews, told us that every single one of them was aware of the ghost stories and that certain officers avoided different areas in the building because of "odd feelings" they would get. Another told us that, a few weeks prior, several officers observed lights turn on in the third floor of Tammen Hall late at night when nobody else was in the building.

A few visitors stopped by with various types of business during the earlier part of the evening. Some were probably just monitoring the project, but word also seemed to have gotten out among the people involved in the Hospital's relocation that the paranormal investigators were in town. One prominent member of city government, District 8 Councilwoman Carla Madison (1956 – 2011) even volunteered to join us and spent some time alone in Shane's room to see if anything might happen.

Once introductions were made, we walked the building with the work crew to determine appropriate locations for monitoring as well as safety policies and

---

27   On the other hand, that we received their testimony second- or third-hand raises questions of credibility in its own right.

considerations and the locations of potential power supplies. Recall the electricity had been shut off in the entire building, so we determined the nearest source of electricity was the front of a newer administrative building (which still stands) and we ran hundreds of feet of heavy duty extension cord from there to our base of operations on the fifth floor (specifically, in Room 560—near but not in Shane's room) and subsequently to our various cameras and other equipment.

Figure 1.8. Our base of operations (photo: Bryan Bonner).

Because the building was so large, we knew there was no way we were going to be able to cover the entire location even with the expanded crew we brought to this investigation. However, we also had an opportunity we didn't ordinarily have: because the building was completely evacuated and the only other people in or out were police officers and contractors with whom we were in contact, we were able to leave a lot of our equipment mounted even when we weren't on site, allowing us to conduct a multi-day investigation without losing additional time breaking down and setting up each day. Over the years, we've become quite efficient at setting up and tearing down large amounts of equipment faster than you'd think possible, but it still takes some time, so this was a great opportunity for us.

We chose several key areas to monitor. Of course the most important of those, given the ghost stories we'd read and had been told, was Room 562, also known as Shane's room. In that location, we mounted three video cameras, two mounted toward the rear of the room and facing the door and one near the patient restroom and pointed toward where the patient bed once would have been (it had been removed prior to our investigation). Additionally, this room was fitted with two microphones and a seismometer to determine if there were any unusual vibrations. Finally, we placed four "control objects" consisting of three foam balls and a stuffed animal in the room under the constant surveillance of one additional video camera.[28]

---

28   So-called "control objects" are any objects placed under surveillance in a supposedly haunted location. Particularly when the ghost stories involve children who might want toys to play with, the thought is that the spirits might move the objects.

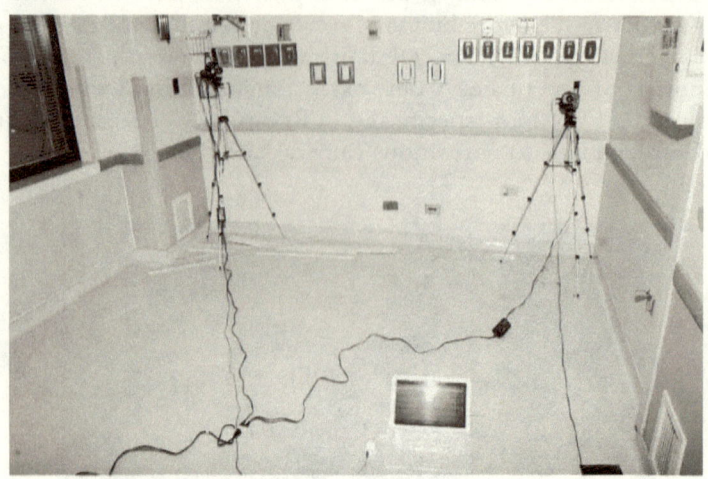

Figure 1.9. Our set-up in Shane's room (photo: Bryan Bonner)

Most of the rest of our monitoring was likewise on the fifth floor. We out-fitted the neighboring room, Room 563, with four video cameras and two control objects consisting of a foam ball and a stuffed animal. One additional camera and one additional microphone were placed in the hallway overlooking these two rooms. In the fifth-floor pharmacy, we placed two video cameras and one stuffed animal control object. Just outside the pharmacy, the hallway was monitored by one video camera. Across from the pharmacy was a patient activity room where we placed one camera and a stuffed dinosaur as a control object. Finally, we chose to monitor Room 555 with a video camera and a control object consisting of a teddy bear specifically because we were unaware of any paranormal claims asso-ciated with that room so we planned to use it as a sort of control case.

The process of setting up was a bit more eventful than usual. In addition to the unusual circumstances involving running extension cords from outside the building to power our equipment, the lack of electricity and heat posed additional challenges and slowed down our work. You try setting up a complete surveil-lance system in a haunted hospital in the middle of the night in the pitch dark in sub-zero temperatures[29] and see how fast you work!

Unusual occurrences also plagued our set-up process. Normally, we're a well-oiled machine, able to get everything in place and operational without any trouble. In this case, one of our members had trouble getting the video feed of our seismograph in Shane's room to display on our monitors. When he went back

In our investigations, we photograph their exact locations and orientations before and after the investigation and also keep them under constant video surveillance so even the slightest of movements can be detected.

29   For reasons we've never quite understood, it seems to have become a tradition that when Rocky Mountain Paranormal investigates, the weather doesn't like to cooperate. On numerous investigations, we've found ourselves working in the middle of a blizzard. In this case, though we weren't snowed in, we did find ourselves in an unheated building during a cold spell.

in the room to check (while another member watched the monitors so they could relay any information back and forth), he discovered that not only was the A/V cable disconnected (which we *could* theoretically write off as an oversight due to our working conditions, but we would also swear we genuinely *thought* we had it connected), but the cable had been carried completely out of the room.

Because of all this trouble, we weren't able to complete our set-up until about 1:00 a.m. Most unusual for us. But one nice thing did come out of that. While we were monkeying with all the equipment, we got the chance to get used to our surroundings and develop a strong "sense" of the building. We also took temperature and EMF[30] readings throughout the set-up process (and, indeed, throughout the entire investigation). Though we kept checking these readings throughout the entire process, we can jump the gun a bit now and point out that EMF levels were uncharacteristically low, consistent with a building whose power had been shut off, and did not fluctuate noticeably throughout the investigation. Temperatures dropped throughout the evening due to decreasing external temperatures and a few missing windows in some of the rooms. Other rooms were reasonably well-insulated and their temperatures only decreased slightly as the investigation wore on.

Figure 1.10. A padded room in an abandoned hospital is decidedly creepy (photo: Bryan Bonner).

Once the investigation started in earnest, we employed our usual method of sitting quietly for most of the time and monitoring from our base of operations. We've explained elsewhere why we use this approach, but it served us particularly well here because we weren't more than an hour into this process before we started hearing a lot of unusual sounds.

30   Electromagnetic fields. See previous volumes for a detailed description of how they relate to both supernatural and natural explanations of paranormal claims. In brief: some believers think paranormal entities give off EMF radiation, while skeptics point out that elevated EMF levels can, in certain individuals predisposed to be sensitive to them, cause hallucinations or feelings of uneasiness (see Chapter 16). Low frequency vibrations can have similar effects, which is part of the reason we used the seismometer.

First, we heard sounds we would describe as akin to banging and footsteps coming from Shane's room (of course when no one was in there). Because of the demolition process, there was a lot of debris on the floor, and footsteps could easily be heard as people would shuffle, snap, and crack their way through a room or down a hallway. The sounds we heard from Shane's room were similar to the sounds we heard when someone was walking around. However, nothing showed up on any of our video feeds of the room during these experiences.

At about 2:00 a.m. (Saturday morning now), we heard similar sounds coming from the hallway just outside our base of operations. A couple of our members stepped out to investigate and saw a human silhouette at the end of the hallway, moving toward the stairs (recall, there was no power so elevators were out). After a moment's wishful thinking that we'd witnessed a ghost, we quickly determined it was one of the police officers making his rounds. That turned out to be just what the doctor ordered, though, because it gave us an opportunity to compare the sounds of his footsteps to the sounds we'd been hearing in Shane's room to see if we could tell any difference between them. Unfortunately, this experiment wasn't quite conclusive. The sounds were similar enough to think they might have a similar cause, but they were not exactly the same and we couldn't be certain about much more than that.

About another hour later, one of our members was volunteered to sit in Shane's room in person.[31] We wanted to see if he would hear the same sounds "live" in the room as the rest of us were hearing through our headphones back at base camp. Five minutes in, and he was already reporting lots of sounds from all different areas of the room, but there didn't seem to be much of a pattern to them. Some but not all of the sounds he reported were audible in our headphones.

Field investigation is tricky business. We always approach an investigation with a solid game plan before we start. Usually, as we've mentioned, that involves a lot of sitting in silence and monitoring. But because we are out in the field, we have to be ready to modify those plans and adapt to changing circumstances. Here, we knew we needed to quickly come up with an experiment we could perform that might help us get to the bottom of the sounds we were hearing. It wouldn't do to come back another night with additional equipment—we might do that anyway, but there would be no guarantee the phenomena in question would continue on another evening, so we had to act now.

Our new experiment was simplicity itself: two members of our team would *both* sit silently in Shane's room, on opposite ends of the room but both in full view of our cameras. If either of them heard a sound, they were to point in the direction they thought it came from. Buildings, especially old ones, do have a tendency to expand, contract, and to pop and groan during times of changing temperatures (which this evening certainly was) so part of our goal was to use these two investigators to triangulate the source of sounds and determine whether we were hearing footsteps from somewhere else in the building, sounds of the build-

---

31   A standard practice we've mentioned before: when someone gets tired and needs a rest during an investigation, our standard practice is to make that person "bait." He or she has to go rest in the most haunted location, sit still and in silence while the rest of us monitor on video and audio, and see if anything happens.

ing itself settling, or something completely different and potentially paranormal. Throughout the experiment, both individuals heard numerous sounds, but they didn't report the same sounds or coming from the same locations even though they were only feet from one another. The mystery continued.

For the remainder of this narrative, the reader should be advised that this investigation continued until Sunday morning of February 24; that is, this investigation took place over two long nights. From now on, we're condensing both evenings' happenings into a single narrative for ease of reading.

At one point, another member revisited the same room to continue experimenting solo, but the sounds stopped during his occupancy. Disappointing from an investigative standpoint but it did allow for a refreshing nap.

Later, after the naps and experiments were completed, we heard sounds again, this time coming from Room 563—the neighboring room to Shane's—and the adjoining restroom. Additional sounds of footsteps came from the hallway. These continued in fits and starts throughout the entire duration of our investigation and we never did get to the bottom of them.

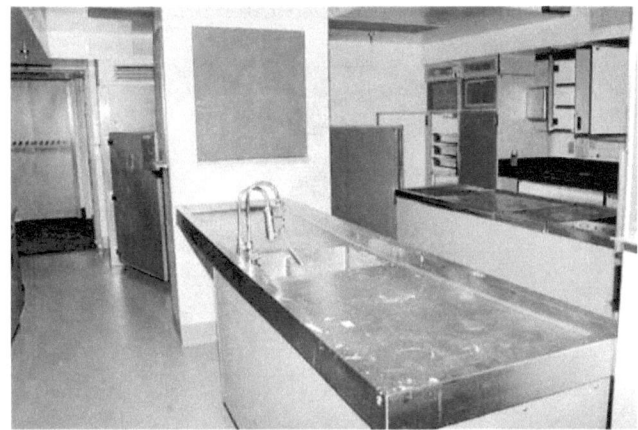

Figure 1.11. The Children's Hospital morgue (photo: Bryan Bonner).

While returning from the nearest functional restroom (in another building), one of our investigators thought the rest of the team played a practical joke on him. Behind him, someone—or, indeed, some*thing*—made quite a ruckus, stomping about with what sounded like heavy footsteps and perhaps something crashing about. Every team member, though, was accounted for, the room in question was completely empty (it wasn't one of the ones under direct video surveillance but the hallways were and we double-checked the room in person to make sure no one got in there undetected somehow), and the sound remains completely unexplained.

Toward the end of the investigation, one of our volunteers went to Room 563 to see if we'd have any better luck identifying the sounds coming from that room. Once again, the sounds continued but couldn't be pinpointed or explained. At one point, another investigator joined and repeated the same experiment we described above. And once again, the two individuals signaled sounds at different times and from different locations. The monitors at our base of operations heard all of the sounds either individual reported but we couldn't gain any additional information about their potential source(s).

Twice throughout the investigation (once on each evening), our team members all left for a quick meal break, welcome both as an opportunity to rest and regain some energy and as another investigative opportunity. While our people were off site stuffing their faces, we left the now *completely* abandoned rooms under video and audio surveillance to see if anything might happen in our absence. Video footage showed a whole lot of nothing, but the audio revealed the same sounds continued even when the building was completely empty.

Something that's important to understand about this investigation is that in some ways, it really ticks all the boxes for our inner horror *and* history nerds. On the latter point, we've already discussed the great honor we felt at being tasked with documenting the location's final days and the heartfelt messages written on the walls, so we did spend a fair amount of our time photographing all of these "graffiti" (albeit legal and approved graffiti). But on the former point, what the reader needs to realize is that conducing this investigation was like stepping onto the set of any number of our favorite horror movies.

Figure 1.12. The abandoned Children's Hospital looked like the setting of a horror movie (photo: Bryan Bonner).

Ghost stories are fun in and of themselves. Being in the location where they take place and trying to solve the mysteries is even better. Best of all, though, is being in a freezing cold, pitch dark abandoned and allegedly haunted hospital in the middle of the night! If you've ever seen a horror movie set in some abandoned haunted hospital, this place really did look exactly like what you've seen in the movies. Debris littered the floors. Windows were missing from their frames. Discarded medical equipment lay strewn about in the rooms.[32] The only light

32   As a matter of fact, with a few limitations like not touching any of the copper which they planned to reclaim and recycle, we were told we could take plenty

we could see in the entire building came from our own shop lights and lanterns. Walking those dark hallways with just a beam of self-provided light in front of you, headed into some dark and unknown abyss is the kind of experience we all live for…but it's also the kind of experience that puts one in a certain frame of mind that could easily color how one interprets events.

Usually, we're pretty good at keeping our thinking straight even in these kinds of situations. Decades of experience and plenty of training (formal and informal) have seen to that. But there was one experience toward the end of this investigation that remains one of the most frightening experiences we've ever had on a case.

Cold temperatures and long hours meant we had to take periodic breaks to stretch our legs and warm ourselves up. During one of these breaks toward the end of the investigation, the entire team decided to go at once to walk around the fifth floor, taking a few measurements along the way, and then rest for a moment in a neighboring building that had functioning heaters and coffee machines. As we walked down a hall which looked very much like the one depicted in Figure 1.12 (though this shot was taken at a different moment), many of our thoughts turned to some of those old horror movies we'd seen. In other words, we were psychologically primed to give ourselves any kind of scare at the slightest provocation.

Provocation came in the form of a loud "crunch" sound coming from the end of the hallway ahead, still shrouded in darkness. Everyone present heard it and all agreed that it sounded like someone had stepped on some glass lying on the floor. We all got a bit excited because we knew our entire team was together so our first thought was that maybe some ghost was making noise at the end of the hall. Maybe or maybe not would that be your own first thought, but recall the conditions under which we were working and you'll understand it. So of course we did what any paranormal investigators would do: we continued our march at a slightly quicker pace. Just as we rounded the corner, we again heard the sounds, this time almost unmistakable as footsteps, rounding the next corner at the end of this new hallway. Again we pursued.

Several hallways and several corners later, almost all at once, we stopped in our tracks. Seemingly at the same moment, we'd all remembered the same thing. Earlier in the evening, a police officer had mentioned being freaked out on the fifth floor and threatened that if he saw anything, he would throw lead. This is not the situation we wanted to be in. Maybe there was a ghost wandering the halls and we were following it. On the other hand, maybe there was a freaked out police officer walking the halls who now thought *we* were the ghost and was getting ready to defend himself with force!

People often ask us about the most frightening thing we've encountered on our investigations. Generally speaking, we give disappointing answers because even when we find things that could be interpreted as paranormal or ghostly, we're not the sort of people who really fear the ghosts. But we discovered that day that we do fear being mistaken for the ghost by a potentially trigger-happy

of souvenirs. It is to our eternal regret that we neglected to take one of the child-sized autopsy tables they offered us because it would be a great piece in our "museum," but at the time we couldn't think of a way to transport it or a place to store it.

officer of the law.

As such, we quietly backtracked and made our way back to our base of operations. We figured we'd sit there for a while until the officer—if, indeed, that's who we'd been hearing—had a chance to finish his rounds and go back downstairs. At that point we'd go down and ask if anyone had recently been up to the fifth floor and have our answer.

A bit later, we went back down and did exactly that. We asked if there'd recently been anyone making rounds up on fifth.

"We just hand a shift change," they said. And so they had no idea, because anyone who might have been up there had already left for the day.

"Oh well," we figured. "We're coming back in two weeks. We can ask then." Surely if that had been one of the officers, he was just as frightened by the whole thing as we were and would remember it a couple weeks later. So we left for the weekend and planned to inquire on our next investigation.

Life had other plans. So did the Federal Government.

While we were preparing for the follow-up investigation, we received a call from the contractors we'd been working with and told the entire building was now a pile of rubble. It seems the Department of Homeland Security had different plans than we did. They also had learned the old hospital was abandoned and scheduled for demolition so they intervened to use it as site for a training exercise. We're not privy to the full details of their operation, of course, but apparently it involved demolishing the entire building and then allowing their agents and canines to train on exploring the rubble. Regardless of their exact purpose, there was no building to which we could return.

For our purposes, that means we had to leave a lot of questions unanswered. We never learned whether it was an officer we'd heard in the hallways and we never discovered the source or sources of the sounds we heard throughout the duration of the investigation. While we'd normally therefore consider this still an open case in our files because of those outstanding mysteries, the Hospital's destruction means the case must be closed and filed away as permanently unsolved and now probably unsolvable.

References & Further Reading

Beaton, G. M. (2012). *Colorado Women: A History*. Denver, CO: University Press of Colorado

Bruinius, H. (2007). *Better for All the World: The Secret History of Forced Sterilization and America's Quest for Racial Purity*. New York: Knopf Doubleday

Children's Hospital Colorado (n.d.). Our History. *Children's Hospital Colorado*. <https://www.childrenscolorado.org/about/history/>

City and County of Denver (2023). Denver Individual Landmarks. *City and County of Denver*. <https://www.denvergov.org/files/assets/public/v/4/community-planning-and-development/documents/landmark-preservation/individual_landmarks_list.pdf>

Denver Public Library (n.d.). Denver Obituary Project. *Denver Public Library Special Collections and Archives*. <https://history.denverlibrary.org/research-tool/denver-obituary-project>

Denver Terrors (2021). Denver Children's Hospital. *Denver Terrors*. <https://denverterrors.com/denver-childrens-hospital/>

Engaging Data (n.d.). US Baby Name Popularity Visualizer. *Engaging Data.* <https://engaging-data.com/baby-name-visualizer/>

Goldberg, R. A. (1981). *Hooded Empire: The Ku Klux Klan in Colorado.* Urbana, IL: University of Illinois Press

Goodstein, P. (1996, 2001). *The Ghosts of Denver: Capitol Hill.* Denver, CO: New Social Publications

Goodstein, P. (2006). *Denver from the Bottom Up Vol. 3: In the Shadow of the Klan: When the KKK Ruled Denver, 1920-1926.* Denver, CO: New Social Publications

Hendricks, R. L. & Foster, M. S. (1994). *For a Child's Sake: History of the Children's Hospital, Denver, Colorado, 1910-1990.* Denver, CO: University Press of Colorado

Love, M. C. T. (1906). A Children's Hospital. *Denver Medical Times, XXVI*(1): 27-28

Packard, C. S. (1934). *A Little Story of the Children's Hospital of Denver.* Denver, CO: Children's Hospital Association

State Historical and Natural History Society of Colorado (1927). *History of Colorado.* Denver, CO: Linderman Co

Tomasso, D. W. (2019). Tammen Hall: National Register of Historic Places Inventory – Nomination Form. [Register No. 100004612]. National Register of Historic Places. <https://npgallery.nps.gov/GetAsset/8e8414ff-eb50-46ad-b91b-d645e6daa0af>

2
Restoring History: Elitch Theatre

Almost anyone who grew up in or around Denver has some memory of Elitch Gardens (commonly known as "Elitch's"). Depending on an individual's age, they might remember its old location in northwest Denver or the new (and current, as of this writing) location downtown. Either way, most know it only as an amusement park and might be surprised to learn that its centerpiece was once the historic Elitch Theatre (whose address is 4600 West 37th Place in Denver), a renowned center of arts and culture whose history is at least as fascinating as its ghost stories. But, as any old theatre ought, it does indeed have its ghost stories and so Rocky Mountain Paranormal was honored to investigate during a period of renovation.

Figure 2.1. Elitch Theatre (photo: Bryan Bonner).

The History

We all know Elitch Gardens (or Elitch's) if we grew up in or around Denver. Most of us have fond memories of the rollercoasters. Younger people might have grown up with the water park. And as horror nerds, of course we have a special place in our hearts for the haunted houses they construct around the Halloween season. Some of us personally remember with great fondness their first "dark ride" haunted house, which was built sometime around 1960 and demolished to make way for a Skee-Ball arcade in the 1980s.[1] But before amusement parks were the ride-centric places we all know and love today, they went through an evolution of different forms of entertainment (see also Volume 2, Chapter 11). Elitch Gardens was no exception. Indeed, its name hearkens back to a time when it was much more of an open garden space than a home to thrill rides and roller coasters. But it was much more than that, and at its heart was the Elitch Theatre.

In order to have a full understanding of the park and the Theatre, it's important to know a little something about the people who made it all come to life. Our story will begin with Mr. John Elitch, Jr. (1850 – 1891), though it was his wife, Mary Elitch Long (1856 – 1936) who unexpectedly became the center of the tale.

Mr. Elitch was born on April 10, 1850 in Mobile, Alabama to John Elitch, Sr. and Hulda May Hamilton Elitch. An interesting historical side note (albeit one with absolutely nothing to do with our story here) is that he was a direct descendant of Stephen Hopkins (1707 – 1785) who, as a delegate to the Continental Congress from Rhode Island, was a signatory to the Declaration of Independence.[2] The family relocated to California where Mr. Elitch attended Santa Clara college and assisted his father as a restauranteur. Shortly thereafter, he met Mary Elizabeth Hauck in church and fell in love. He courted her through letters and eventually proposed marriage by the same medium; knowing her father would not consent to the arrangement, the couple eloped to San Jose in 1872 when John was twenty-two and Mary was sixteen.[3]

For her part, Mary was born in Philadelphia in 1856 but relocated with her

---

1   DRDb (n.d.) Haunted House: Elitch Gardens. *Dark Ride Database.* <https://darkridedatabase.com/rides/haunted-house-6/> (accessed August 27, 2025).

2   Borrillo, T. A. (2012). *Denver's Historic Elitch Theatre: A Nostalgic Journey.* Denver, CO: Self-published.
Dier, C. L. (1932, 2023). *The Lady of the Gardens: Mary Elitch Long.* Denver, CO: Historic Elitch Theatre.

As an interesting side note, Hopkins suffered terrible palsy in his hands and had to use both of his hands to sign. Upon doing so, he said "My hand trembles, but my heart does not." Of interest to theatre fans, appropriate to this particular chapter, his role in America's founding is humorously immortalized in *1776*, the musical by Sherman Edwards and Peter Stone, and the 1972 film adaptation of the same title by Peter H. Hunt and Peter Stone.

3   Borrillo, T. A. (2012). *Denver's Historic Elitch Theatre: A Nostalgic Journey.* Denver, CO: Self-published.

family to California where she spent most of her childhood assisting her parents with their fruit farming business.[4] Little else is known of her life prior to her marriage to John, though her own fame ultimately overshadowed his some years later.

Upon their marriage, the young couple settled for a time in San Francisco, where John used his experience having assisted in his father's restaurant to manage the California Theatre's restaurant. Perhaps in part because of his work in that setting and all the various entertainment professionals who crossed his path (and whose acquaintance would come in handy years later), he developed a lifelong passion for the theatre. He and Mary developed a dream of someday owning and managing a resort featuring a theatre and a zoo.[5]

Eventually, the couple would relocate to Colorado. First they arrived in Denver, but eventually established a restaurant in Durango. After a time, they sold this property to establish a new restaurant in Denver. This allowed John to cultivate relationships with the movers and shakers in town and he went on to co-found the Denver Athletic Club in 1884 and then to open the Elitch Palace Dining Room, then the largest restaurant in Denver, in 1886.[6] The following year, John and Mary purchased the 16-acre Chilcott farm, planning to use it to supply fresh produce for their restaurant. The farm was located in what was then the town of Highland, to the west of Denver; now Highland is a neighborhood within Denver proper. Of their new neighborhood, Mary said, "The Highlands was something of a wilderness, for few streets were in common use. Mr. Elitch and I would drive from the gates of our ranch diagonally across the plains and down the hill, across the Platte River into Denver. A visit to 'the city' was a day's event to us."[7]

Despite plans to use the farm to support the restaurant, the couple's lifelong dream of becoming zookeepers and theatre owners was not forgotten. Just a year later, they sold their restaurant and put all their energy and resources into creating a cultural center which they intended to feature gardens, a zoo, musicians, theatre, and more. Elitch Zoological Gardens was born and opened for business on May 1, 1890, with such dignitaries as the Mayor of Denver, P. T. Barnum, and Mr. and Mrs. Tom Thumb—all friends of Mr. Elitch—in attendance.[8]

The word "zoological" in the name suggests something rather different from the theatre we're discussing in this chapter or the theme park into which the establishment eventually developed, but it was in line with the Elitches' thoughts about creating a space for both natural and cultural beauty. In fact this "park" (for want of a better word) was home to one of the first American zoos west of Chicago[9] and the Elitch zoo would eventually branch off to form the Denver Zoo

4   Colorado Women's Hall of Fame (n.d.). Mary Elitch Long. *Colorado Women's Hall of Fame*. <https://www.cogreatwomen.org/project/mary-elitch-long/> (accessed June 28, 2025).
5   Borrillo, T. A. (2012). *Denver's Historic Elitch Theatre: A Nostalgic Journey*. Denver, CO: Self-published.
6   *Ibid.*
7   Quoted in: Dier, C. L. (1932, 2023). *The Lady of the Gardens: Mary Elitch Long*. Denver, CO: Historic Elitch Theatre.
8   Elitch Gardens (n.d.). Park History. *Elitch Gardens Theme & Water Park*. <https://elitchgardens.com/plan-a-visit/park-history/> (accessed June 28, 2025).
9   *Ibid.*

which still exists today.[10] That wasn't the only first. Over the years, it would also be home to Denver's first symphony orchestra, Denver's first botanical gardens, Colorado's first children's museum and activity center, Denver's first motion picture theater,[11] and more—all the more impressive when one realizes Denver itself was only thirty years old at the time![12]

The Elitch Theatre itself was built according to Mr. Elitch's specifications. He wanted something reminiscent of London's Globe Theatre, itself built by William Shakespeare's "Lord Chamberlain's Men," but while the Globe was a round open-air theatre, the Elitch was an octagonal indoor one, so either Elitch had some details wrong or made some artistic changes of his own.[13]

Tragedy, unfortunately, was brewing despite these undeniable successes. After closing the doors following the first theatrical season, John Elitch assembled a touring company of performers and headed west with a vaudeville act. But when he reached San Francisco, he fell ill and ultimately died of pneumonia on March 10, 1891, with his wife Mary at his side, less than a year after opening his grand vision for Elitch Gardens.[14]

At this point, Mary steps into the spotlight. She wasn't about to let her and her late husband's dreams die so easily. She came back to Colorado and, despite having no business background of her own, saw to the reopening of their facility for the next season and single-handedly managed it for the next twenty-six years.[15] She spent the rest of her life living on the grounds (even after it had been sold—see below) and became known throughout Denver as "the Gracious Lady of the Gardens."[16]

It wasn't an easy task. The Mary Elitch who returned to Denver in 1891 was a thirty-four-year-old heartbroken widow, short on cash and without any background or experience running a business—much less one of this size—on her own. But she made it work. She sold an interest in the business to local capitalists but remained in charge and eventually repurchased complete ownership in the company in 1894, which she grew to new heights and successes she and John probably could only have dreamed of just a few years before. For this reason, she is considered something of a feminist icon among Colorado historians and is featured in the Colorado Women's Hall of Fame.[17]

Though our history is primarily interested in the Elitch Theatre specifical-

10   Liban, D. [director] (2007). *The Ghosts of Elitch Theatre* [DVD]. Tinyfist Films.
11   For ease of distinction, we use the spelling "theatre" for live theatre and "theater" for motion picture theaters.
12   Elitch Gardens (n.d.). Park History. *Elitch Gardens Theme & Water Park*. <https://elitchgardens.com/plan-a-visit/park-history/> (accessed June 28, 2025).
13   Liban, D. [director] (2007). *The Ghosts of Elitch Theatre* [DVD]. Tinyfist Films.
14   Borrillo, T. A. (2012). *Denver's Historic Elitch Theatre: A Nostalgic Journey*. Denver, CO: Self-published.
15   Liban, D. [director] (2007). *The Ghosts of Elitch Theatre* [DVD]. Tinyfist Films.
16   Elitch Gardens (n.d.). Park History. *Elitch Gardens Theme & Water Park*. <https://elitchgardens.com/plan-a-visit/park-history/> (accessed June 28, 2025).
17   Colorado Women's Hall of Fame (n.d.). Mary Elitch Long. *Colorado Women's Hall of Fame*. <https://www.cogreatwomen.org/project/mary-elitch-long/> (accessed June 28, 2025).

ly, it's worth mentioning a few additional highlights in the history of the larger park. A popular dance hall called the Trocadero Ballroom opened in 1917 and remained until the waning popularity of dance halls forced its closure and demolition in 1975. The carousel was added in 1928, requiring three skilled artisans to meticulously carve each figure by hand. A Ferris wheel was erected in 1936. KiddieLand, a region with attractions for the smaller children, was added in 1954, followed in 1965 by "Mister Twister," a roller coaster with "not a foot of straight track in it."[18] The park closed at its historic location in 1994 and reopened in 1995 in central downtown Denver with a more modern feel. A water park was added in 1997. In 1998, Premier Parks, then-owner of the park, acquired Six Flags and renamed the park Six Flags Elitch Gardens. In 2007, another new owner purchased the park and the name revered to Elitch Gardens (dropping the Six Flags branding).[19]

But the Elitch Theatre has a lot of history of its own. From the start, it became known as one of the best "go-to" places for live theatre. No less a figure than Cecil B. DeMille (the great American actor and filmmaker known for such classics as *The Ten Commandments* and *Cleopatra*) called it "the cradle of American theatre."[20] It would go on to become the oldest standing summer stock theatre in the United States.

Just a few of the major stars to grace its stage over the years included: Patty Duke, Edward G. Robinson, Douglas Fairbanks, Sr. (who got his start there at age fourteen when he was hired as a stagehand), Douglas Fairbanks, Jr., Vincent Price (royalty to us horror fans, of course), Sarah Bernhardt, Grace Kelly, Raymond Burr, Darren McGaven, Cesar Romero, Barbara Bel Geddes, Kitty Carlisle, Morey Amsterdam, Dick Van Patten, Sid Caesar, Jose Ferrer, John Astin, Steve Allen, Ginger Rogers, William Shatner, David McCallum, Tyrone Power, Sr., Tyrone Power, Jr., Nancy Walker, Julie Newmar, Helen Bonfils (of local Denver fame), Judd Hirsch, Mickey Rooney, Victor Borge, and many more.[21]

Theatre lovers might not know that the very person for whom the prestigious Tony Awards are named likewise got her start right there at the Elitch Theatre. According to a recollection by Mary Elitch herself, Antoinette "Tony" Perry's first on-stage role was a walk-on role at the Elitch at the age of eleven years (with her first credited role about five years later).[22] She would go on to have an internationally-successful career in the theatre as a performer, director, producer, and patron and was ultimately honored by being made the namesake of the Antoinette Perry Award for Excellence in Broadway Theatre (best known now simply as the "Tony Award") given by the American Theatre Wing and The

18   Elitch Gardens (n.d.). Park History. *Elitch Gardens Theme & Water Park.* <https://elitchgardens.com/plan-a-visit/park-history/> (accessed June 28, 2025).

A replica, known as "Twister II" still operates at the park's new location.
19   *Ibid.*
20   Liban, D. [director] (2007). *The Ghosts of Elitch Theatre* [DVD]. Tinyfist Films.
21   Elitch Theatre (n.d.). A Century of Stars. *Historic Elitch Theatre.* <https://historicelitchtheatre.org/history/stars/> (accessed June 28, 2025).
22   Borrillo, T. A. (2012). *Denver's Historic Elitch Theatre: A Nostalgic Journey.* Denver, CO: Self-published.

Broadway League.[23] Her own biography is also fascinating reading but sadly beyond the scope of this volume.

Returning to our main history, Mary Elitch married her business manager, Thomas D. Long, in 1900, becoming Mary Elitch Long. The new couple continued to grow the Elitch Gardens business over the following years, but there's some controversy regarding what happened next. According to the *Colorado Women's Hall of Fame*, "With the death of Long's second husband, Thomas Long, in 1906, J. K. Mullen and Jim Gerger began running the non-theat[re] operations and helping with financial decisions."[24] However, the *Colorado Encyclopedia* has the couple working together on the business for at least a decade following their marriage.[25] Unreliable sources around the Internet (without citations) claimed Thomas Long died in an automobile accident in 1920.

To settle the debate, we dug into the newspaper archives looking for obituaries and found the real answer. First, the September 26, 1914 edition of the *Rocky Mountain News* contained this one-paragraph notice: "Thomas D. Long, vice president and general manager of the Elitch-Long Amusement company, is in a serious condition at St. Anthony's hospital from a nervous breakdown. He has been a patient there for three weeks. He may be moved to a sanitarium."[26] That at least proved he survived longer than 1906. But then we found the most important piece of information, in the September 14, 1920 edition of the same newspaper: "Thomas D. Long, former manager of Elitch's gardens, was instantly killed late yesterday afternoon when an automobile he was driving plunged down a thirty-foot embankment on the Denver-Colorado Springs highway three miles north of Colorado Springs. Mr. Long was for several years connected with the North Denver amusement park, but left there some time ago. His wife, Mrs. Mary Elitch Long, had not heard of the fatal accident until notified by The News."[27]

So that settles that debate. We're not sure where the confusion came from, but it's clear that Mr. Long sadly passed in 1920. The report of his illness in 1914 also sheds some further light on why the Elitch Gardens and Elitch Theatre properties changed hands over the years. Clearly as early as 1906, Mullen and Berger were brought in to manage many of the company's operations, but we also knew that in 1916, Mary Elitch Long sold the property to Denver business interests—first at auction to Oscar L. Malo and subsequently to John Mulvihill[28]—but it wasn't immediately clear why. It seems likely that this sale was necessitated by

23   Murphy, J. E. (2017). *Colorado Myths & Legends: The True Stories Behind History's Mysteries*. 2nd ed. Guilford, CT: Twodot.

24   Colorado Women's Hall of Fame (n.d.). Mary Elitch Long. *Colorado Women's Hall of Fame*. <https://www.cogreatwomen.org/project/mary-elitch-long/> (accessed June 28, 2025).

25   Carr, S (n.d.). Mary Hauck Elitch Long. *Colorado Encyclopedia*. <https://coloradoencyclopedia.org/article/mary-hauck-elitch-long> (accessed June 28, 2025).

26   Rocky Mountain News (1914, September 26). Thomas D. Long Is Ill. *The Rocky Mountain News*.

27   Rocky Mountain News (1920, September 14). Former Manager of Elitch's Killed In Auto Accident. *The Rocky Mountain News*.

28   Carr, S (n.d.). Mary Hauck Elitch Long. *Colorado Encyclopedia*. <https://coloradoencyclopedia.org/article/mary-hauck-elitch-long> (accessed June 28, 2025).

Mr. Long's illness beginning two years earlier in 1914.

Whatever the reason, the sale was made in 1916. Some historians have suggested that perhaps Mary was a little naïve when she sold the property and thought the businessmen were going to provide capital and assistance while leaving her in charge,[29] but whether or not she expected it, that was not the case. It was a complete and total sale, with just a few stipulations: the name would not be changed, Mary would keep her beloved box at the Elitch Theatre, and she would be allowed to continue to live on the premises.[30] These promises were kept: the Elitch Theatre remains the Elitch Theatre (and, for that matter, Elitch Gardens remains Elitch Gardens even though it's no longer connected to the same property), Mary kept seeing plays from her box, and she kept living in her cottage on the grounds until her health necessitated she move in with her sister-in-law in 1932.[31]

Mary Elitch Long died of a heart attack on July 16, 1936 and is buried in Fairmount Cemetery alongside her first husband (though he'd died in California, she'd had his remains moved back to Colorado where she planned for them to be buried together). She was inducted into the Colorado Women's Hall of Fame in 1996 and the Colorado Business Hall of Fame in 1998.[32] She's remembered both for her contributions to local business and for her love of the theatre.

Throughout its more than a century of history, the Elitch Theatre was a major cultural attraction. Both in the theatre world and in the city at large, it was the place to go for fine entertainment. Its longevity came as something of a surprise to everyone, though. This wasn't the kind of construction that was intended to last for more than a century. Few buildings of its era were. One expert described its foundation as "masonry on dirt."[33] But even as its structure started to fail, it was beloved by the community both locally and nationally. It was designated a Denver Individual Landmark in 1995.[34] Perhaps more impressively, it was also listed on the National Register of Historic Places in 1978.[35] Clearly, it needed to be preserved.

Others agreed. In 2002, a foundation called the Historic Elitch Gardens Theatre Foundation (HEGTF) was organized and began raising funds toward a complete renovation and restoration project aiming to bring the building up to modern standards and keep it standing while also retaining its historic charm and appearance. In 2005, the Elitch Theatre received a "Save America's Trea-

29   Liban, D. [director] (2007). *The Ghosts of Elitch Theatre* [DVD]. Tinyfist Films.
30   *Ibid.*
31   Carr, S (n.d.). Mary Hauck Elitch Long. *Colorado Encyclopedia.* <https://coloradoencyclopedia.org/article/mary-hauck-elitch-long> (accessed June 28, 2025).
32   *Ibid.*
33   Liban, D. [director] (2007). *The Ghosts of Elitch Theatre* [DVD]. Tinyfist Films.
34   City and County of Denver (2023). Denver Individual Landmarks. *City and County of Denver.* <https://www.denvergov.org/files/assets/public/v/4/community-planning-and-development/documents/landmark-preservation/individual_landmarks_list.pdf> (accessed June 12, 2025)
35   Norgren, B. (1978). Elitch Theatre: National Register of Historic Places Inventory – Nomination Form. [Register No 78000844.]. *National Register of Historic Places.* <https://s3.amazonaws.com/NARAprodstorage/lz/electronic-records/rg-079/NPS_CO/78000844.pdf> (accessed June 28, 2025).

sures" grant from the Federal government in the amount of $300,000 toward restoration of the old building.[36] That wasn't going to be nearly enough, though. But the money came pouring in—some $5 million in all, from a combination of federal, state, city, and private grants and donations—and the exterior reconstruction project (phase one of the full project) began in 2006.[37] A firm called OZ Architecture was brought in to complete the project, taking care to preserve all the charm, quirks, and oddities of the original building while also making it fit for continued use; at the same time, documentarian David Liban was brought in to document the process and Rocky Mountain Paranormal was called to investigate the paranormal claims—but we'll get to that part in our next section.[38]

Phase two of the reconstruction began in 2012 and was completed in 2014 and involved repairing or replacing most of the interior fixtures including lighting, safety upgrades, and a modern fire suppression system. Phase three began in 2020 after a 2018 storm caused $800,000 in damages and involved a repaired roof, new restrooms, painting, and landscaping.[39] The theatre is now once again open for business and hosting both performances and historic tours, though they continue fundraising for additional improvements.

Paranormal Claims

Any old theatre worth its salt has some ghost stories. It comes with the territory. Whether it's because theatre people have a superstitious streak[40] or because any longstanding theatre has inevitably seen its share of tragedies or simply because there's something magical and otherworldly of the theatre experience itself—or, likely, a combination of all three—it seems just about every playhouse has a ghost. There's even an old tradition we find particularly charming that theatres must leave a "ghost light" illuminated on the stage after everyone leaves for the evening so the ghosts still have some light.[41] Elitch Theatre, of course, is no exception.

Unfortunately, when it comes to specifics, there aren't a ton we can report.

---

36   Craine, K., Dana, C., & Fletcher, R. (n.d.). Save America's Treasures: Preserving the Legacy of Our National Experience. *President's Committee on the Arts and the Humanities*. <https://www.govinfo.gov/content/pkg/GOVPUB-PR-PURL-LPS79036/pdf/GOVPUB-PR-PURL-LPS79036.pdf> (accessed June 28, 2025).

37   Westword Staff (2017). Historic Elitch Theatre Hosts First Major Concert in Decades. *Westword*. <https://www.westword.com/music/historic-elitch-theatre-hosts-first-major-concert-in-decades-9399307> (accessed June 28, 2025).

38   Liban, D. [director] (2007). *The Ghosts of Elitch Theatre* [DVD]. Tinyfist Films.

39   Elitch Theatre (n.d.). Theatre Timeline. *Historic Elitch Theatre*. <https://historicelitchtheatre.org/history/building-timeline/> (accessed June 28, 2025).

40   Longstanding tradition, for instance, prohibits uttering the name of Shakespeare's play *Macbeth* inside a theatre lest one be cursed. We've done so and remain, as far as we can tell, curse free.

41   This may have originated as a safety measure rather than a superstitious one, but we prefer to think theatre people really are leaving the lights on for the ghosts; either way, the ghostly lore has attached itself inextricably to the tradition.

It seems just about everyone who's ever worked at (and many who've just visited) the Theatre have a ghost story to tell, but most follow a common pattern, so this is going to be a fairly short section in the book.

The most common ghost associated with Elitch Theatre, appropriately enough, is Mary Elitch Long herself. It makes sense, given her love of this specific Theatre and of theatre in general. She's thought to watch over the location and continue keeping it safe.

Indeed, many people even credit Mary's ghost with the Theatre's fortuitous but unexpected longevity.[42] Honestly, though we have no way to prove it, we think that's a charming enough explanation we endorse it as the best one, at least on a metaphorical level (we have to maintain our neutral skepticism with regard to whether it's *literally* true).

Proving once again that in the world of the paranormal, every unusual claim is followed by an even stranger one, John Hickenlooper, then-Mayor of Denver (later Governor of Colorado, and as of this writing, United States Senator from Colorado) offered a different but no less paranormal possibility: "Or there could be a highlands anti-gravitational field that somehow is focused and concentrated there that actually reduces the effect of gravity in that very specific area."[43] He said it with a smile on his face, so we'll give him the benefit of the doubt and assume he was joking.[44] But there are people in the world who believe in such things.

According to an oft-repeated rumor, legendary actress Shelley Winters, while doing rehearsals for a play at the Theatre, stopped what she was doing, pointed toward one of the boxes, and said (words to the effect of), "who's the broad in the boa?" There was nobody in the box. But, the story goes, she'd pointed specifically to Mary's own box. And Mary was known to frequently wear a long gown and a feather boa.[45]

Figure 2.2. Elitch Theatre interior with Mary's private box at far left (photo: Bryan Bonner).

---

42   Liban, D. [director] (2007). *The Ghosts of Elitch Theatre* [DVD]. Tinyfist Films.
43   *Ibid.*
44   We certainly *hope* he was joking.
45   *Ibid.*

One thing that sets the haunting of Elitch Theatre apart from many of the places we've looked into is that most of the reports—not all, but certainly far more than average—involve visible manifestations of the alleged spirit. More people than we can count have said they've actually seen Mary's ghost, always wearing her gown and boa. Most often, she's reported as being seen in her box but sometimes she seems to wander over to the balcony or the backstage area. Typical haunting stories don't involve so many visual reports. They're not unheard of, but people often claim instead to hear things (footsteps, doors slamming, whispering voices), to feel things (cold spots or bad emotional feelings), or to experience other symptoms like items moving about when no one else is there to move them.

One staff member reported that whenever he would enter the building with his dog tagging along, the canine would, immediately upon setting foot in the Theatre, bolt off to the balcony.[46] Whether that's due to paranormal prompting or not, we can't say, but he certainly considered it unusual behavior.

Another staff member working as an archivist told us that one afternoon, she was working in the Theatre to remove some historic photos and bring them to the archive collection. She'd load up a box and carry it down the aisle and set it downstage left where they'd be gathered to eventually be transferred to the archives. While carrying one such load, she felt like she'd run into a soft, warm, unseen wall. She described it as if she were being hugged by some invisible person in front of her, preventing her from moving forward. After a moment, the feeling passed and she carried on with her work, only to discover later that her experience took place directly below Mary's box.

Our Investigation

Our own investigation into the alleged hauntings at Elitch Theatre coincided with the renovation project. Everyone involved was interested in the ghost stories and we were selected to check them out. A documentarian was also brought in to film the entire process and tell the story. That documentary ultimately became *The Ghosts of Elitch Theatre*. Being followed by a film crew isn't the way we normally operate but it also wasn't a new experience for us. For the purpose of this book, we'll just present our story the same way we would any other, but with the caveat that we had some extra people with us and took periodic breaks to explain to the camera what we were doing and what we'd found.

On the evening of the investigation, we arrived at about 7:45 p.m. and immediately started setting up. Based on the stories we'd been told, we knew the main areas we wanted to monitor would be Mary's box, the balcony (left, center, and right), and the stage. Because there was no curtain separating the stage from backstage at the time, our cameras and other monitors on the stage itself were also able to get a view of the backstage area. We wanted to look at more of the rest of the building, but it was determined that some areas were too dangerous for us to access due to the condition of the building and the ongoing construction project.

---

46   *Ibid.*

Figure 2.3. The stage and backstage at Elitch Theatre (photo: Bryan Bonner).

As is our way, in addition to constant video and audio monitoring, we scheduled hourly readings for both temperature and EMF throughout the building. While we were setting everything up and taking our initial (baseline) EMF readings, we discovered some unexpected guests would be joining us on our investigation: a family of bats had taken up residence in the Theatre.[47] Bats needn't be frightening creatures, but that did seem to set the mood for an evening of looking for ghosts, so we couldn't have been happier.

Throughout our investigation, we never caught anything unusual on video, but there were a few interesting moments.

First, we discovered unusually elevated (and sometimes fluctuating) EMF readings, particularly in Mary's box, though we picked up "spikes" and fluctuations throughout the entire Theatre. Despite these few spikes, most readings were consistent throughout the evening, and their elevated levels were fully consistent with the amount of old wiring in the building. We can't conclusively explain every spike we detected without setting up controlled experiments in which the entire electrical system is under our control, but we consider everything we observed to be consistent with the building's electrical system.

47   While strictly unrelated to paranormal investigation, we have a word of advice as lovers of bats. If you ever find any living in your attic or around your house, don't panic. They mean you no harm. Rather, they're likely to serve you well by eating lots of mosquitoes which might otherwise do you actual harm. While we understand that you probably don't want them living in your house, there are several things you can do to help them find their way to a place they and you will be happier with. The best thing to do if there are a lot of them in your attic is to call a wildlife relocator (not an exterminator) or a bat rescue. Perhaps consider erecting a bat house to give them a more hospitable home. If you ever find a single bat, it's important not to touch it with your hands (unsafe for both you and the bat) and you should instead contain it and/or make sure it's well before you either release it outside or contact a rescue professional.

Temperatures were mostly consistent throughout the evening, decreasing steadily as the evening progressed due to lowering external temperatures and the building's poor insulation and numerous missing windows. Interestingly enough, there was a bit of a "cold spot" for a time in Mary's box, which many paranormal enthusiasts might take for evidence of paranormal activity. We don't think so. In this case, the temperature fluctuation was entirely explainable as the result of environmental factors.

That does raise an important point about "cold spots" in general, though. We'll explain our thinking more fully when we publish our "how-to" book on paranormal investigation, but in brief, the theory in paranormal lore is that cold spots are caused by spirits or other supernatural entities drawing energy from the environment in order to manifest. Though unproven, there's a certain logic to that. But if you remember your elementary thermodynamics, any exchange of energy must necessarily produce heat as a byproduct. That's why the back of your fridge gets hot and why air conditioners need to vent to a window. So if paranormal entities do indeed produce cold spots (and for the reason suggested), we should also be looking for a coincident "hot spot," which never seems to come up in the paranormal lore.

Several exciting moments occurred throughout the investigation, presented here in chronological order.

At about 11:55 p.m., we heard a loud screaming sound. This understandably caused much excitement. Alas, it turned out to be a stray cat across the street.

Around 2:40 a.m., a smoke alarm went off in the pump room. There was no smoke and no fire. This was welcome news because any fire that started in the Theatre before the renovations were completed probably would have destroyed the whole thing. One of the workers told us the Fire Department considered the building a "five minute structure," meaning if a fire started, the entire building would be on fire within that amount of time.

We heard a woman singing at 2:58 a.m. Of course we all immediately had thoughts about the ghosts of singers who performed at the Theatre in the past. Quick investigation, however, revealed it was coming from a passerby outside the Theatre.

Finally, we started breaking down our equipment a little after 4:00 a.m. At about 4:15, we heard an odd banging sound coming from the north stairs and went to investigate. At first, we couldn't find anyone and thought we might have to set everything back up and start monitoring again. Shortly thereafter, though, we traced the sounds to a family of foxes who'd taken up residence below the stairs. Mystery solved.

All in all, it was an eventful evening, but not in the paranormal sense. We had to leave this one much as we found it: a truly magnificent building (indeed, one of "America's Treasures") full of spectacular ghost stories. But despite our best efforts, neither Mary Elitch nor any other ghosts decided to manifest for us during our investigation, so we can neither confirm nor debunk any of the various paranormal claims. This case has to remain open, and we'll hope to be invited back in for a follow-up some day in the future now that the building is in better repair and we might be able to safely conduct a more thorough investigation.

Addendum

Just as we were preparing this book for press, further news came to our attention which we think merits a brief mention. Between the months of April and July, 2025, the Elitch Theatre has been the victim of a string of six break-ins in the wee hours of the morning. The perpetrators appear to be youths who are just exploring the theatre and taking photos, but the building's caretakers have said both that they're concerned about safety during these after-hours illegal explorations and that each break-in has cost the company thousands of dollars in repairs and additional security expenses.[48]

Urban exploration and ghost hunting alike are common pastimes, and they often overlap, so we have no idea whether these youths are motivated more by historical curiosity, by the prospect of looking for ghosts in an allegedly haunted theatre, or simply by the thrill of breaking the law. Regardless, we condemn such illegal entries in the strongest possible terms. We like exploring places after hours, too, but the *only* legitimate way to do so is with the proper permission of the property owner(s).

As of this writing, the culprits have not been identified.

References & Further Reading

Borrillo, T. A. (2012). *Denver's Historic Elitch Theatre: A Nostalgic Journey*. Denver, CO: Self-published

Brick, C. (2018). *Haunted America: Ghost & Legends of Colorado's Front Range*. Charleston, SC: The History Press

City and County of Denver (2023). Denver Individual Landmarks. *City and County of Denver*. <https://www.denvergov.org/files/assets/public/v/4/community-planning-and-development/documents/landmark-preservation/individual_landmarks_list.pdf>

Carr, S (n.d.). Mary Hauck Elitch Long. *Colorado Encyclopedia*. <https://coloradoencyclopedia.org/article/mary-hauck-elitch-long>

Colorado Women's Hall of Fame (n.d.). Mary Elitch Long. *Colorado Women's Hall of Fame*. <https://www.cogreatwomen.org/project/mary-elitch-long/>

Craine, K., Dana, C., & Fletcher, R. (n.d.). Save America's Treasures: Preserving the Legacy of Our National Experience. *President's Committee on the Arts and the Humanities*. <https://www.govinfo.gov/content/pkg/GOVPUB-PR-PURL-LPS79036/pdf/GOVPUB-PR-PURL-LPS79036.pdf>

Dier, C. L. (1932, 2023). *The Lady of the Gardens: Mary Elitch Long*. Denver, CO: Historic Elitch Theatre

DRDb (n.d.) Haunted House: Elitch Gardens. *Dark Ride Database*. <https://darkridedatabase.com/rides/haunted-house-6/>

Elitch Gardens (n.d.). Park History. *Elitch Gardens Theme & Water Park*. <https://elitchgardens.com/plan-a-visit/park-history/>

Elitch Theatre (n.d.). A Century of Stars. *Historic Elitch Theatre*. <https://

---

48   Sims, A. (2025). Historic Elitch Theatre break in caught on camera: 6th break in since early April. *KDVR*. <https://kdvr.com/news/local/historic-elitch-theatre-break-in-caught-on-camera-6th-break-in-since-early-april/?fbclid=IwY2xjawL2bIxle HRuA2FlbQIxMQABHlnALqPe0meMy1fbNbijo-JHxrtHFQCceRXn9lGMfv_B5_FrMVLaaQFV1g6D_aem_GOH84hODoUVWw8qkiCAg3Q> (accessed July 29, 2025).

historicelitchtheatre.org/history/stars/>

Elitch Theatre (n.d.). Theatre Timeline. *Historic Elitch Theatre*. <https://historicelitchtheatre. org/history/building-timeline/>

Liban, D. [director] (2007). *The Ghosts of Elitch Theatre* [DVD]. Tinyfist Films

Murphy, J. E. (2017). *Colorado Myths & Legends: The True Stories Behind History's Mysteries*. 2nd ed. Guilford, CT: Twodot

Norgren, B. (1978). Elitch Theatre: National Register of Historic Places Inventory – Nomination Form. [Register No 78000844.]. *National Register of Historic Places*. <https://s3.amazonaws.com/NARAprodstorage/lz/electronic-records/rg-079/ NPS_CO/78000844.pdf>

Rocky Mountain News (1914, September 26). Thomas D. Long Is Ill. *The Rocky Mountain News*

Rocky Mountain News (1920, September 14). Former Manager of Elitch's Killed In Auto Accident. *The Rocky Mountain News*

Sims, A. (2025). Historic Elitch Theatre break in caught on camera: 6th break in since early April. *KDVR*. <https://kdvr.com/news/local/historic-elitch-theatre-break-in-caught-on-camera-6th-break-in-since-early-april/?fbclid=IwY2xjawL2bIxleHR uA2FlbQIxMQABHlnALqPe0meMy1fbNbijo-JHxrtHFQCceRXn9lGMfv_B5_ FrMVLaaQFV1g6D_aem_GOH84hODoUVWw8qkiCAg3Q>

3

Spices and Spirits: The Yak and Yeti

The Yak and Yeti is a wonderful Indian-Nepalese restaurant and brewhouse now with multiple locations in the greater Denver area, but the focus of our investigation (and of this chapter) is located in a historic former single-family home at 7803 Ralston Road in Arvada. As with many restaurants *and* historic homes, it has collected its share of ghost stories over the years and was the subject of a long series of investigations by Rocky Mountain Paranormal. For a time, it was also home to our series of paranormal lectures and theatrical séance productions. It also serves as one of our favorite examples of the importance of dedicated investigation and follow-up because some of its mysteries took years to unravel.

Figure 3.1. The Yak and Yeti (photo: Bryan Bonner).

The History

Though it's now a public facing venue offering delicious South Asian cuisine and craft beers (brewed on site), the building that would become the Yak and Yeti (at least the one in Arvada—as of this writing, there are now locations in Arvada, Denver, Westminster, and Wheat Ridge[1]) began its life as a single-family home (albeit a large and delightfully-designed one). As such, historic records concerning the properly are necessarily more limited than for some of the other locations we've discussed throughout this series.

The story begins with Arvada's second postmaster, Eli Allen (1829 – 1906). He served as postmaster for the town from 1882 to 1894 as the successor to the first postmaster, Benjamin F. Wadsworth (1827 – 1893), one of the town's founding fathers, for whom the well-traveled Wadsworth Boulevard is named.[2] Though only tangentially related to our story of the home which would become an allegedly haunted restaurant, it seems interesting enough to bear mentioning that Mr. Allen's own daughter, Ida Allen (1859 – 1909) would succeed her father and serve as Arvada's third postmaster for the fifteen years following Eli Allen's service.[3]

Mr. Allen built his home as a single-story frame house measuring approximately twenty-four by twenty-seven feet with a wooden floor and cellar under the house in 1864. Though the building has seen numerous additions, partial destructions, and renovation over the years, we've been told the original cellar still exists under the east side of the present building. Allen's purpose for the house was both to be a good place to raise his family and to maintain stables for racehorses on the property. At the time, Arvada was still mostly farmland with just about thirty homesteads operating in the area.[4] By all accounts, Mr. Allen was quite successful in all of these ventures—the post office thrived, one of his own children succeeded him in his office, and he maintained a large stable of horses on the property.

For a man like Mr. Allen, though, personal success wasn't his only goal. He had interest in community projects to build what would become the now-bustling suburb of Arvada. Beginning in 1859 (and with water flowing by 1960), a ditch was built to carry water between Golden and the Arapahoe bar. It's important to realize that much of Colorado's climate is quite arid, particularly in the Denver area, so irrigation ditches like these were necessary to keep homesteads, farms, and businesses running. In 1864, none other than Walter Scott Cheesman (see Volume 1, Chapter 5 for much more on Mr. Cheesman and the Denver park which now bears his name—thus proving once again that once one starts digging

1   Yak and Yeti (n.d.). *The Yak and Yeti*. <https://theyakandyeti.com/> (accessed July 13, 2025).

2   Barton, B. J., Walker, R., Jr., Lutz, M. R., Gorrell, L., & Gorrell L. (1985). *Arvada: Just Between You and Me: 1904 – 1941*. Arvada, CO: Arvada Historical Society.

3   Find a Grave (n.d.). Eli Allen. *Find a Grave*. <https://www.findagrave.com/memorial/19191742/eli-allen> (accessed July 13, 2025).

4   Briggs, A. (2014). Yak and Yeti in Arvada celebrates 150th birthday of historic building. *The Denver Post*. <https://www.denverpost.com/2014/06/24/yak-and-yeti-in-arvada-celebrates-150th-birthday-of-historic-building/> (accessed July 13, 2025).

into history, ghost stories, or both, the same names just keep popping up all the time) purchased an interest in the ditch, lengthening it to supply water to the farms of numerous individuals including Mr. Eli Allen. In turn, Mr. Allen and others incorporated the canal as the Golden City and Arapahoe Ditch, which eventually took on the name "Farmers' High Line Canal" in 1885.[5] Though essential for irrigating farms in the area and allowing the city to develop, this canal should not be confused with the more-famous Denver High Line Canal which begins at the South Platte River, runs northeast for a length of over seventy miles, and is now a popular hiking trail for nature-lovers.[6]

The house's own story took a different path from Eli Allen's in 1891 when it was taken over by Allen's son, Charles Erwin (sometimes misspelled "Irwin" in publications) Allen (1864 – 1892), who either modified or replaced the home with a brick structure in preparation for his upcoming wedding to Jane Anne "Jeannie" Christie (1866 – 1936; later Jane Anne Hutchings).[7] Unfortunately, the marriage was not to last long. Charles E. Allen passed away only about a year later at the age of twenty-seven or twenty-eight years. Because tragedy never knows when to leave a family alone, his only child, Christobel E. Allen, died just a few years later, aged two or three.[8] The house, therefore, was left in the care and ownership of the newlywed and newly-widowed Jeannie. By this time, the home took on some of the appearance we would recognize today, though numerous other modifications followed.

In 1894, an independently wealthy man named Elias William Van Voorhis (c. 1857 – ?) moved to Colorado for health reasons. This was not an uncommon occurrence at the time, and though we don't know the particular health reasons invoked in this case, we remind readers of the story of how Mr. F. O. Stanley came to Colorado as part of his tuberculosis treatment (Volume 1, Chapter 14). Regardless of the reason, he purchased the home from Mrs. Allen in 1894 (the same year young Christobel passed away, perhaps supplying a psychological motivation for the sale on her part). Mr. Van Voorhis spotted the house on a buggy ride along Ralston Road with his own young bride Cora Van Voorhis (née Rockwell; 1860 – ?[9]), fell in love with it, and arranged the purchase.

During the Van Voorhis ownership of the property, they hired a Denver ar-

5   Genealogy Trails History Group (n.d.). Arvada. *Genealogy Trails History Group*. <https://genealogytrails.com/colo/jefferson/history_Arvada.html> (accessed July 13, 2025).

6   Many canals share the name "High Line" or "Highline" because they're named for the "high line" engineering principle in which the canals are designed to follow the naturally occurring highest possible elevation, allowing water to flow driven only by gravity rather than by any kind of expensive and complicated pumping mechanism.

7   Find a Grave (n.d.) Jane Anne "Jeannie" Christie Hutchings. *Find a Grave*. <https://www.findagrave.com/memorial/8601530/jane_anne-hutchings> (accessed July 13, 2025).

8   Find a Grave (n.d.) Charles Erwin Allen. *Find a Grave*. <https://www.findagrave.com/memorial/19191784/charles_erwin-allen> (accessed July 13, 2025).

9   Family Search (n.d.) Cora Bell Rockwell. *Family Search*. <https://www.familysearch.org/en/tree/person/details/K4X7-9WC> (accessed July 13, 2025).

chitecture firm to design additions to the property which were completed in 1895. The new addition joined the original building at the tower, giving the building its current appearance.

Elias and Cora had twin daughters but only one survived: Elizabeth W. "Betty" Van Voorhis (1897 – ?), who married Milton Francis "Nic" Nicholson (1895 – 1970) who was part of a well-known family of Masons who built many of the churches in the area.

As for the occupancy of the home, Elias and Cora welcomed, in addition to their own child, Cora's sister, mother, and uncle to live at the house. Uncle Ned was said to take refuge from all the women in the house by hiding in the tower room, which he called his "sanctuary."

In 1929, the home caught fire. It was caused by—we kid you not—"baby mice playing with wooden matches." At least that's what the fire marshal said about it. Cora acted quickly to save her two grandchildren—also living in the house at this time—by wrapping them in wet blankets and shepherding them down the stairs and out of the burning building through the smoke and flames.[10] Never underestimate a grandmother's protective spirit.

During their own residency at the house, Nic and Betty would celebrate their birthdays (September 1 and 2, respectively) by throwing a grand party beginning with a feast of stuffed pork chops in the upstairs dining room at home and culminating with dancing until midnight at the Trocadero Ballroom at Elitch Gardens (see Chapter 2 and once again marvel at the coincidences), sometimes followed by date cake back at home.

Cora died on a date we have been unable to determine sometime in the 1940s from a fall down the stairs at home. She was known to suffer fainting spells, and it seems she collapsed at the top of the stairs and fell all the way down, fracturing her head and neck, ultimately dying at home within a few days.[11] We should note the stairs at the building currently (which become relevant in the ghost stories to follow) are a replacement. The current staircase is a mirror image of where the old ones used to be. Rumor says that when Cora fell down the stairs, she hit her head on every step. If we can be forgiven for jumping ahead to our investigation a bit, we did determine experimentally that this would have been impossible unless she were dragged down the stairs by her feet, which did not happen.

At this point, the home underwent a huge alteration. Following her mother's death, Betty divorced Nic, cut down many of the trees on the property, and sub-divided the thirty-three-acre homestead into lots she could sell. She retained the home itself but turned it into apartments, expanding the usable area by moving the chicken coop from across the property and attaching it to the house. It remains there to this day and serves as the Yak and Yeti's kitchen and restrooms.

Nic and Betty's daughter Katherine (known as Katie) lived in an apartment in the old barn until her graduation from college.

10   Briggs, A. (2014). Yak and Yeti in Arvada celebrates 150th birthday of historic building. *The Denver Post.* <https://www.denverpost.com/2014/06/24/yak-and-yeti-in-arvada-celebrates-150th-birthday-of-historic-building/> (accessed July 13, 2025).

11   Austin. S. (2010). Ghost of Yak and Yeti. *Colorado Community Media.* <https://coloradocommunitymedia.com/2010/10/28/ghost-of-yak-and-yeti/> (accessed July 13, 2025).

Despite remaining in the hands of only two families for many decades (apartment tenants notwithstanding), the home eventually fell into disrepair, but money was too tight for the owners and a process began of gutting the insides and preparing the majestic old home for demolition. Enter, in 1997, a group of investor/contractors led by British expatriate Geoffrey Bruce. They spent the next three years employing their own labor and capital to renovate the building and convert it into a brewpub. They called it The Cheshire Cat. It was Bruce's desire to start brewing quality beers after suffering what he considered to be the inferior quality of American beer for too long.[12]

Finally, the building entered its current stage of development in 2008 when Dol Bhattarai, a Nepalese expatriate restauranteur, purchased it to transform it into the second of his Yak and Yeti restaurants.[13] It was at this stage (and shortly after it became the Yak and Yeti) that Rocky Mountain Paranormal became involved in the story. But before we get there, we need to turn back the clock, because the Yak and Yeti's new owner and staff were not the first people to experience ghost stories in the building.

Paranormal Claims

This is a bit of an odd one for us, because it seems like just about everybody involved in the restaurant (whether in its current capacity as the Yak and Yeti or in one of its prior incarnations) has a ghost story to tell about it. It should therefore seem like we ought to be able to fill dozens of pages with detailed accounts. However, most of the stories follow a similar format and structure, so to repeat them all would be redundant in the extreme. Instead, we present an overview of the stories we've heard reported over the years.

Most of the stories involve the ghosts either of Cora Van Voorhis or her Uncle Ned. Cora is often seen on the stairs and Uncle Ned in his tower room "sanctuary."

The earliest claims we've come across come from the Van Voorhis family themselves. Cora's own granddaughter Katie (who we met briefly in the history section) reported several ghostly apparitions over the years. Beginning shortly after her grandmother's funeral, she reported ghostly visits from both Cora and Uncle Ned. We've heard reports (but have been unable to confirm them) that as a young boy, Katie's brother, Milton Edward "Nic" Voorhis Nicholson (1927 – 1977) likewise received visits from his grandmother, who would visit him at night and sit at the edge of the bed to carry on long conversations. This room, now a dining area, is still (or at least was still at the time of our investigations) known as "Nic's Room."

In the years since, numerous staff and visitors alike have reported seeing a

12   Briggs, A. (2014). Yak and Yeti in Arvada celebrates 150th birthday of historic building. *The Denver Post.* <https://www.denverpost.com/2014/06/24/yak-and-yeti-in-arvada-celebrates-150th-birthday-of-historic-building/> (accessed July 13, 2025).

13   Austin. S. (2010). Ghost of Yak and Yeti. *Colorado Community Media.* <https://coloradocommunitymedia.com/2010/10/28/ghost-of-yak-and-yeti/> (accessed July 13, 2025).

woman fall down the stairs. In some cases, they either run to investigate or to alert staff as to the emergency only to find that there was never really anyone there.

During the renovations between 1997 and 2000, contractors reported several ghostly encounters. Often, they complained that hammers and other tools they'd been using mysteriously moved to other parts of the building. Some even claimed that Cora liked to trip people on the third stair, though no one ever got hurt.[14]

Figure 3.2. The haunted staircase (photo: Bryan Bonner).

---

14   *Ibid.*

When the building was still The Cheshire Cat, one story involves a customer who handed her credit card to a blonde waitress to pay her tab during a large party. Only trouble was: they did not employ any blonde servers at the time. Everyone freaked out, thinking the woman's credit card had either been lost or stolen, but she came back later reporting the same blonde waitress returned the card to her. No one ever figured out who the mystery waitress was, nor whether she was a prankster or perhaps one of the ghosts.[15]

Both as The Cheshire Cat and as the Yak and Yeti, it's been the reported site of what some people might call poltergeist activity. People report seeing shadows and hearing sounds throughout the building even when they're the only ones there. Objects and bottles have been said to move—either visibly or to disappear and then reappear later somewhere else.

One individual reported to us that he'd been working in the building late at night as a contractor and listening to his rock 'n roll music while he worked. When he was working around the bar, he said, someone or something scratched him but he couldn't see any culprit. But when he changed his loud rock music to something more peaceful, the room "felt" better and he had no further incidents. He took this as a sign that one of the ghosts disapproved of his choice in music.

Some of the paranormal claims have been a bit more unusual and not the kind of stuff we read about in every report of a haunting. One customer saw two versions of her own cousin—an apparent doppelganger of sorts, wearing exactly the same clothes—in two places at exactly the same time. Another employee cut himself and immediately found a bandage laid out for him on the table. During a Halloween radio broadcast from the restaurant, two women witnessed a water pitcher move several inches even though no one was near it.[16]

Several ghost hunting teams preceded Rocky Mountain Paranormal to the site over the years. A common report from several of them was a "vortex" of EMF energy near the bottom of the stairs. "Vortex" is one of those words we hear a lot in paranormal literature but it's often ill-defined. In science, it's defined as location of whirling or rotating matter (fluid or air). But in paranormal lore, it's often thought to be a location of concentrated paranormal or spiritual energy (though the word "energy" itself is often poorly defined in the paranormal literature, so it's entirely unclear how one might detect or measure one of these vortices.[17] Paranormal photography enthusiasts often use the label "vortex" for streak-like luminous manifestations in their photographs, which seems to be an unrelated definition; we handled that sort of vortex in greater detail in Volume 1, Chapter 24 of this series. In this case, the prior teams were of the opinion that the room near the bottom of the stairs containing this vortex was the point of ingress and egress for spirits haunting the restaurant. They seemed to view the ghosts as not haunting the location permanently but traveling between the building and the spirit world and using this vortex as a means of travel.

---

15   *Ibid.*

16   *Ibid.*

17   Often "vortexes," but we use the correct pluralization.

Figure 3.3. The site of the reported "vortex" (photo: Bryan Bonner).

Shortly after taking over the building, the owner of the new Yak and Yeti, who had until then been completely unaware of his new property's haunted reputation, had an experience of his own. The first such experience he reported was hearing a door open and close on its own when no one else was around. It was after that experience that he inquired about it of his manager (who had also worked there when it was The Cheshire Cat) and first learned of the ghost stories. After that, he witnessed such remarkable things as teapots and dishes moving across the bar on their own.[18]

It wasn't long after that they called in Rocky Mountain Paranormal to conduct our own investigation. It is to that we now turn our attention.

Our Investigation

Our own work at the restaurant began shortly after it became the Yak and Yeti in 2008. Following this initial investigation, though, we returned several times for follow-up inquiries. Additionally, it became one of our "home bases" of sorts. For many years, in addition to our investigative work, we partnered with the restaurant to offer educational and entertainment offerings to the public, including our popular "Ghosts of Colorado" lecture series as well as our always-fun

18   Austin. S. (2010). Ghost of Yak and Yeti. *Colorado Community Media.* <https://coloradocommunitymedia.com/2010/10/28/ghost-of-yak-and-yeti/> (accessed July 13, 2025).

theatrical séances, in which we used theatrical scripts and magicians' tricks to recreate the appearance of a successful classic séance. For our purposes here, we won't focus on those public-facing events but instead describe our investigation into the reported hauntings.

On the first night of the investigation, we arrived at about 9:00 in the evening, right around closing time after the dinner service. Following our usual protocol, we started by setting up all of our monitoring equipment, taking baseline readings for EMF, temperature, and other measurable phenomena, and began actively monitoring the building at approximately 11:00 p.m.

Our first order of business was to address the "vortex" reported by other paranormal teams who'd looked into the location. They had reported a room near the bottom of the stairs as having abnormal electromagnetic activity which they interpreted as a sort of portal between the Yak and Yeti and the spirit world. To begin our investigation, we started by taking EMF readings.

Sure enough! There was indeed a strong EMF reading in this area. Rather than accept that as proof-positive the building was haunted, though, we tried to figure out if there could be a natural explanation for the anomalous reading, and it didn't take us long to find one. Part of the building's alarm system included a large amount of unshielded wiring right in that area. We figured it was likely the culprit. To confirm it, we temporarily disabled the alarm. The EMF "vortex" disappeared. When the alarm was on, the readings were abnormally high. When it was off, they were indistinguishable from those in the rest of the building.

In that way, we partially confirmed and partially debunked the prior teams' claim. We confirmed that they had indeed detected an EMF anomaly. But we debunked the paranormal interpretation as to its cause. Not bad for a start, but solving one piece of the mystery does not solve the entire mystery.

Shortly thereafter, one of our team members went upstairs to the room which had belonged to the younger Nic (the one who'd supposedly seen his grandmother's ghost in that room) to spend some time monitoring the room alone. While he was there, he heard a strange sound. It was, he said, like the sound of "someone thumping the light fixture" or flicking it with a fingernail.

He got up to investigate, but as he approached the light fixture from which he'd heard the sound, the noise moved. Now it sounded like it was coming from a different light fixture. Again he followed it and again it moved to yet *another* fixture. He followed the weird thumping noise to three different lights, but never figured out what caused the noise. No one else was with him on the upper floor at the time, so it shouldn't have been caused by movements of any of our other team members, but we can't say much more than that. Unfortunately, some mysteries have to remain mysterious and, as we always remind people, we have to be comfortable with "I don't know" as the only honest response to many questions or puzzles in this line of work.

While he was still in the room alone and observing, he heard an unusual sound that could only be described as a voice at a distance. While trying to determine where the voice was coming from, he got up and started to wander around the room again. It appeared that no matter where he went in the room the voice would follow. After the attempt to determine what the voice was he yelled down to the rest of the team at our base of operations to see if maybe one of us was watching videos or talking. We confirmed that nobody was making any noise, so we sent one person up to see if they could find the source. After a few minutes, they both heard the voice again. This was a familiar voice. The additional investigator immediately recognized it. It was the phone in the back pocket of the first investigator's trousers. As he moved around, he was inadvertently pressing the dial and a pre-recorded message was playing. It said, "if you would like to make a call, please hang up and try again."

After a while, he decided to give up the search for the original thumping sound and return to our little base of operations.

While watching the monitors at the base location, we started to go back through the footage by "scrubbing" along the timeline to see if we'd missed anything while watching live. We noticed that a chair in Nic's room appeared to have moved when no one was anywhere near it.

This got us all a little bit excited because it's not every day furniture moves by itself. Staff members told us that all the furniture in the area had to be arranged precisely or Cora's ghost might put things back where they belonged, so we were extra excited that our anomaly matched some of the location's lore. We spent most of the rest of the investigation attempting to study this moving chair. Of course, we continued with our initial monitoring as well, and spent much of the time in silence. But we spent those quiet times thinking of ideas about the chair and spent most of our breaks from silence testing them out.

We had little success. The video and audio monitoring during our quiet times didn't yield much of interest. Neither, though, did our attempts to recreate or explain the moving chair phenomenon.

Few things get us more excited than an unsolved mystery. One of the local news affiliates even ran a story about that moving chair, reporting it as exactly what it was at the time: a deeply puzzling mystery we couldn't figure out.

One thing we always remind people, though, is that just as we have to be comfortable with admitting when we don't have a solution, we should also always add the word "yet" to that statement. Often, we only get one crack at an investigation and so many of our unsolved mysteries do remain unsolved permanently. Some of the buildings in our case files don't even exist anymore. But when we do get the opportunity to follow up on an investigation, we take it.

We were fortunate to work in and with the Yak and Yeti for a period of many years, giving us ample opportunity to chase every lead we could think of. In that time, we never did figure out what our team member had been hearing around the light fixtures upstairs, but we did, after many *years* of work, figure out the moving chair. As it happens, we also got really lucky. All those years later, when we finally figured it out, it happened to be while the cameras were rolling for a story to be broadcast on national television.

The solution, and the reason it took so long to figure it out, was that the chair had to be arranged in such a particular way that our attempts to recreate it simply never hit the right combination of circumstances. One leg of the chair had to be situated on the base of the table while the other legs were on the floor, allowing it to wobble. One of the other legs had to be on a particular loose floorboard, *and* someone had to then step on the far end of that loose floorboard, causing the chair to tilt.

Figure 3.4. The haunted chair configuration (photo: Bryan Bonner).

In a way, this could still be seen as something of a ghost story. The only reason this works at all is because many of the wood floors in the building are still original. It may not have been a literal ghost moving the chair, but the chair only moved when one of us trod exactly the same floorboards as the home's original occupants more than 150 years ago!

Nevertheless, we were able to find this natural explanation. Recreating this without knowing the exact configuration necessary to make it happen would have been all but impossible, which is why it took us so long to figure out. Now that we know the solution, we were able to recreate the same phenomenon many times.

Stories like this contain an important lesson. Just because we haven't yet found a natural explanation for some phenomenon doesn't mean we can declare it proof of a ghost and give up. Paranormal investigation often means chasing moving goalposts. Until we find real incontrovertible proof of a ghost, we have to accept that no matter how strange a phenomenon might be, and no matter how many ways we've tried and failed to find the natural explanation, we have to remain open to the possibility that there is some such explanation. It's our job to keep looking for it.

On the other hand, just because we have solved some of the Yak and Yeti's ghostly mysteries, that doesn't mean the *other* phenomena necessarily do have natural explanations. We have to remain equally open to the possibility of a ghostly explanation. For example, that sound of thumping on the light sconces still remains unexplained. We'll have to keep looking.

For now, though, we have been able to close the book on some but not all of the ghost stories at the Yak and Yeti. We can say with certainty that it's a great spot with wonderful food and magnificent history. Beyond that, it remains an open case in our ledgers.

References & Further Reading

Austin. S. (2010). Ghost of Yak and Yeti. *Colorado Community Media.* <https://coloradocommunitymedia.com/2010/10/28/ghost-of-yak-and-yeti/>

Barton, B. J., Walker, R., Jr., Lutz, M. R., Gorrell, L., & Gorrell L. (1985). *Arvada: Just Between You and Me: 1904 – 1941.* Arvada, CO: Arvada Historical Society

Brick, C. (2018). *Haunted America: Ghost & Legends of Colorado's Front Range.* Charleston, SC: The History Press

Briggs, A. (2014). Yak and Yeti in Arvada celebrates 150th birthday of historic building. *The Denver Post.* <https://www.denverpost.com/2014/06/24/yak-and-yeti-in-arvada-celebrates-150th-birthday-of-historic-building/>

Family Search (n.d.) Cora Bell Rockwell. *Family Search.* <https://www.familysearch.org/en/tree/person/details/K4X7-9WC>

Family Search (n.d.) Elias William Van Voorhis. *Family Search.* <https://www.familysearch.org/en/tree/person/details/LZ2S-8M5>

Find a Grave (n.d.) Charles Erwin Allen. *Find a Grave.* <https://www.findagrave.com/memorial/19191784/charles_erwin-allen>

Find a Grave (n.d.). Eli Allen. *Find a Grave.* <https://www.findagrave.com/memorial/19191742/eli-allen>

Find a Grave (n.d.) Jane Anne "Jeannie" Christie Hutchings. *Find a Grave.* <https://www.findagrave.com/memorial/8601530/jane_anne-hutchings>

Genealogy Trails History Group (n.d.). Arvada. *Genealogy Trails History Group.* <https://genealogytrails.com/colo/jefferson/history_Arvada.html>

Yak and Yeti (n.d.). *The Yak and Yeti.* <https://theyakandyeti.com/>

4
Police and the Paranormal: Belmar Park

This is an unusual one for us because it is either one or two investigations, depending on how you figure it. Belmar Park is an open public park in Lakewood, Colorado, but it's home to one of the most remarkable historic properties we've ever seen. Heritage Lakewood Belmar Park, located at 801 South Yarrow Street, is something of a hybrid museum and park. If you look at the visitor center, it looks like any other small municipal museum. But if you wander through the park, you'll realize that the entire park is a museum, home to (as of this writing) fifteen historic buildings, many of which have been painstakingly relocated to the park from their original locations around town for exhibition and preservation. Our own investigation focused on two of those historic buildings within the park: the Streer-Peterson House, and the Ralston Country Schoolhouse.

Figure 4.1. The Streer-Peterson House (photo: Robert Lewis).

Figure 4.2. The Ralston Country School House (photo: Robert Lewis).

The History

Heritage Lakewood Belmar Park is a large historic property. No, that's not quite right. It *is* a historic property in and of itself, but much more than that, it is a *collection* of historic properties. It was established as a park in the 1970s. Since then, numerous historic buildings, typical of those seen in twentieth century Colorado, have been lifted from their foundations and moved, in their entirety, into the park, creating a sort of open-air museum with fifteen historic buildings and tens of thousands of museum-quality artifacts. People use it as any other public park—which it is—but they also use it as a museum, a place where history seems to come alive in a way quite unlike any other place in Colorado.

Because Heritage Lakewood Belmar Park (henceforth just Belmar Park for convenience, though technically the Heritage Lakewood historic park/museum is only part of the broader Belmar Park) is such an unusual property, it's difficult to know where to begin our history. Should we discuss the history of the Park itself? The individual histories of all the buildings that have been moved there? The histories only of those buildings we have (so far) had the chance to investigate? It's a difficult question. In fact, we initially considered dividing this chapter in two, dedicating each to one of the buildings, but that would seem to lose the opportunity to properly discuss the uniqueness of the Park as a whole.

As such, we're going to do a little bit of all of those options. First, we'll provide a brief history of the entire Belmar Park, because it's a fascinating tale in and of itself. We'll then give you a small overview of the buildings to which it's been made home. Finally, we'll give a more detailed history, to the extent we're able, of the two buildings we investigated individually.

Most of the discussions of the history of Belmar Park begin in 1973 when the City of Lakewood acquired 127 acres of land (later expanded to its current 132 acres) and formally established it as a public park. For our own history, though, we think it's important to go a little further back to see exactly how the property became what it now is. Without that context, we wouldn't even be able to understand why it's called "Belmar."

Our story begins before Lakewood was a city. In the early 1900s, several Jefferson County communities incorporated as cities, but the area that would become Lakewood did not until, following the Second World War, several Denver mayors developed an appetite for expansion and annexation into Jefferson County. Attempts to incorporate Lakewood began as early as 1947, but it wasn't until June 24, 1969, facing increased threats from then-Mayor of Denver Bill McNichols to annex much or all of east Jefferson County, that the city finally incorporated. Initially it was called Jefferson City, but its name changed to Lakewood just nine weeks later.[1]

Anyone from the Denver area has probably heard the name "Bonfils" (pronounced BON-Feez) plenty of times. The Bonfils family were among the wealth-

---

1   MacPhail, S. (2019). *The True Story of How Belmar Park Came into Existence: A Case Study of How Citizen Activism Can Work.* Archived online at <https://savebelmarpark.com/download/Belmar%20Park%20History%20%20-%20A%20Story%20of%20How%20It%20Came%20To%20Existence.pdf> (accessed July 13, 2025).

iest and most powerful people in Denver, and their influence, through a variety of philanthropic donations, is still felt throughout the Denver area today. Because a portion of the Bonfils estate is central to our story, it's worth taking a moment to provide some brief biographical treatment of a few key players.

Frederick Gilmer Bonfils (1860 – 1933) was one of the main movers and shakers of his day. In 1895, along with Harry Heye Tammen (see Chapter 1), he arranged to purchase the struggling *Denver Evening Post* for the sum of $12,500 (equivalent to about $478,000 in 2025 dollars), and together they renamed it *The Denver Post* and turned it into one of the region's most successful newspapers (and one of the few still in operation even to this day).[2] Though not strictly related to our story here, it's an interesting historic sidenote that Bonfils and Tammen took substantial bribes of $250,000 and 320 acres of valuable land to refrain from reporting on the Teapot Dome Scandal, which itself became relevant in one of our other paranormal cases (see Volume 2, Chapter 9) after being among the first to uncover the story.[3]

Another interesting side story occurred in conjunction with the case of Alferd (sometimes spelled Alfred) Packer (1842 – 1907), famous as the "Colorado Cannibal," subject of the fictionalized *Cannibal! The Musical* (itself an early work of the minds who would go on to create *South Park*), and namesake of the student dining hall at the University of Colorado Boulder,[4] in addition to countless paranormal stories and ghostly reports we have not yet found the opportunity to investigate. During Packer's trial for murder, *The Denver Post* ran a series of articles by Polly Pry, the pen name of Leonel Ross Campbell (1857 – 1938), accusing Packer of cannibalism, though they would eventually go on to assert that while he may have been guilty of survival cannibalism, he was innocent of murder.[5] In a case proving that both newspaper reporting and the practice of law were quite different at the turn of the last century, Packer's own attorney, W. W. Anderson (1845 – ?), having objected to some of Pry's articles, entered the offices of *The Denver Post* and shot both Mr. Bonfils and Mr. Tammen (neither fatally, though Bonfils' own injuries were such that his survival was unexpected and quite fortunate; some might even say miraculous) until he was restrained by Pry herself, subsequently arrested, and ultimately acquitted of the charges.[6]

2   Leavitt, C., & Noel, T. J. (2016). *Herndon Davis: Painting Colorado History, 1901-1962*. Boulder. CO: University Press of Colorado.

3   McElroy, J. (2024). Essay: A century ago, Denver newsmen helped unearth nation's biggest political scandal. *The Colorado Sun*. <https://coloradosun.com/2024/06/30/teapot-dome-scandal-civil-discourse-jack-mcelroy-carl-magee/> (accessed July 13, 2025).

4   A local historian told us another interesting side note about the dining hall. When it was established, the first sign announced it as the "Alfred E. Packer Dining Hall." The initial "E" was added in either accidental or comical confusion for Alfred E. Neuman of *Mad Magazine* fame.

5   Denver Public Library Special Collections and Archives (n.d.). Polly Pry (1857 – 1938). *Denver Public Library*. <https://history.denverlibrary.org/colorado-biographies/polly-pry-1857-1938> (accessed July 14, 2025).

6   Whitehead, C. (1942). A Glimpse of the Old Time Bar. *Denver Law Review, 19*(6): 147-148.

Following Mr. Tammen's death, much of his estate went into the preservation and expansion of the Children's Hospital (see Chapter 1), leaving the Bonfils family in charge of the *Post* and its operations.[7]

Mr. Bonfils and his wife Belle Barton Bonfils (1967 – 1935) had two daughters, Helen Gilmer Bonfils (1889 – 1972) and Mary Madeline "May" Bonfils (1883 – 1962), whose contributions to Denver history somehow managed to overshadow even such a storied legacy as their father's. Both girls had a strict Catholic upbringing and went on to contribute in a variety of ways to the development of business and culture in the Denver area, but after May decided to elope at the age of twenty-one years with a non-Catholic sheet music salesman named Clyde V. Berryman (1881 – 1959), the younger Helen became the "favored daughter" of the family, a development which will become quite relevant to our story later.[8]

For her part, Helen Bonfils assumed management of *The Denver Post* following her father's death in 1933, attempting to bring more journalistic integrity to the paper (rather than her father's more sensationalistic approach), acted at the Elitch Theatre (see Chapter 2) and on Broadway, and, lacking heirs of her own, made a legacy for herself and her family through a wide variety of philanthropic and cultural grants and investments both privately and through the Frederick G. Bonfils Foundation, which she ran after her father's death.[9]

Elder daughter May Bonfils, though elder siblings traditionally inherit the lion's share of the family assets and responsibilities, never achieved quite the same degree of notoriety, almost certainly as a direct result of family fallout from her aforementioned elopement with Mr. Berryman, of which her family did not approve. Nevertheless, she achieved numerous great things of her own. Though she disliked being in the spotlight, she was an accomplished musician and, like her sister lacking heirs of her own, dedicated the bulk of her fortune both during her life and posthumously to a variety of philanthropic causes including endowing the University of Colorado Medical Center with its Clinic of Ophthalmology, the Loretto Heights College with its library and auditorium, the Denver Museum of Natural History (now the Denver Museum of Nature and Science) with the Bonfils Wing, the United States Air Force Academy in Colorado Springs with its Catholic Chapel and many more, for which she was inducted into the Colorado Women's Hall of Fame in 1985.[10]

It was May Bonflis' estate that ultimately became Belmar Park, but to understand that is to understand complicated legal and real estate maneuverings that

7   Leavitt, C., & Noel, T. J. (2016). *Herndon Davis: Painting Colorado History, 1901-1962*. Boulder. CO: University Press of Colorado.

8   Varnell, J. (1999). *Women of Consequence: The Colorado Women's Hall of Fame*. Boulder, CO: Johnson Books.

9   *Ibid*.

10   Colorado Women's Hall of Fame (n.d.). May Bonfils Stanton. *Colorado Women's Hall of Fame*. <https://www.cogreatwomen.org/project/may-bonfils-stanton/> (accessed July 14, 2025).

Riley, M. G. (2006). *High Altitude Attitudes: Six Savvy Colorado Women*. Boulder, CO: Johnson Books.

Varnell, J. (1999). *Women of Consequence: The Colorado Women's Hall of Fame*. Boulder, CO: Johnson Books.

we'll try to summarize here even though the full story is quite beyond the scope of this volume.

The estate in question was a roughly 250-acre parcel of land in what eventually became Lakewood. A large parcel of that land originally belonged to founder of Denver's First National Bank Charles Brewer Kountze (1844 – 1911) for whom Kountze Lake (now part of Belmar Park) is named. He either sold or leased a large parcel of his land, including the lake, to Frederick Bonfils, who combined that parcel with other acquisitions such that at the time of his death, he owned the land in question.[11]

Upon his death in 1933, Mr. Bonfils left an estate worth just north of $14 million (which we have adjusted for inflation to be worth just shy of $350 million in 2025 currency). Perhaps without realizing it, he also left a legal battle that would dominate his daughters' lives for the following years and cause them to completely disown each other. Because of his disapproval of May's marriage to Mr. Berryman, the entire estate went to Helen except for an annual allowance of $12,000 per year (just under $300,000 in 2025), increasing to $25,000 should she divorce, though a court ruled that the will's encouragement of divorce was unenforceable and awarded her the full amount. Following Belle Bonfils' death in 1935, an additional $10.5 million was added to the estate, also all going to Helen. Subsequent legal battles between the sisters resulted in the courts awarding May $5 million in cash from Mrs. Bonfils' estate, a small cash award from Mr. Bonfils' estate, 15% of *The Denver Post* stock, and 10 acres of the estate ultimately destined to become Belmar Park.[12] Contradictory reports suggest she may have won all 250 acres of what we'll call the Belmar Estate (though that name came later), but it is clear that either through inheritance and lawsuit or through purchase, she eventually acquired the entire parcel and added an additional 500 acres, bringing the grand total of the estate to a whopping 750 acres of prime real estate.[13]

The grand irony of the whole matter is that the marriage which caused a rift in the family was essentially over long before these legal battles. May Bonflis and Mr. Berryman had separated as early as 1916, though they remained legally married and she continued to pay him a small allowance before finally managing to secure a divorce (and restoration of her maiden name) in the mid-1940s.[14] Had

11   MacPhail, S. (2019). *The True Story of How Belmar Park Came into Existence: A Case Study of How Citizen Activism Can Work.* Archived online at <https://savebelmarpark.com/download/Belmar%20Park%20History%20%20-%20A%20 Story%20of%20How%20It%20Came%20To%20Existence.pdf> (accessed July 13, 2025).

12   Varnell, J. (1999). *Women of Consequence: The Colorado Women's Hall of Fame.* Boulder, CO: Johnson Books.
Wood, R. E. (2005). *Here Lies Colorado: Fascinating Figures in Colorado History.* Helena, MT: Farcountry Press.

13   MacPhail, S. (2019). *The True Story of How Belmar Park Came into Existence: A Case Study of How Citizen Activism Can Work.* Archived online at <https://savebelmarpark.com/download/Belmar%20Park%20History%20%20-%20A%20 Story%20of%20How%20It%20Came%20To%20Existence.pdf> (accessed July 13, 2025).

14   *Ibid.*

May been a bit more aggressive in divorce proceedings or had Mr. and Mrs. Bonfils survived another decade to see the divorce finalized, it seems quite likely the
entire legal battle may have been avoided and the Bonfils family unity preserved,
though this is of course only blind speculation on our part.

Once the legal battles had been sufficiently settled, May set about turning her
land into a secluded paradise and began construction in 1936 on a twenty-room
mansion modeled on Marie Antoinette's *Petit Trianon chateau* in Versailles, France,
bringing a piece of European decadence to what would become Lakewood. The
white terra-cotta mansion was completed in 1937 by Denver architect Jacques
Benedict and included a chapel where local priests could read private masses
as well as a dental clinic.[15] She named the entire estate "Belmar" combining her
mother's first name (Belle) with her own given first name of Mary.[16]

At some point, May Bonfils met Charles Edwin Stanton (1909 – 1987) and
engaged him to install an elevator in her mansion. This meeting would ultimately
lead to their marriage in 1956 when she was seventy-three and he was forty-six
years of age. Reports seem to suggest that though they may have had affection
for one another, the marriage was in large part an arrangement to secure the management of the Belmar estate and to ensure Helen Bonfils would not be able to
inherit any of it. Though their marriage was a civil one, they renewed their vows
in Rome in 1961 and received a blessing by Pope John XXIII.[17]

Upon her death in 1962, May Bonfils Stanton left about half of her estate
to Mr. Stanton. In 1970, following his late wife's wishes, he donated the mansion
and ten acres of the estate to the Archdiocese of Denver but continued working
on managing the remainder of the estate and administering philanthropic causes,
separately from Helen's own charitable work, through the Bonfils-Stanton Foundation until his own death in 1987. The donation to the Archdiocese mandated
that the property could only be used for religious purposes and, if sold, the sale
must mandate the destruction of the mansion so it could not be used for any
other purpose.[18]

15   Varnell, J. (1999). *Women of Consequence: The Colorado Women's Hall of Fame.*
Boulder, CO: Johnson Books.
Riley, M. G. (2006). *High Altitude Attitudes: Six Savvy Colorado Women.* Boulder, CO:
Johnson Books.
16   MacPhail, S. (2019). *The True Story of How Belmar Park Came into Existence:
A Case Study of How Citizen Activism Can Work.* Archived online at <https://
savebelmarpark.com/download/Belmar%20Park%20History%20%20-%20A%20
Story%20of%20How%20It%20Came%20To%20Existence.pdf> (accessed July 13,
2025).
17   Varnell, J. (1999). *Women of Consequence: The Colorado Women's Hall of Fame.*
Boulder, CO: Johnson Books.
Riley, M. G. (2006). *High Altitude Attitudes: Six Savvy Colorado Women.* Boulder, CO:
Johnson Books.
18   MacPhail, S. (2019). *The True Story of How Belmar Park Came into Existence:
A Case Study of How Citizen Activism Can Work.* Archived online at <https://
savebelmarpark.com/download/Belmar%20Park%20History%20%20-%20A%20
Story%20of%20How%20It%20Came%20To%20Existence.pdf> (accessed July 13,
2025).

Enter another major player in the development of Belmar Park. Local developer Gerri Von Frellick (1916 – 1933), working in conjunction with Mr. Stanton (and in accordance with Bonfils' vision for the land) built the Villa Italia indoor shopping center on Bonfils-Stanton land in 1966 (this shopping center would be remodeled several times before finally closing in 2001, but that's a side note to our story). Both because he liked the idea of a park and because he was concerned about commercial development in the area damaging the value of his Villa Italia property, he joined forces with the local effort to create a Belmar Park, which had begun to spring up in the 1960s as citizens thought it would be the best use for the estate.[19]

By the 1970s, those efforts were becoming more organized. The Ladies Auxiliary of the Knights of Columbus (the Catholic fraternal organization) offered tours of the Belmar estate as fundraisers in 1971, but then the Archdiocese determined they couldn't find sufficient religious purposes for the property and put it up for sale. Meanwhile, a student-led effort to save Kountze Lake sprung up following observations that water levels were depleting (which itself sparked an administrative controversy surrounding whether the body of water was a "lake" or a "catch basin," because different authorities would have power and responsibility to tend to different types of body of water. It turned out that the Archdiocese had sold the land to a company called Craddock Development, and just about everyone except a group of citizens, students, and Mr. Von Frillick were interested in (and probably draining the lake to begin the process of) turning the property into commercial development. But they'd require City Council to approve new zoning ordinances for the land.[20]

The resulting legal, administrative, and public relations battles are complicated and convoluted in a way only municipal governments could possibly manage, and frankly probably boring to anyone who doesn't practice real estate law as a profession. But the long and short of it was that by 1972, students, citizens, historians, and those interested in preserving both the magnificent old mansion and the beautiful open space were engaged in all-out war with the City Council, Archdiocese, developers, and commercial interests over the fate of the Belmar Estate. In March of that year, the City Council approved the Craddock-proposed rezoning, prompting a lawsuit to overturn the decision. Press coverage and comments from certain members of the public, in turn, caused Craddock to file a defamation suit against several of those interested in the estate's preservation. Appraisals were done and redone. Issues were raised, revised, and dismissed. The whole process became something of a legal and political quagmire until it finally reached the voters in January of 1973, offering a two-part question, the first of which would authorize the creation of a park and the second a means to fund its

Varnell, J. (1999). *Women of Consequence: The Colorado Women's Hall of Fame.* Boulder, CO: Johnson Books.

19   MacPhail, S. (2019). *The True Story of How Belmar Park Came into Existence: A Case Study of How Citizen Activism Can Work.* Archived online at <https://savebelmarpark.com/download/Belmar%20Park%20History%20%20-%20A%20Story%20of%20How%20It%20Came%20To%20Existence.pdf> (accessed July 13, 2025).

20   *Ibid.*

acquisition. It was approved by a two-to-one majority of voters and Belmar Park was established, though it would take some years before Kountze lake was refilled (and then, only to about half its original size).[21]

Unfortunately for lovers of architecture, though the citizens groups got their wish and the park was established, the mansion itself was lost. It was demolished in 1971 to make way for "Irongate Executive Plaza," an office park named for the mansion's original iron gate and only surviving member.[22]

Heritage Lakewood Belmar Park, the historic portion of the broader park, began its history with the opening of the Lakewood Heritage Center at 801 South Yarrow Street in 1986 on the 100th anniversary of Colorado's statehood. Initially it was planned to be a museum of twentieth century Colorado history but plans were quickly expanded to include the relocation of historic buildings from around the Denver area, thus preserving historic buildings (albeit not in their original locations) which would otherwise be lost either to disrepair or to make way for further developments. In a sense, we can view the loss of May Bonfils Stanton's Belmar mansion as the impetus for an enterprise which has gone on to preserve numerous other buildings which otherwise likely would by now have been destroyed.

It is to those buildings we now turn our attention, focusing first on the two we had the opportunity to investigate, and with greater detail on the history of those, followed by brief mentions of the other features in the Park.

We begin with the Streer-Peterson House, built sometime in the 1870s or 1880s. It originally stood where Hole #1 of the Fox Hollow Golf Course is today, and was picked up from its foundation and moved to Belmar Park in 1986, where it's staged for visitors with furnishings from the 1920s.

Our story here begins with Morris Streer (1892 – 1955), sometimes spelled "Strear," though historic documents show he spelled it "Streer," a Russian Jewish immigrant to the United States who married Sophia Dinner Streer (1896 – 1978), another Jewish immigrant from Russia, in Colorado in 1913; they had at least four children we know about: Ruth (1914 – ?), Dorothy (1915 – 2006), Sidney (1917 – 1988), and Leonard (1924 – ?).[23] They weren't the first owners of the house, but they are the first whose names we've been able to identify, though we've heard (through correspondence with Belmar Park employees and tour guides) the home housed eight families over the years, and some of the Streers' predecessors included at least one who was flooded out, one perhaps forced out by the 1893 silver panic (the largest economic depression in American history until the Great Depression), and others who may have relocated due to drought. The Streers maintained the property from 1918 to 1921 and also operated the Morrison Farm and Dairy.

Prior to our investigation, we were told a story which we've so far been

21   *Ibid.*

22   Noel, T. (2018). May Bonfils and Her Lost Belmar Mansion: A Lavish Lakewood Estate House a Wealth of Benevolence. *History Colorado.* <https://www.historycolorado.org/story/articles-print/2018/10/19/may-bonfils-and-her-lost-belmar-mansion-lavish-lakewood-estate> (accessed July 14, 2025).

23   Family Search (n.d.) Morris Strear. *Family Search.* <https://ancestors.familysearch.org/en/LT98-1PV/morris-strear-1892-1955> (accessed June 14, 2025).

unable to confirm through public records but repeat here because it has become a part of the building's lore. We were told that during the Streers' residence of the house, they allowed friends (identified as Milton Goldblatt and Harry Katz, though again we have not found sufficient records to confirm these identities) to use one of the outbuildings to operate an illegal still. Colorado established prohibition even before the United States did, so alcohol was illegal in Colorado between 1916 and 1933. According to this legend, their bootlegging operation was so successful it was generating incomes of up to $1,000 per day (equivalent to about $16,000 per day in 2025). Further, the legend says their building was the subject of law enforcement raids in 1919 and 1921, one of which was the largest still bust ever in Jefferson County history.

Given such a large operation, we expected to be able to find more information on the raids and busts, but record-keeping in those days was not as good as today; furthermore, not all historic records have yet been digitized and made available for public inquiry, so it may be a true story whose details are buried in archives we've not yet had the opportunity to visit.

Figure 4.3. The house was staged with items appropriate to a 1920s Jewish family; note the prayer shawl and yarmulke (photo: Robert Lewis).

Much of the home's staging in its current location is consistent with a 1920s Jewish immigrant family, though the items and artifacts were not the specific ones that belonged to the Streers. At the front door is a *mezuzah* holding a small scroll of Jewish scriptures or teachings meant to bless those who entered the house. The front room is equipped with a *menorah*, and a dresser on the second floor is a *yarmulke* and prayer shawl.

Additional furnishings outside the building include farmyard equipment appropriate to the period but sourced from different locations. Many Americans farmed their own produce in their back yards at this time due to rations during the First World War, so though the objects were sourced from elsewhere, they would have been familiar to the Streers. One of the most prominent features is a windmill Belmar Park's management sourced from a Lakewood property which was manufactured by the Dempster Mill Manufacturing Company in 1922. Park tour guides told us the windmill is still operational but not connected to a well so it's not in use.

One of the later sets of occupants before the building stood empty for a while was the Peterson family, headed by Arthur Peterson (dates unknown). We haven't found much information about these people except that they owned the house between 1939 or 1940 and 1974 or 1975. One of the tour guides told us they still have descendants living in the area who occasionally visit the house. It seems they may not have occupied the home for the entire duration of their ownership. One report, again from a tour guide, said the home was vacant for several years before being purchased by the Army Corps of Engineers in 1975.

Regardless, we do know its original location is now part of a golf course and that it was preserved and relocated to Belmar Park in 1986.

Figure 4.4. The Streer-Peterson House windmill (photo: Robert Lewis).

What we know of the Ralston Country School House, the second of the locations to which we were given access for our investigations, likewise comes from Belmar Park staff and tour guides.

It was built in the 1860s and originally served as a Methodist-Episcopal church in Arvada but was converted into a one-room schoolhouse shortly thereafter in 1871 when the church's congregation split.

Education in those days was not the industrialized process our public education system is today. Modern education is based on the Prussian model of education in which students are grouped by age and pushed through public schools with a goal of creating quality employees for industry or military service. Though this model was introduced into American education in limited regions as early as the 1840s, rural or semi-rural education in the 1870s featured these one-room schoolhouses serving students of varying age under what's sometimes known as the Bell-Lancaster method in which students would largely teach each other with teachers serving as guides or monitors. Compulsory education was uncommon until the twentieth century.

A second room was added to the building in 1919, allowing students to be divided into two groups (one for grades one through four, and the other for grades five through eight). In 1956, the building was relocated to the grounds of Vivian Elementary School in Lakewood where it was used as a temporary school building. It was slated for demolition in 1982 but the school district had the wisdom to preserve it by donating it to Heritage Lakewood. It was moved to Belmar Park in 1983. Two outhouses were added in 1998, constructed by a local troop of Eagle Scouts.

Figure 4.5. Ralston Country School interior (photo: Robert Lewis).

The interior of the schoolhouse is decorated much like a schoolhouse from that era would have been. The main building consists of one large room with just a few offshoot rooms serving as restrooms, storage, and a kitchen. Bookcases line some of the walls, blackboards line others, and there's an organ in one corner suitable for teaching a music class. Next to the organ is an American flag. Careful observers might note that this flag only has forty-eight stars in its design. That's because it's a period-accurate flag. Each star represents one of the States in the Union, but Alaska and Hawaii weren't admitted to the Union until 1959 (several months apart).

The second room, added in 1919, is used currently as storage for all of the desks and chairs, keeping the main room as one large empty space.

While those were the only two locations we investigated (at least for now; we'd welcome a return invitation to look at some of the other properties at some point in the future), we should also mention some of the other buildings preserved in Belmar Park.

The Estes Motel, built in 1948 or 1949, is preserved as a window into life in the 1950s. It was built and operated by Clifford and Christine Estes who originally used it as a home but expanded it into a motel some years later, responding to increased traffic at its original location on West Colfax Avenue.[24]

24   Incidentally, we could write an entire book just on the ghost stories of Colfax. Local residents know it as a strange stretch of road crossing a good portion of the Denver area. Though it was originally home to elite residential districts, most

Figure 4.6. Ralston Country School organ and 48-star flag (photo: Robert Lewis).

Ethel's Beauty Salon, also known as Gil and Ethel's, was built in 1948 and originally served as a laundry. In the 1950s, it was used as a "five and dime" store, a type of store popular in the early to middle twentieth century in the United States where people could buy a variety of goods for either a nickel or a dime. They were the spiritual forerunners of the dollar stores many of us might be familiar with today (even though many of them, due to high rates of inflation, no longer charge only a dollar for their goods). It was then purchased in 1961 by Gil and Ethel Gomez who operated it until 1996 as a beauty salon. It was moved to Belmar Park in 1998, with the front room staged to look like the five and dime store and the back room staged to look like Ethel's Beauty Salon.

The White Way Grill, also built in 1948, was another East Colfax property originally located just east of the historic Aurora Fox Theater. Diners of the era were built as prefabricated constructions (some were repurposed railway dining cars). This one was built by the Valentine Diner Company in Wichita, Kansas and took its name from the "great white way," a reference to the bright lights in theatre districts. It served its original function until it was moved to Belmar Park

of its length was subsequently rezoned for commercial purposes and unfortunately many stretches of its length are now known as sketchy neighborhoods, though current revitalization efforts are underway in several areas. It's known as the longest continuous commercial street in the United States and many of its historic buildings are reputed to be haunted; we have only just begun to scratch the surface of all of them in our investigations.

in 2003. A neon sign in the front window is a reproduction of the original, which would signal to hungry passers-by when the diner was open.

The Calving Barn, Machinery Sheds, Auction House, and Caretaker's Cottage were all part of the original Belmar Estate. During May Bonfils' life, she used a portion of the land to house deer and cattle, and Mr. Stanton continued maintaining her prized flock of animals between her death and his donation of the Belmar estate in 1970. Though the mansion was torn down, these buildings still stand as part of the historic Belmar Park. All of them except the Auction House are in their original locations—the Auction House once stood where the Ralston Country School House now stands, so it's been moved a bit but still stands on the overall property. Most of these buildings were used for exactly what their names imply. A calving barn is a structure designed to house cattle throughout the birthing and nursing process. The auction house was used for livestock auctions. The machinery sheds were housing and storage for farm equipment, and the caretaker's cottage was a home for an on-site permanent caretaker.

The Farmhouse and Turkey Coop has a somewhat uncertain early history. According to oral history, it began its life sometime in the late 1800s as a carriage house on the Hallack Estate located at South Pierce Street and West Alameda Avenue, but insufficient records exist to prove or disprove this story. The tale continues that sometime in the 1910s, the building was hoisted onto logs and by that means rolled, intact, to the northeast corner of West Ohio Avenue and South Wadsworth Boulevard. This is certainly possible. Obviously with modern technologies, numerous buildings have been relocated in one piece to Belmar Park. But given the time period, it would have been a remarkable—but certainly plausible—achievement. It's thought based on structural evidence that at least two new additions were made to the home at this location. More direct evidence substantiates the remainder of its history. It changed hands several times, serving as a fish farm, a sheep farm, and in the 1940s, the Maplecrest Turkey Farm.

In 1956, the powers that be decided they needed to widen Wadsworth Boulevard and it was slated for destruction. By coincidence, none other than May Bonfils Stanton owned it at this time and saved it by selling it to Lucy Webber who had it moved to West Yale Avenue and Zuni Street. Apparently this home was cursed with the bad luck to always be in the way of road projects because in 1980, Zuni Street was in need of widening and the house was again targeted, but was given to Belmar Park where it has now found its permanent home.

Paranormal Claims

Despite the high concentration of historic buildings from around the greater Denver area located in Belmar Park, it's not one of the most famously haunted places in the state. We haven't seen it on any "most haunted" lists or anything like that. Nevertheless, many of its buildings have at least one or two ghost stories volunteers, staff, and guests have reported witnessing over the years. Park staff have dutifully collected and curated these stories, and they provided them to us at the beginning of our own investigation.

As we did in the history section above, we'd like to present not only the ghost stories connected to the two buildings we had the opportunity to investigate but all of the ghost stories from Belmar Park we're aware of. Contrary to the history section, we will begin here with the stories we did *not* have the chance to investigate, and conclude this section with the Streer-Peterson House and the Ralston Country School in preparation for our own investigation in the following section.

One thing to keep in mind as we begin these ghost stories is that several people believe May Bonfils Stanton continues to haunt numerous locations around the entire Belmar property.

At the Estes Motel, no specific ghosts have been identified, but some visitors report feeling as if they're in the presence of "energy" from some of the thousands of guests who stayed at the Motel when it was in operation. Two guest rooms in the Motel have been staged and preserved as if they were still occupied (perhaps with the guests just off doing tourist business for the day) and some people think ghosts of prior guests might enjoy that sense of familiarity.

Of the Belmar Park-original buildings (Auction House, Calving Barn, Caretaker's Cottage, and Machinery Shed), the only one specifically identified with any ghost stories we've heard of is the Calving Barn, where an employee has reported—on more than one occasion—hearing the sound of footsteps in the barn when no one else was there, sometimes accompanied by the sound of creaking steps on the staircase.

Even some of the modern additions to Belmar Park have ghost stories of their own. The museum and visitor center isn't a historic building itself, but it is home to a wide range of historic artifacts, and some believe ghosts may have attached themselves not so much to the property but to some of the items contained therein. Staff working at the gift shop have reported arriving to work in the morning only to discover books and other items have been dislodged from their shelves and strewn about the floor or shuffled around to incorrect locations.

Doors are said to open and close of their own accord throughout the building. People hear footsteps and creaking floors when they're alone in the building. Some have even seen strange shadows, again when they were alone.

One employee reports working alone in the building in the evening (after closing time but not late at night). She said she heard someone call her name. Upon investigation, though, there was no one else in the place.

Another modern building is the Conservation and Preservation barn, where the museum's permanent collections are stored when under study, in for repair, or otherwise not out on display. One former employee reported working on the first floor of this building one day when he heard, coming from upstairs, the unmistakable sound of footsteps along with something being dragged across the floor. As has been the case with most of these stories, he was alone at the time but went upstairs to investigate. Of course, no one was there. He did, however, find the source of the dragging sound: a pitchfork. It was one of the objects in the museum's collection but out of its proper place, and it is said it was being dragged across the floor by some ghostly presence—perhaps the spirit of a long deceased farmer.

Finally, we come to the ghost stories of the two buildings we had the chance to investigate. Ghost stories are, unfortunately, fairly limited about these locations but even one good tale is enough to catch our interest, and there are at least a couple good tales.

The Streer-Peterson house only has one real ghost story, but it's been independently repeated by at least two visitors to the Park who had no knowledge of each other. In both cases, the witnesses reported that when they were walking through the park on a moonlit night, they happened to look up to the second floor window at the front of the house where they say a woman in farm clothes standing by the window, looking out over the sidewalk. Of course, these events took place after hours when the house was locked up and no one was inside.

That brings us to the Ralston Country School, where a few different stories have been reported. The first was given to us firsthand by the witness, who was one of the staff members with whom we worked on our investigation. It happened during the winter season and he was cross-country skiing through the park alone one evening. Drawn for some unexplainable reason to the School, he stopped to take a photo with his smartphone. On the photograph, he saw a mysterious mist he couldn't see in person, and part of that mist took the shape of a mysterious—and, to him, terrifying—human face.

On a separate occasion, a painter was working in the building doing some routine touchups to the paint. While working, he heard the sound of someone sweeping the floor. As has been the case with most of these stories, he was alone at the time and upon searching the building found no one else had joined him. He thinks it could have been the ghost of a former teacher sweeping the school to keep it clean and ready for students.

Speaking of the students, there are also several reports of ghostly students in the school. People don't seem to have ever actually *seen* any of them, but multiple witnesses say they've heard the sound of marbles rolling across the floor and attributed it to the ghosts of former students playing marbles in their old schoolhouse.

Armed with those ghost stories, we were ready to begin our investigation.

Our Investigation

The lead-up to our investigation came in the form of an inquiry as to our public educational offerings. Around the Halloween season, staff and volunteers at Belmar Park had started offering "haunted" tours of the property, sharing their knowledge of the history combined with some of the ghost stories they'd collected. Events like those were a smash hit. They wondered about bringing us in to give a similar presentation, either as a tour or as a lecture, for interested parties. We said of course we'd be happy to, but that it would be even better if we could be given access to one or two of the historic buildings in advance so we could conduct a proper investigation of our own (and *on* our own) before bringing the stories to the public. They agreed and after a bit of discussion we settled on the Streer-Peterson House and the Ralston Country School House as the sites of our investigations, to be conducted on separate evenings. After that, we'd give a public lecture to share our results with their patrons and Park staff would be on hand to help people tour some of the buildings after the lecture. The plan was in motion!

The first building under investigation was the Streer-Peterson House. We knew the only real ghost story that house had was the woman in the window, so we planned to dedicate the bulk of our investigative efforts to making sure that window was under constant surveillance. However, we're nothing if not thorough, so we brought in all the equipment and planned to monitor as much of the building as possible. Fortunately, it's a small house so it wasn't a terribly complicated set-up.

When we arrived at about 6:00 in the evening, we took note of several things. First, it was a ghastly hot afternoon, and the old building had no air conditioning, so we left the front and back doors open while we set up to hopefully create a cross-breeze. It helped a bit, but not enough for real comfort. When we entered, we also noticed the entire house—particularly the kitchen and entryway where we were to establish our base of operations—was permeated by a truly foul stench.

Figure 4.7. The source of the foul stench (photo: Robert Lewis).

We quickly found the source of the unpleasant aroma—there was a dead and decaying rabbit under the front porch, and several dead rats scattered about the property near the house. Several rodent traps had been placed around the property, so we weren't surprised to discover a rodent problem—it was, after all, an open space park with a bunch of historic buildings scattered throughout it. To find several dead specimens all at once and all clustered around this one house, though, was unusual. We're not saying it's paranormal, but we just hoped that wasn't going to be an omen of things to come for our evening. To his great credit and with our eternal gratitude, our staff host for the evening took it upon himself to remove the carcasses while we finished setting up our equipment, and the smell faded rapidly.

On a side note concerning the animals in the Park, we did notice later in the evening there were several coyotes about, presumably looking to snack on some of the rodents we'd seen. We detected them first by sound and then saw a couple of them lurking in the distance. They didn't give us any trouble and we left them alone, but we did take note of their presence both for safety and in case their presence might explain any unusual sights or sounds we might detect throughout the investigation.

As we started setting up, we went through our usual routine of thoroughly mapping out the entire building and its surrounding land and documenting everything with photographs. This was the first investigation on which we used spherical (or "360-degree") cameras to ensure we didn't miss a single inch of the property. Video files on those cameras are huge, but we think it's worth the more time-intensive data processing and larger file sizes to get a more complete picture of the location, though we still use good old fashioned still and video cameras as well.

The building is a small two-floor house, plus an underground cellar with an access door in the back yard. The cellar was being used, at least as of our investigation, as storage for Christmas decorations. It's delightfully creepy, as all cellars tend to be, but we didn't choose to focus too much on it because as an underground cellar, it was part of the new foundation rather than part of the original house from its original location. It didn't become relevant to our story until, oddly enough, another evening when we were on our investigation of the nearby schoolhouse, but we'll get to that in a moment.

For this one, we used a four-camera surveillance set-up, with two cameras in the upstairs master bedroom (the one with the window where people reported seeing a ghost overlooking the sidewalk)—one camera near the door with a view of the window and one on a dresser near the window looking over the bed, where we also placed an air quality monitor within the camera's view. Another camera was placed in a child's bedroom also on the upper floor, and one additional camera was in the downstairs living room or parlor. Two microphones were sufficient to hear anything in the house, and we placed one in the same downstairs living room/parlor and one near the door between the two rooms upstairs. Our base camp was in the kitchen on the downstairs floor, between the front and back doors. We also placed "cat balls"—small lightweight balls that light up when jostled or disturbed—in front of two of the cameras and a stationary EMF meter in front of one camera, just in case.

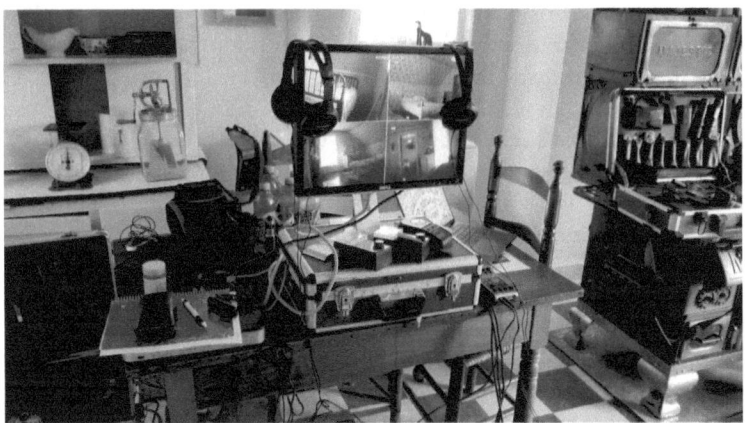

Figure 4.8. Our base of operations (photo: Robert Lewis).

Set-up took us about two hours. That was slower than usual for us, especially for such a small property, but the delays were caused by the search for the foul smell (and subsequent removal of the rotting carcasses) and a few interruptions by passers-by who were just walking through the park and wondered what we were doing.

When all of that was done, we took a round of baseline measurements, monitoring temperature at two locations in the house, and EMF and seismic activity at five various locations throughout the building. Throughout the evening, we would take such measurements hourly. At this point, we can jump ahead to the conclusions at least with regard to those measurements and say we did not find any anomalies. EMF levels were low and consistent throughout the evening, we found no measurable seismic activity, temperatures decreased only slightly as ambient temperatures fell outside, and air quality was consistently clean. We did take the opportunity to test for lead paint in the building's interior because long-term lead exposure can sometimes lead to a variety of mental conditions which might be implicated in a paranormal investigation. Indeed the building does contain lead paint; however since no one is living there and the only reports of ghostly activity we've heard have been from witnesses walking outside, we only note that finding parenthetically. It doesn't seem to have any relevance to the paranormal claims.

As we monitored the video and audio throughout the investigation, we found the place was surprisingly noisy. Nothing was happening on our cameras, but every few minutes we heard another sound that needed to be identified. Most of those were quickly and easily explainable as people walking through the park or the sounds of traffic passing on a nearby busy road. On the former point, we need to point out that though the buildings close and are locked after business hours, the park itself is open to the public all night, so we did notice a small number of late-night walkers.

Things got more interesting at about 10:30 that evening, though, when we heard a strange beeping sound. It sounded like the chirp of an alarm system. We knew it shouldn't be coming from any of our own devices, but we figured perhaps it was part of a security alarm or fire alarm in the building. Our host wasn't sitting with us all night, but was still elsewhere on the property doing some of his work, so we just filed it away for later, figuring we'd ask him when we regrouped later on.

We thought we got lucky a little later. We heard a knock at the front door. We'd asked our host to send a text message before he returned so he wouldn't accidentally interrupt any critical experiments or monitoring but figured he just forgot and was checking in on us (as he promised to do a few times throughout the evening). But when we answered the door, no one was there. We quickly stepped outside with our flashlights and wandered a bit up and down the street and around the house to see if anyone was there. No one was. We have no idea who—or what—made that noise that sounded to all the world like a knock at the front door.

Monitoring continued and every few minutes, in between sounds we could quickly identify as discussed above, we heard sounds that reminded us of footsteps and, perhaps even stranger, of furniture being moved around. Recall, though, that we had cameras all through the house, and could tell that no furniture was actually moving. Nor was anyone else in the building—we were the only ones there and we were sitting on our butts gathered around our monitors exchanging confused looks with each other as we listened.

It got stranger still. About two hours later, we heard another beeping sound, but this one wasn't just one or two chirps. It was loud and persistent, as if someone were playing a note from a tone generator. Later investigation, when we brought our recordings back home and ran them through audio analysis software, revealed it was indeed a pure tone at a frequency just shy of 1700 Hz. We spent a few minutes double-checking all of our equipment to make sure nothing was playing a tone or alarm from some setting we'd never heard before—none of our equipment was doing anything strange. As far as we could tell, none of the electronics in the house were doing anything either.

When he returned at the end of the night, we asked our host if there were any kind of alarm or anything and told him about the sound we'd heard. He said no alarms went off in that building—or anywhere in the Park for that matter—and the sound we described was quite unlike anything he'd heard in the building or from any of the Park's security systems before.

We wrapped up this investigation in the wee hours of the morning, a bit excited but also frustrated by the mystery. We hoped that when we returned about a week later to look at the School House, even though we'd be in a different building, we might experience some of the same phenomena and be able to pinpoint their source.

Life had other plans for us. The second investigation ended up being another eventful night, but for entirely different reasons. Just to offer a final note on the Streer-Peterson house before we move on to the School, we can say we never were able to explain the knocking, footsteps, sounds of furniture moving, nor the strange tone. No mysterious ghostly woman ever materialized in that second floor window like the ghost stories promised, but we got some new mysteries of our own to consider. As for the woman in the window, we remind the reader that just because we didn't see anything during our investigation doesn't mean there never was such a ghost. But on the other hand, just because we weren't able to explain by natural means all the things we experience doesn't necessarily mean they were caused by a ghost, either. These have to be filed away as unsolved mysteries and perhaps someday the subject of a follow-up investigation.

But now, on to the Ralston Country School House!

Our plan for this investigation followed a similar protocol to its predecessor the week before. Because the bulk of the building consisted of one large room with just a few smaller rooms as offshoots, we figured we would dedicate the bulk of our monitoring to the main room, and established our base of operations in the hallway near the front door and kitchen. We placed three cameras in the main room, strategically located to give us a view of the entire room, and a fourth in the second room used for desk storage. We placed a stationary EMF meter in that second room, and cat balls throughout the main room to determine if anything was moving.

Three microphones were placed around the building and a fourth just outside the front door to determine if any anomalous sounds might be coming from outside. We remembered the strange sounds we heard at the Streer-Peterson House the week before and wanted to make sure we'd be better equipped to triangulate the source of any of the sounds by determining which microphones picked up more or less volume of any noises detected.

Because one of the stories involved people hearing the sound of ghostly children playing with marbles, we staged a small collection of marbles and little "jingle bells" near one of the microphones by the teacher's desk. We figured if any ghosts happened to move those around, even if the movement was too subtle for the cameras to pick it up, we should hear the jingling of the bells.

Figure 4.9. The teacher's desk (photo: Robert Lewis).

As before, we concluded our setup with baseline measurements of EMF, temperature, seismic activity, and other environmental conditions. This building tested negative for lead paint and air quality remained clean throughout the evening. Everything was set-up by about 7:00 in the evening, though we had some guests from the Belmar Park staff observing us for the first hour so we didn't "go quiet" until about 8:00.

In terms of results, we'll begin by talking about the hourly EMF, temperature, and seismic monitoring. For the most part, these readings were consistent, with one exception we'll get to in a moment. Temperatures were largely consistent, again just decreasing slightly as outdoor air temperature cooled as the evening wore on. We also didn't detect much of anything on the seismometer, nor most of the EMF readings.

In order to understand the one exception, we need to first clarify (or remind readers of our prior volumes) that we measure two kinds of electromagnetic radiation during our investigations. Alternating current (AC) fields are primarily generated by manmade electrical equipment such as lights, computers, security systems, and so forth; direct current (DC) fields are more often naturally-occurring. We measure for both, but we have to be very careful with the latter because even small movements of a human body in a building with our sensitive meters can cause false positives.

In the case of the Ralston Country School House, we never detected any anomalous DC readings. Similarly, AC readings were very low in most of the locations and consistent with nothing other than the building's own lights or our own equipment. However, when we took measurements in the back room where the desks were stored, our readings would occasionally jump from about one milligauss (consistent with readings in the rest of the building) to three or four milligauss. While these are still not *strong* fields—nothing anyone should ever worry about—we couldn't figure out what was causing those spikes. Nothing was being turned on or off, nothing was moving around the building, and we couldn't find any major electrical boxes or hubs near that room (either inside or on the building's exterior). We suspect these EMF spikes may have been caused by some electrical device inside the building power-cycling or transmitting data, but we never did find the source despite hours of searching. That's one topic we'd like to follow-up on if we're ever invited back.

Other than curiosity regarding those EMF readings, most of the early part of the evening proceeded without much excitement. We did hear noises throughout the evening similar to those in the Streer-Peterson House—most notably noises that sounded like footsteps. The first of these occurred at about 8:37 p.m. and as was the case in the other building, all our team members were accounted for, nothing showed up on video, and we still have no idea what caused the noise. A potential naturalistic explanation might be that these were just the sounds of the old building settling—any old building does have characteristic pops, creaks, and groans—but even we have to admit they sounded eerily like footsteps. Absent in this building, though, were the noises that sounded like furniture moving or that bizarre pure tone sound.

The greatest excitement in the evening occurred at about 10:00. While we were silently listening, we heard a string of popping sounds. One of our team members, who happened to have his headphones off at that moment, immediately knew what it was, but several others took their headphones off to try to listen "naturally" and verify whether it was what we thought it was. Sure enough, the sound repeated, and sure enough, it was *exactly* what those of our team members experienced with such things thought it was.

Gunshots.

They weren't particularly nearby so we didn't feel threatened. They were coming from somewhere across the street, outside the Park. As near as we could tell, they sounded like they were coming from what looked like an apartment building. We noted the anomaly in our notebooks with the appropriate time stamp so we wouldn't get confused when listening back to our recordings, took the opportunity (since we'd broken our silence anyway) to conduct one of our hourly walkaround and measurement sessions, and went to go back to monitoring as before.

A few minutes later, we saw flashlights approaching from across the park. Then flashing red and blue lights. What looked like it must have been about half of the Lakewood Police Department's night shift was approaching. Knowing paranormal investigations probably look suspicious to any law enforcement officer who didn't know what was going on, a couple of us non-threateningly stepped onto the front porch to greet the two officers who were walking up (the rest of them were spreading out seemingly through the entire park).

"Let me guess," one of us said. "You heard some gunshots?"

The officers confirmed that's what they were responding to. We mentioned we'd heard the same—and indeed, recorded the sound, should that happen ever to become relevant—and pointed in the direction we thought we'd heard the sounds coming from. After a bit of conversation, they mentioned they'd had similar reports from that apartment complex before, but they'd responded to the Park instead because someone had seen a bunch of suspicious-looking people lurking in a normally-closed historic building on a security camera and they thought something was going on.

After a bit of a pause, one of them then asked the million-dollar question: "So…what *are* you guys doing?"

We mentioned back in Chapter 1 that police and other first responders often have a bit of a superstitious streak. We weren't kidding. The minute we mentioned the word "ghost" and explained that we were—with permission from Park staff—conducing a paranormal investigation, the senior officer of the duo immediately wanted to be anywhere else but standing there talking to us. His younger partner asked a couple questions about the ghost stories and we said we hadn't seen too much of interest yet but shared the stories we'd been told (in abbreviated format) and that satisfied his curiosity enough to allow him to be dragged off by his older and visibly unsettled colleague. That we accidentally recorded this entire conversation on the microphones we'd stationed around the building is a source of endless amusement to us.

They quickly went on their way, presumably headed across the street, and we went back to work as if nothing had happened. Paranormal investigation is a funny business. Even when nothing ghostly is going on, when you put yourself in unusual circumstances, unusual things tend to happen. Indeed, we could write an entirely new book just of all these kinds of weird stories that have nothing to do with the paranormal but are nevertheless weird experiences we've had on investigations.

This night wasn't yet done with unusual but non-paranormal happenings, though. Right around midnight, we noticed a suspicious looking person lurking around some of the nearby buildings while a couple of us were returning from a short break to take a walk. We told the other team members to stay inside and lock the door while we investigated (though we had no authority over the Park so we didn't actually want to confront anyone, regardless of what he was up to). As far as we could tell, he was a homeless person just visiting the park. No harm done, though a bit later, we observed him trying to break into the cellar of the Streer-Peterson House. We reported this to the staff member who was hosting us, but by that time, the man had given up his breaking and entering attempts and instead settled for a nap on a bench, so he decided to just leave well enough alone.

About an hour later, we thought we heard the sound of voices, but we determined those were most likely distant voices from someone walking through the park. A little later still, we heard some unusual thumping sounds that seemed to come from near the teacher's desk. One of our team sat there for a time while the rest of us listened, but we didn't hear any more of the sounds while trying to conduct that experiment.

We concluded our work and starting packing everything up at around 2:00 in the morning.

Police and homeless visitors aside, though, we didn't even notice the weirdest thing until a few days later when we were listening back to all our recordings. Clear as a bell in our audio recordings, though we didn't hear anything at all in person, was a bizarre screeching sound. It sounded animal, but yet "wrong," somehow, and at the very least not something we could easily associate with any of the nocturnal wildlife (such as coyotes) we might expect to see or hear in Belmar Park. The nearest we can describe the sound, since obviously we can't recreate it in print, is to imagine trying to scream while inhaling instead of exhaling. That's not *exactly* what we heard, but it's pretty close.

Once again, we've run that recording through audio analytic software and asked plenty of people what they make of it, and so far we have absolutely no idea what it could possibly be, nor why it showed up in our recording even though not one of our team members heard it in person. Stranger still, when we tried to triangulate its position by determining which of the microphones picked it up most clearly, it seemed to occur with equal volume on all four of our audio tracks. That is: it seemed to come from everywhere at once.

Like the Streer-Peterson House, then, we didn't actually witness any of the particular stories that had been reported to us, but we did experience some unusual and unexplained audio, planting this case firmly in the "still open and unsolved" category. As always, we're not saying it is or isn't a ghost (or ghosts); we're just saying we had an unusual experience we can't (yet) explain.

References & Further Reading

Colorado Women's Hall of Fame (n.d.). May Bonfils Stanton. *Colorado Women's Hall of Fame.* <https://www.cogreatwomen.org/project/may-bonfils-stanton/>

Denver Public Library Special Collections and Archives (n.d.). Polly Pry (1857 – 1938). *Denver Public Library.* <https://history.denverlibrary.org/colorado-biographies/polly-pry-1857-1938>

Family Search (n.d.) Morris Strear. *Family Search.* <https://ancestors.familysearch.org/en/LT98-1PV/morris-strear-1892-1955>

Family Search (n.d.) Sophia Dinner. *Family Search.* < https://ancestors.familysearch.org/en/LT98-1PP/sophia-dinner-1896-1978>

Leavitt, C., & Noel, T. J. (2016). *Herndon Davis: Painting Colorado History, 1901-1962.* Boulder. CO: University Press of Colorado

MacPhail, S. (2019). *The True Story of How Belmar Park Came into Existence: A Case Study of How Citizen Activism Can Work.* Archived online at <https://savebelmarpark.com/download/Belmar%20Park%20History%20%20-%20A%20Story%20of%20How%20It%20Came%20To%20Existence.pdf>

McElroy, J. (2024). Essay: A century ago, Denver newsmen helped unearth nation's biggest political scandal. *The Colorado Sun.* <https://coloradosun.com/2024/06/30/teapot-dome-scandal-civil-discourse-jack-mcelroy-carl-magee/>

Noel, T. (2018). May Bonfils and Her Lost Belmar Mansion: A Lavish Lakewood Estate House a Wealth of Benevolence. *History Colorado.* <https://www.historycolorado.org/story/articles-print/2018/10/19/may-bonfils-and-her-lost-belmar-mansion-lavish-lakewood-estate>

Riley, M. G. (2006). *High Altitude Attitudes: Six Savvy Colorado Women*. Boulder, CO: Johnson Books

Varnell, J. (1999). *Women of Consequence: The Colorado Women's Hall of Fame*. Boulder, CO: Johnson Books

Whitehead, C. (1942). A Glimpse of the Old Time Bar. *Denver Law Review, 19*(6): 147-148

Wood, R. E. (2005). *Here Lies Colorado: Fascinating Figures in Colorado History*. Helena, MT: Farcountry Press

5

Hotel of Celebrities and Statesmen: The Brown Palace

Located at the intersections of 17th Street, Tremont Place, and Broadway (and with a formal address of 321 17th Street), the iconic triangular-shaped Brown Palace hotel in the heart of Denver is impossible to miss. Its very appearance is connected to Denver's unique layout and telegraphs to all passersby that important and interesting things must be happening inside. Indeed, throughout more than a century of its history, it has been home and resting place to some of the world's greatest movers and shakers. It's been the site of both triumph and tragedy, hotel to the rich and famous, and, if certain stories are to be believed, a favorite haunt of perhaps more than its share of ghosts.

Figure 5.1. The Brown Palace hotel (photo: Bryan Bonner).

The History

In many ways, the history of the Brown Palace is tied to the history of Denver itself. One of the first things just about anyone would notice upon seeing the grand hotel is its unusual triangular shape. It has that shape, of course, because it fills basically the entire triangular lot of land formed by three streets: 17th Street, Tremont Place, and Broadway. Given that most cities are laid out on a rectangular grid (at least for the most part), one might be moved to wonder why Denver has such an unusual arrangement of roadways. Indeed, if you look at a map of Denver, you'll see that broad stretches of the city are laid out as one would expect, with a gridwork of roadways running north to south and east to west. But just north of Colfax Avenue and east of Broadway,[1] suddenly all the roads (and the buildings between them) shift by about forty-five degrees. Why?

The reason Denver's streets are so famously confusing basically comes down to a combination of the city's rapid growth in the mid-to-late 1800s and geographical features. Specifically, Denver's streets follow not one but *two* grid systems. A city called Auraria[2] ran parallel to Cherry Creek and the city of Denver ran parallel to the South Platte River. The end result was two towns operating on distinct grid systems. But as the populations grew, the towns merged into a single City of Denver in 1860. It was, though, too late to completely normalize everything to a single grid system (either following the cardinal directions or following the geographic features) and so Denver's streets remain a confusion to tourists probably forevermore.[3]

Thus did the history of Denver mold the very shape of one of its oldest hotels.[4] In order to understand the history of the Brown Palace itself, we need to know a little bit about its founder, Henry C. Brown (1820 – 1906). In addition to founding the hotel (which is our main point of interest, of course), he also had a hand in the development of the State Capitol itself, which turned out to be quite

---

1   The boundaries marking this shift are different in different neighborhoods, but the overall pattern is the same throughout most of Denver.

2   No longer a separate city, but whose name lives on in the Auraria Campus, home to the Metropolitan State University of Denver, the University of Colorado Denver, and the Community College of Denver. The campus still preserves some historic homes as offices spaces despite developing modern buildings across the bulk of their land.

3   Williams, B. (2024). Why are streets in downtown Denver crooked? History can explain. *KDVR*. <https://kdvr.com/denver-guide/why-are-streets-in-downtown-denver-crooked-history-can-explain/> (accessed July 29, 2025).

4   The Oxford Hotel was built a year before the Brown Palace. Some historians will claim the Brown Palace as the oldest continuously operating hotel in Denver. Technically this is true, as the Oxford did not serve as an active hotel for a few years during its history. Therefore, though we're mincing words here, technically the correct thing to say is that the Brown Palace is the oldest *continuously operating* hotel in Denver, while the Oxford is the oldest hotel in Denver *still in operation*. Regardless, both are remarkable historic properties and their ages are only one year apart anyway, so it's of little real-world concern.

the convoluted process.

Henry Cordes Brown was born on November 12, 1820 in Ohio, to Polly Newkirk and Samuel Brown, who had served in the Revolutionary War and sired no fewer than nineteen children during the course of his two marriages. Both of Henry's parents had died by the time he was seven years old, so he spent his adolescence working on farms until he learned carpentry and relocated to West Virginia in search of work in his new trade.[5] He married his first wife, Anna Louise Inskeep, in 1841. They would have two children together before Mr. Brown spent the next several years traveling the world. When his wife died in 1854, he left the children in the care of their maternal grandparents (their only living grandparents, recall), and eventually married Quaker schoolteacher Jane Cary Thompson in 1858, with whom he traveled west during the Pikes Peak Gold Rush, eventually settling in Denver.[6]

Mr. Brown established a boarding house and carpentry shop as his business and set about establishing himself as one of the great pioneers of the city, even building Denver's first church (in the Methodist denomination).[7] In 1863, he applied for a land grant and ultimately purchased 160 acres of property he called Brown's Addition, part of which he farmed and part of which he divided into lots separated by streets and then sold to other interests.[8] Despite these massive successes in his life, not all was well, as he lost both his shop and the church he'd helped to established in a large 1864 Cherry Creek flood.[9] Nevertheless, he persevered and became, partly through his hard work and partly through historic misadventure, one of the greatest developers in Denver's early history.

At around this time, there was much controversy concerning the location of Colorado's State Capitol. The Territorial Legislature originally met in Denver in 1861, then what is now known as Colorado Springs later that year, Denver again in 1862, and Golden in 1863, while the seat of the Governor remained in Denver. Various cities, particularly those with some animosity toward Denver, wanted the Capitol in their own boundaries. But Denver was destined to become the Capitol, and in 1867, Governor Alexander Cameron Hunt convinced the legislature to pass a bill moving the Capitol back to Denver (and establishing it there on a more permanent basis) provided the City of Denver would donate to the then-territory (now state) ten acres dedicated to the purpose. Mr. Brown saw an opportunity

5   Hall, F. (1895). *History of the State of Colorado, Embracing Accounts of the Prehistoric Races and Their Remains*. United States: Blakely Printing Company.

6   Faulkner, D. B. (2010). *Ladies of the Brown: A Women's History of Denver's Most Elegant Hotel*. Charleston, SC: The History Press.

7   McGrath, M. D. (1934). *Real Pioneers of Colorado Volume 1*. Archived by the Denver Library at <https://history.denverlibrary.org/sites/history/files/RealPioneersColorado.pdf> (accessed July 29, 2025).

8   Hall, F. (1895). *History of the State of Colorado, Embracing Accounts of the Prehistoric Races and Their Remains*. United States: Blakely Printing Company.

9   Faulkner, D. B. (2010). *Ladies of the Brown: A Women's History of Denver's Most Elegant Hotel*. Charleston, SC: The History Press.
McGrath, M. D. (1934). *Real Pioneers of Colorado Volume 1*. Archived by the Denver Library at <https://history.denverlibrary.org/sites/history/files/RealPioneersColorado.pdf> (accessed July 29, 2025).

to turn his property, which was at the time not particularly attractive, into prime real estate at the cost of just ten acres, which he promptly donated for the purpose of building the Capitol. Thus did Brown's Addition become the Capitol Hill neighborhood.[10]

An interesting side note in all of these real estate speculation stories has become the subject of local legend. Many people interested in Denver's early history will ask about "the cow," having heard rumors that some cow was responsible for derailing or moving real estate deals that, in todays money, would have been worth millions to billions of dollars. As always, there's a lot of rumor and innuendo and a lot of the legends get the stories wrong, but there is truth at the heart of it all. One such cow belonged to Mr. Walter Scott Cheesman (see Volume 1, Chapter 5), who grazed a single cow on prime Capitol Hill real estate so that it could be taxed as farmland instead of commercial real estate until such time as he could capitalize on his investment.[11] Another such cow belonged to Mr. Brown himself, which he grazed on the odd triangular property which would later become the Brown Palace, presumably for similar reasons.[12]

The payoff of this story and all of the real estate speculation in historic Denver was that indeed, the Colorado State Capitol would be built on property donated by Mr. Brown. In a flash, his property suddenly became some of the most valuable real estate in the city—and, indeed, it has remained some of the prime real estate in the country, though of course every city's real estate markets has its various ups and downs. With the Capitol moving in and with businesses in the area expanding, the need for a grand and elegant hotel with accommodations suitable to the world's great dignitaries became immediately clear, and Mr. Brown was just the person to make it happen.

Sparing no expense on his project, Mr. Brown hired architect Frank E. Edbrooke (who also built the Oxford Hotel, which opened a year before the Brown Palace) and construction began in 1888. Edbrooke designed the hotel in the Italian Renaissance style and used Colorado red granite and Arizona sandstone for the exterior. The interior was designed In a grand atrium style (quite the impressive structure, especially for its day, and no less beautiful now) so guests could stand on balconies overlooking the elegant common area below. Remarkably, they chose to eschew the more common wood construction, even for much of the interior, instead building the floors and walls from terracotta blocks, which allowed it to be only the second building in the entire nation to be certified as "fireproof." The grand total for the construction was a mindboggling $1.6 million, with additional costs for furnishing the rooms and common areas bringing the total cost to approximately an even $2 million (equivalent to more than $71 million in 2025). It featured 400 guest rooms upon its opening in 1893 (now down to 241) which could be reserved for a princely sum ranging from $3 to $5 per evening.[13]

10   Goodstein, P. (1996, 2001). *The Ghosts of Denver: Capitol Hill.* Denver, CO: New Social Publications.

11   *Ibid.*

12   Brown Palace (n.d.). Rich Tradition with a Twist at the Historic Brown Palace. *The Brown Palace Hotel.* <https://www.brownpalace.com/our-hotel/history/> (accessed July 29, 2025).

13   *Ibid.*

The hotel was built atop a 720-foot deep artesian well (fed by the same water source—the Arapahoe Aquifer (Denver Basin)—which supplies the Deep Rock Bottling Company), supplying water to the building and also contributing to its fireproof reputation.

Figure 5.2. The Brown Palace atrium (photo: Bryan Bonner).

All of these elements—the grand atrium, the fireproof design, the elegance—combined with the fact that it was at the time Denver's tallest building earned it a write-up both in an 1892 *Scientific American* cover story and in just about every architectural history of Colorado ever since.[14]

But before we move on to the rest of the Hotel's history, spanning more than a century, we need to return to Mr. Brown himself and close out his own biography. In 1893, the same year the Brown Palace opened, Mr. Brown's second wife Jane died after thirty years of marriage. Though growing advanced in years, Mr. Brown would marry once more, this time to Mary Helen Matthews. At the time of their wedding, she was nineteen years of age, compared to Mr. Brown's seventy-four years. Their marriage lasted only six years and ended in divorce. She would go on to marry John Douglas Campbell and return most of Mr. Brown's expensive gifts.[15]

Mr. Brown's life came to an end during a trip to San Diego, California on March 6, 1906 when he was eighty-six years old. As an honor, and likely in gratitude to his donation of the land on which the Capitol was built, his body lay in

14   Noel. T. J. (1997). *Buildings of Colorado.* New York: Oxford University Press. Scientific American (1892). Iron and Steel in Large Buildings. *Scientific American, 66*(21) [May 21, 1892]: 325.

15   Faulkner, D. B. (2010). *Ladies of the Brown: A Women's History of Denver's Most Elegant Hotel.* Charleston, SC: The History Press.

state at the State Capitol in Denver, where he received approximately 2,000 visitors before being buried with his second wife (from his longest marriage) Jane at Denver's Fairmount Cemetery.[16]

In 1959, an annex building was added to the Brown Palace in the form of a 22-story tower across Tremont Place and connected to the main Brown Palace building by both a skybridge and an underground service tunnel.[17] This building was originally operated under the Brown Palace brand but offering more budget-friendly rooms than the palatial main hotel, but was subsequently rebranded as a Comfort Inn and most recently as a Holiday Inn Express.

There is also a tunnel connecting the Brown Palace to the Navarre Building. Built in 1880, this remarkable structure became a gambling house and brothel in 1889, and the tunnel would have been used to move Denver's elite to and from the more illicit accommodations unseen by prying eyes. The Navarre is no longer used for this purpose, of course, but it still stands.[18] Our understanding is that the tunnel in question still exists but that it has been sealed off. For more about Denver brothel history, see Chapter 14.

Though accommodations have been modernized and rooms have been renovated several times throughout the years, the overall character and function of the Brown Palace have remained consistent throughout its entire history. Today, stepping through those doors into the grand atrium offers a glimpse of what it must have been like to enjoy all the luxuries of the past, but with all the comforts of the present. Because it has remained so consistent in its overall appearance and in its function, though, the most interesting history of the Brown Palace has been the guests who've stayed there, sometimes just taking a room for a night or for a week, and sometimes spending years or decades in full-time residence. Indeed, during the Great Depression, the eighth floor was split and converted into two floors of apartments, whose rents helped the hotel stay in business even as many others were forced to close due to the struggling economy.

A common misconception about the Brown Palace is that it was named for the famous "Unsinkable" Molly Brown (1897 – 1932). Of course, as we've seen, this is not the case, and it was named for its original owner. However, the famous socialite and Titanic survivor was a longtime Denver resident[19] and was a regular guest at the Brown Palace, often using the hotel's spaces for her various social engagements and fundraisers. Whenever she chose to stay overnight, she always

16   Windsor Beacon (1906). Body Lay in State – Henry C. Brown. *The Windsor Beacon* (March 17, 1906).

17   Indeed, Denver has numerous tunnels running from building to building that most visitors—and even most residents—might never know about. Though this one is clearly a service tunnel for the hotel, many were used to move contraband and call girls to some of Denver's business and political elite away from the prying eyes which lined the public streets.

18   It's also reputedly haunted. We have not yet investigated but would like to do so if ever invited.

19   The Molly Brown House, her actual residence for most of her adult life, is located on Pennsylvania Street and now operates as a historic museum dedicated to telling the story of Molly Brown's life. It is also famously reputed to be haunted but, as of this writing, we have not yet had a chance to investigate there.

insisted upon staying in Room 629, where it's said that she stayed only a week after the Titanic sank in 1912.

But she wasn't the only one. Jefferson Randolph "Soapy" Smith (1860 – 1898) was a regular guest at the Brown Palace during his years in Denver. Unfortunately his full life story, colorful and fascinating as it is, is well beyond the scope of this book, but in brief: he was an infamous confidence man, gambler, crime boss, gangster, saloon proprietor, and political kingmaker whose life and (mis) adventures have inspired works of literature, film, television, stage, and more for over a century now.

Of international historic significance, Dr. Sun Yat-sen (1866 – 1925) stayed at the Brown Palace shortly before becoming the first Provisional President of the Republic of China following the 1911 Revolution, of which he was a major leader and which ended China's imperial dynasty; in that way, he can be considered the founder of the Republic of China (not to be confused with the People's Republic of China, the nation's current political identity following Mao's Cultural Revolution). Another international leader to grace its rooms was Marie (1875 – 1938), the last Queen of Romania and wife of King Ferdinand I (1865 – 1927).

Several Presidents of the United States have stayed at the Brown Palace, including:

- William Howard Taft (1957 – 1930), our 27th President,
- Warren G. Harding (1965 – 1923), our 29th President,[20]
- Harry S. Truman (1884 – 1972), our 33rd President,
- Dwight D. Eisenhower (1890 – 1969), our 34th President,
- Ronald Reagan (1911 – 2004), our 40th President, and
- Bill Clinton (b. 1946), our 42nd President.

Dignitaries from the entertainment world who've stayed at the Brown Palace include both John Wayne (1907 – 1979) and The Beatles.

As with any hotel, not everyone who stayed there should be considered a heroic character (obviously, readers' positions regarding whether some of the individuals mentioned merit that honor will vary and this is neither the time nor the place to get into historic political arguments, as amusing as that might be to all of us). Back in Chapter 1, we pointed out that Colorado's early history is uncomfortably connected with that of the Ku Klux Klan (about which, there will also be more in Chapter 7). While some of that was likely less malevolent than modern eyes would think (that is, a lot of members likely joined more to be part of "the boys' club" rather than out of any truly deep racial animosity), some characters are undeniably connected with the white supremacist movement, and one of those was a noteworthy guest at the Brown Palace. His name was John Galen Locke (1871 – 1935) and he was a "Grand Dragon"[21] and founder of the

20   For more on the Harding Administration and an unfortunate scandal therein, see Volume 2, Chapter 9 of this series.

21   The highest rank in a regional "Realm" and subject under the authority of the "Imperial Wizard." KKK nomenclature grew increasingly convoluted and, frankly, embarrassing, over the years. In 1946, *The Adventures of Superman* radio serial broadcast a series of episodes in which the Man of Steel exposed the actual codewords used by Klansmen (based on research done by Stetson Kennedy), which some historians have credited with the Klan's subsequent steep decline in membership. Whether

Colorado chapter of the KKK who helped establish the Klan's influence in State politics. We wouldn't ordinarily dedicate much attention to such a character in what's meant to be a book of fun ghost stories, except that he didn't only stay at the hotel but actually died of a heart attack during a political meeting held at the Brown Palace on April 2, 1935.[22] At the risk of spoiling our next section, we can mention now that we're unaware of any particular ghost stories related to Locke's demise.

And then there was Louise Sneed Hill (1862 – 1955), a noted Denver socialite, wife of noted Denver businessman and civic leader Crawford Hill (1862 – 1922) and daughter in law of United States Senator Nathaniel Peter Hill (1832 – 1900). Louise Sneed was born in North Carolina in 1862, relocated to Tennessee to live with sisters after her parents died, and eventually found her way to Denver in 1893, where she met Crawford Hill; the two married in 1895, though her new mother in law would say she was "sick" about the marriage, thinking Louise was marrying for power and influence rather than out of a true emotional connection.[23]

Whether or not her mother in law was correct about the motivation behind the marriage is a matter the players in the drama have all taken to their graves, but she was certainly correct about one thing: Louise Sneed Hill was interested in social influence and was disappointed by what she considered to be Denver's poor quality of social life at the time. Her life as a socialite began in earnest when she started inviting those she considered the highest of high society to play card games at her and Crawford's mansion. The group of thirty-six elite women came to be known as the Sacred 36. When the group snubbed Molly Brown, who desperately wanted to get into such an elite club (her own social ambition has become the stuff of legend), she (Brown) called Hill the "snobbiest woman in Denver," though she would eventually gain admittance to Hill's card tables after she became famous for surviving the sinking of the Titanic.[24]

In 1908, Hill founded the *Who's Who in Denver Society* (later the *Blue Book* of Denver Society listings),[25] which lasted for more than a century until (probably due to the growth of the Internet) it finally shut down in 2011.[26]

Sometime in or around 1914, she met Bulkeley Wells[27] (1872 – 1931), a businessman with mining interests who became a good friend to both Louise and Crawford, even though he would carry on a torrid affair with the former for

---

Superman gets all the credit or whether social attitudes had simply shifted enough to lead to their downfall, we like the story.

22   Daily Sentinel (1935). Dr. John Galen Locke is Dead, Former Klan Head and Political Power, Dead. *The Daily Sentinel* (April 2, 1935).
23   Riley, M. G. (2006). *High Altitude Attitudes: Six Savvy Colorado Women.* Boulder, CO: Johnson Books.
24   Faulkner, D. B. (2010). *Ladies of the Brown: A Women's History of Denver's Most Elegant Hotel.* Charleston, SC: The History Press.
25   *Ibid.*
26   Simon, S. (2011). Denver's Society Bible Hits the Skids. *The Wall Street Journal.* <https://www.wsj.com/articles/SB10001424052970204450804576623273813179898> (accessed July 29, 2025).
27   Sometimes spelled Buckeley Wells.

some years. Eventually Wells' wife, Grace Livermore Wells, divorced him in 1918 and then Crawford Hill died in 1922, leaving Louise to think he would stand by her side; instead he eloped with a younger woman and eventually lost his mining empire, dying by suicide, nearly destitute, in 1931.[28]

For her part, Louise Hill remained at her mansion until the combination of a stroke and the uncertainty resulting from World War II led her to take up a permanent residence at the Brown Palace, where she spent approximately the last decade of her life in Room 904. She died on May 28, 1955. Though it was primarily health that led her to change residence to the hotel, subsequent rumors have connected her abandonment of the mansion and move into the Brown Palace to her affair with Wells.

There have also been some murders.

The second and less famous murder at the Brown Palace took place in 1946, and our narrative of these events is based on the case as described in the Colorado Supreme Court decision on the matter.[29]

Ronald Frederick Smith was a police officer serving in Grand Junction, Colorado and a veteran of World War II. On September 30, 1946, he went to Denver for a medical examination by the Veterans Bureau, which he concluded at about 2:30 in the afternoon, after which he visited several bars including the Ship Tavern at the Brown Palace, to which he returned several times, telling other bar patrons he'd missed his train. At about 5:30 in the evening, one Mr. Phillip Van Slyck took a seat at the bar and a conversation with Mr. Smith followed. Witnesses describe Mr. Smith's demeanor throughout this encounter as a "little bit high," loud and boisterous, but not particularly combative or aggressive.

At one point, Mr. Smith seemed to confuse Mr. Van Slyck for some wanted fugitive and a bizarre conversation ensued.

"You are wanted by the law in Grand Junction, aren't you?" Smith asked.

Mr. Van Slyck said he was not. Mr. Smith then said he'd seen Van Slyck in Grand Junction or some other place before that evening, which the latter denied, insisting he'd never met Mr. Smith before. Then Smith directed his attention to a table occupied by some young women and suggested Van Slyck should "ask that blonde dame for a date" because "you shouldn't have any trouble getting a date." They then turned, apparently quite hastily as Smith seemed to be leading the conversation in various directions all at once, to the subject of their respective ages (twenty-three years in the case of Smith and twenty-six for Van Slyck). Smith then said they were going to have a "big time tonight" and offered to by his interlocutor a drink, which Van Slyck declined, saying he needed to leave shortly for a prior engagement.

Apparently offended by Van Slyck's refusal of his offer to pay for a drink, Smith became more hostile.

"You are just like all the rest of them," he said. "You would turn me in in a minute if you got the chance. Go ahead and turn me in."

28   Riley, M. G. (2006). *High Altitude Attitudes: Six Savvy Colorado Women*. Boulder, CO: Johnson Books.

Faulkner, D. B. (2010). *Ladies of the Brown: A Women's History of Denver's Most Elegant Hotel*. Charleston, SC: The History Press.

29   *Smith v. People*, 120 Colo. 39 (1949).

Van Slyck didn't know what he was talking about. "Whom do you want me to turn you in to?"

"You know who—that fellow over there you just gave the 'hi' sign to."

Van Slyck would later testify he hadn't been paying attention to anyone else at the bar and had no idea who Smith was talking about.

At that point, Mr. Smith produced a firearm and offered it to Mr. Van Slyck. "Go ahead and shoot him," Smith said.

"Shoot whom?"

"You know who."

Trying to play it all off as some kind of joke—and (we assume) probably also desperately looking for any opportunity to safely extract himself from this increasingly troublesome conversation—Van Slyck laughed and returned the firearm to Mr. Smith.

"You'd better put it away before somebody gets hurt," he said.

At this time, Smith pointed the gun at Van Slyck, saying, "you son of a bitch! You haven't got any guts!"

They argued over the gun for a moment, during which time Smith broke the cylinder of the revolver to demonstrate that it was loaded. Smith then began speaking rapidly, threateningly, and somewhat incoherently: "This is all a dream. Isn't it nice that way? I could shoot you right now and nobody would know it. Nobody would even remember. I am going to do it, too. You're a nice guy. I like you and you like me, but you're going to die tonight."

At this time, Smith again pointed the revolver directly into Mr. Van Slyck's midsection. This time, the latter reacted to defend himself. In the ensuing struggle, the firearm went off five times. None of the bullets struck Mr. Van Slyck, but one of them hit Dr. James Mullen, a physician who'd been at the bar and who had interacted with Smith earlier in the evening. Court records indicate the bullet penetrated Dr. Mullen's left lung through the chest, fractured his left clavicle and fifth cervical vertebra, and caused him to die before the police surgeon arrived on the scene.

Witnesses subdued Mr. Smith and he was eventually arrested and questioned by the police. During questioning and at trial, he maintained that he remembered getting into a conversation with Van Slyck but didn't remember anything else until he woke up strapped into a hospital bed in police custody. He pleaded not guilty by reason of insanity but was found guilty of first degree murder. He was initially sentenced to life imprisonment but later transferred to the Colorado State Hospital due to his mental condition.

Though horrifying enough, and surely the stuff of psychological thriller novels (though we're unaware of any works of fiction directly inspired by this case), Smith's murder of Dr. Mullen has been overshadowed in the history books by another infamous murder that took place in the Brown Palace's Marble Bar in 1911.

This more famous murder is the stuff of tabloid newspapers and operas! It involves high society, infidelity, a love triangle (actually more of a love quadrilateral, as we'll see), betrayal, intrigue, and all the stuff the masses love in a true crime story. For the sake of space, we'll only summarize the story briefly here, but readers interested in a highly detailed deep dive into the subject would do well to consider journalist Dick Kreck's 2003 book on the topic entitled *Murder at the Brown Palace: A True Story of Seduction & Betrayal.*

Though he ended up being only a relatively minor player in the events that would follow, our story rightly begins with John W. Springer (1859 – 1945), a noted social figure at the time, as well as an attorney, banker, businessman, and owner of a 10,000-acre ranch on which sat the Highlands Ranch Mansion.[30] But it was his personal life rather than his business ventures that become relevant to our story. His first marriage was to Eliza Clifton Hughes, who died in 1904, leading to his second marriage to Isabel Patterson Folck (1880 – 1917) in 1907. This was also her second marriage, undertaken shortly after she divorced her first husband, a salesman named John E. Folck.

This was not destined to be the happiest of marriages. Isabel was twenty years John's junior and apparently unprepared to give up her youthful indiscretions and enjoyment of the night life. Perhaps to allow her easier access to society than from their ranch, the Springers rented a suite at the Brown Palace consisting of Rooms 600 and 602, which of course led directly to the rest of our story. The trouble was, Isabel maintained flirtatious communications outside of her marriage. It's not clear whether these friendships ever progressed beyond verbal and written communication, but they certainly progressed further than mere friendly conversation. Two suitors in particular are relevant to our story.

Sylvester Louis "Tony" von Phul (1878 – 1911) was a bit of the wild and perhaps reckless sort who passed his time hot air ballooning, for which he developed a reputation as one of the most accomplished professionals. He also apparently had a reputation as a bit of a ladies' man, and had a particular affection for Isabel Folck prior to her marriage to John Springer, which he never gave up even as Isabel Folck became Isabel Springer. Apparently the feelings were mutual and she wrote him numerous letters. Of particular interest were letters written in early 1911 pleading with him to join her in Denver, which he eventually did.

Also in 1911, the Springers met Harold Francis "Frank" Henwood (1876/7 – 1929), a well-to-do businessman who began a business partnership with John Springer and a friendship with both John and Isabel. But of course, this relationship, too, quickly began to seem somewhat more than merely friendly.

Thus were four people involved in what we're calling a love quadrilateral. And in 1911, they all found themselves in the same place at the Brown Palace. Isabel seemed rather indecisive about her men. She was married to John Springer, wrote Frank Henwood on May 12, 1911 asking him to help her get back some letters she'd written to Tony von Phul, but still wrote affectionate letters to von Phul as late as May 20. Behind Mr. Springer's back, Henwood and von Phul began to quarrel, each probably thinking on the one hand that he was defending Isabel's honor and on the other hand that he had some personal interest in removing from her life another competitor for her attention and affections. On May 23, Isabel told Henwood that von Phul had threatened to show her letters to her husband if she broke off their (Isabel and von Phul's) relationship and dispatched Henwood to retrieve the letters and prevent the scandal.

On the same day, von Phul also checked into the Brown Palace where he re-

30   The Mansion is also reputed to be haunted but has not yet been the subject of one of our own investigations.

quested Room 603 to be near the Springers' suite. Almost immediately, Henwood confronted him. The two would exchange words about the letters in question and Isabel's affections multiple times throughout this and the following day, the full details of which are spelled out in Kreck's aforementioned book. For our purposes, what we need to realize is that both thought the other was meddling in "his" relationship with Isabel. Neither seemed to give much thought to Mr. Springer, and neither seemed to consider that Isabel's affections might lie elsewhere. For her part, Isabel seemed inclined to maintain these feelings in both men.

The fateful confrontation occurred late in the evening of May 24, 1911. Both Henwood and von Phul had been attending *Follies of 1910* at the nearby Broadway Theater. The show let out at about 11:15 and Henwood first made his way to the Brown Palace's Marble Bar. A few minutes later, von Phul followed suit, prompting Henwood to leave and return to the theatre. But he didn't stay there for long. About five minutes later, he returned to the Marble Bar, where he confronted von Phul about their prior bickering. By all accounts, he was polite at first, but the tension between them must have been almost unbearable, and von Phul responded by threatening to go upstairs and attack Mr. Springer. Other patrons, noticing there was about to be a fight, started to rearrange themselves or to leave the bar.

The precise details of what happened next remain somewhat controversial and were the subject of not one but multiple trials and appeals. What everyone agrees on is that von Phul struck Henwood first, sending the latter to the ground. When he came back up, he'd drawn his revolver. Henwood maintained that von Phul was still facing him and moving to attack. Other witnesses suggested von Phul had by then turned back to the bar. It's not entirely clear what happened except that Henwood fired multiple shots before bystanders were able to restrain him and wrestle the firearm from his hands. Three people were hit.

One was Tony von Phul, who died the following day at St. Luke's Hospital.

Another was George E. Copeland, a forty-three year old businessman who'd just been a bystander during the whole affair. He died of his wounds several days later on June 1.

Finally, Henwood's bullets also struck James W. Atkinson, inflicting serious but non-deadly injuries.

Henwood never denied firing the gun. It would have been useless given how many people saw him do it. Nevertheless, he maintained his innocence, arguing that he'd acted in self-defense. Police and prosecutors disagreed and charged him with murder. At an initial trial in 1911, he was found guilty but was later granted a new trial in 1913, which again returned a guilty verdict. He was initially given the death penalty but it was later commuted to life imprisonment. He died in 1929.

Just to wrap things up, John Springer divorced Isabel later in 1911. He would go on to a third marriage, this time to Janette Elizabeth Orr Muir Lotave, in 1915. He died in 1945. For her part, Isabel died at the age of thirty-seven years in April of 1919. The *Rocky Mountain News* ran an obituary calling her a "vampire" and pointing out that "the wages of sin is death."[31] Not necessarily the way we'd like to be memorialized, we must say!

---

31   Quoted in: Kreck, D. (2003). *Murder at the Brown Palace: A True Story of Seduction & Betrayal.* Golden, CO: Fulcrum Publishing.

That about does it for our history. We conclude with just one final fact and point of honor for the Brown Palace: in 1970, it was recognized on the National Register of Historic Places.[32]

Paranormal Claims

Well, now! After hearing all those tales of intrigue and murder, we suspect you must be ready to hear all the ghost stories that sprang from those stories. Alas, we're sorry to disappoint, but those murders, believe it or not, don't actually seem to have led to any ghost stories we've been able to discover. Sure, the stories of the murders get told by people alongside ghost stories all the time, but none of the claims ever seem to involve the characters from those chapters in Brown Palace history. Nothing, of course, proves ghosts related to those murders *don't* haunt the hotel, but we've never heard specific claims along those lines.

Be that as it may, there are plenty of ghost stories at the Brown Palace related to other events and characters, and it is to those that we now turn our attention.

Recall from the History section above that Louise Sneed Hill spent her last years in Room 904 at the Brown Palace. During the early 2000s, a historic tour of the hotel included her room as one of its stops. Guides would tell tourists about Hill's affair with Bulkeley Wells and that she'd died in that very room. For a period of several months, Room 904 was removed from the tour due to a remodeling program. Employees at the front desk at the time of this remodel reported that they'd receive phone calls from the vacant Room 904, but that no one was ever on the other end when they answered and that whoever called would just hang up without saying anything. Weird enough, right? It gets weirder. Because the room was being remodeled, the telephone had been removed from Room 904. These phone calls were coming from a room without a phone in it! When the tour guide learned of this, she stopped telling the story of the affair when tours resumed some months later, and the phone calls stopped. Believers think this may have been the ghost of Louise Sneed Hill trying to get people to stop airing her dirty laundry. If that's the case, it certainly seems to have worked.[33]

The murders we mentioned earlier might not have led to many ghost stories, but other deaths have: suicides. Around the turn of the last century, there was a disturbing trend in which individuals intending to commit suicide would check into grand luxurious hotels to commit the act. Better to be found by strangers working as hotel staff, apparently, than to burden one's own family with the task of discovering and cleaning up one's mortal remains. In a horribly twisted sort of way, we can even understand how people would come to that conclusion, though we'd also like to offer a word on behalf of the poor hotel cleaning staff who'd be unfairly burdened with something that was never meant to be their problem. Another reason might simply have been to end one's life in conditions of elegance

---

32   Fink, R. (1970). Brown Palace Hotel: National Register of Historic Places Inventory – Nomination Form. [Register No. 70000157]. *National Register of Historic Places.* <https://npgallery.nps.gov/GetAsset/9b648c8c-45dd-4c67-a9f7-df6a00c05ceb> (accessed July 29, 2025).

33   Of course, now we've resurrected the story. Whoops.

and luxury rather than in one's ordinary conditions at home.

Shortly after 1900, one such woman visited the Brown Palace. Sometime during her stay at the hotel, she went to the second-floor landing of the grand staircase and threw herself down the stairs, fatally breaking her neck.

Patrons and employees of the hotel alike have witnessed a ghostly repetition of this event. Minor details vary depending on who tells the story, but it usually goes something like this. Some hotel guest would frantically run to the front desk to report they just saw someone fall down the stairs. Whenever anyone would investigate, though, no one would be found. Apparently the front desk staff have sometimes told witnesses not to worry because "it's just the ghost" and calmly returned to their normal routine.

Paranormal lore suggests there may be different types of haunting. An "intelligent haunting" is thought to be what we'd ordinarily think of as a ghost. It's a soul doomed to wander the earth or something like that. Details may vary, but these are thought to be sentient beings with actual agency and which can interact with the world in some intelligent manner. "Residual hauntings," conversely, are thought to be something like playing back a recording. These "ghosts" aren't intelligent or conscious souls. Rather, they manifest simply by replaying some event (often a horrifying, traumatic, or violent event) from the past.

The story of the ghost on the Brown Palace staircase has the hallmarks of the latter type of haunting. For our part, we remain neutral on the subject of whether it's even meaningful to talk about classifications of haunt. Our goal is to first document any anomalous phenomena, to rule out potential natural explanations, and only then to consider any of these questions of who, what, why, and how.

Hotel suicides aren't only a thing of the past. These things still happen in hotels from time to time and probably every hotel in the world has had it happen. They tend to try to keep such stories under wraps either to protect their business or simply to avoid alarming their other guests, but sometimes we manage to hear things through the grapevine.

In the mid-2000s, another suicide took place at the Brown Palace. Records are sparse, but a credible source connected to the hotel told us that one of the hotel maids decided to commit suicide. She went to one of the upper floors, wrapped herself in a sheet to minimize the mess she'd leave for others to clean up, and flung herself off the balcony to the atrium floor below. Specific ghost stories connected to this suicide haven't turned up yet, but given the hotel is reputedly haunted by other suicide victims, it seems worthy of mention.

Back in the late 1990s, one of the hotel's employees was going about his business cleaning the lobby late at night (or rather, at about 3:00 in the morning). He heard the sound of music coming from the bar and restaurant, but of course no one was supposed to be in there at that hour. He went to investigate. Much to his surprise, he found a quartet of musicians playing right there in the lounge! He asked who they were and what they were doing there and one of them said "we live here." As soon as the employee left the room, the music stopped. When he looked back, the musicians had vanished.[34]

---

34   The idea of a ghostly party in a hotel, bar, or restaurant is not an uncommon one. We've encountered such stories numerous times both in ghostly fiction and on our

Going back further in the hotel's history, this time to the 1980s, a large guest room was converted into the "Club Room," which served as a private gentlemen's club until the late 1980s. This room featured a bar and was accessible only by membership. In the decades since, it has been converted into an events venue for weddings, parties, and banquets. It's also supposed to be one of the most haunted spots in the hotel. Numerous employees have reported seeing people in the room when no one was supposed to be there. One former manager described a time when he entered the room to make sure no one was in there without authorization (apparently a common occurrence even without considering ghost stories, as homeless people often wandered in to find a place to rest). Upon entering, he discovered a man dressed as a formal bartender standing behind the bar. When he asked the man what he was doing there, the ghostly bartender "just walked through the wall." Shades of Stephen King's *The Shining*, if you ask us, though it was of course Estes Park's Stanley Hotel (see Volume 1, Chapter 14) that inspired that particular story, rather than the Brown Palace.

On another occasion, a florist was setting up arrangements for an event in the same Club Room. Someone appeared in the corner of the room wearing historic attire and walked across from the women's restroom to the opposite wall and then proceeded straight through the wall.

Figure 5.3. The Club Room (photo: Bryan Bonner).

Yet another employee on yet another occasion, also while preparing the room for an event (in this case, conducting final inspections before opening it up to host a wedding reception), stepped into the women's restroom. Water was running in the sink faucet. But as she approached to turn it off, she saw the handle move by itself to shut the water off.

Finally, in the same room, a guest reported seeing the reflection of a person standing behind him in the glass of a bookcase. But when he turned around, no one was there.

---

own investigations. A couple noteworthy instances of the latter are found in Volume 1, Chapter 14 and Volume 2, Chapter 14.

Our Investigation

Paranormal investigation is a strange business. Two of the most common questions we're asked when we give lecture or at book signing events are whether we really believe in the ghosts and whatnot given our openminded but skeptical approach, and what's kept us going for well over a quarter of a century at this point. The answers to those are related and what it all really comes down to is: almost every investigation ends up surprising us in one direction or another. On one investigation, we might visit a place with barely any paranormal lore and end up finding some of the most remarkable experiences ever. On another, we might go to a place with more ghost stories than we can count, remarkable history, creepy architecture, the works, and end up finding nothing at all.

Unfortunately, the latter comes close to being the case with our investigation of the Brown Palace.

We were invited to conduct an investigation spanning several days, including some additional informal "investigation" time when we got to look around and investigate a bit while leading ghost tours, one stop of which visited the hotel. During both the formal investigation and the tour stops, we were given access to both the Club Room and Room 904 (Louise Sneed Hill's room), as those were the places we wanted to focus our investigation. Yes, it's true that some of the stories are also focused on the atrium and staircases, but since the hotel wasn't closed down during our investigation, we assumed these public areas would have too many data contaminants so we focused on the areas we thought we'd be able to do the best research.

Figure 5.4. Room 904: Louise Sneed Hill's room (photo: Bryan Bonner).

As usual, most of our investigation was focused on silent monitoring and taking environmental measurements. Toward that end, we set up eight video cameras, three microphones, two seismometers, several still cameras, several AC and DC electromagnetic field detectors, and found...bupkis.

While disappointing, It's important to point out that these kinds of results are to be expected from time to time during paranormal investigations. Seldom do alleged paranormal phenomena occur on a fixed schedule, so it's entirely possible we were just there at the wrong times. Nevertheless, negative results are still results and while they may not prove or disprove anything, we think it's important to report them to the public nonetheless.

During our investigation, we did also try to look into the claims regarding a reflection manifesting in the bookcase glass and the water faucet turning itself on and off in the women's restroom.

For the former, we wanted to see just how reflective the glass was under a variety of circumstances and whether reflection and refraction might combine to result in either a duplicated or distorted image in the glass. Our hypothesis was that the witness who saw a reflection standing behind him might have seen some kind of distortion of his own reflection. We were easily able to capture reflections of our investigators in the glass under certain lighting conditions, but none of our attempts resulted in distorted or duplicated images. We therefore don't think he saw a duplicated or distorted copy of his own reflection. Whether he got spooked by his own ordinary reflection, saw someone else who quickly left the room before he turned around, or saw a ghost reflected in the glass remains an unsolved and potentially unsolvable mystery.

Regarding the water faucet, we did everything we could think of to recreate the phenomenon of it turning on and off seemingly by itself. We checked to see if it were the sort on a light sensor (it wasn't) and then stomped about the room and moved all the doors and generally did anything we could think of to get the handle to move either from the off to the on position or vice versa. Nothing worked. Since we weren't present when the original phenomenon was reported, there's little else we can say on that note. Just as an interesting side project, we did manage to recreate the phenomenon using magicians' trickery, but we don't think that's a plausible explanation for the original claim.[35]

Separate from our investigations, one of our team members did find the opportunity to visit the Brown Palace on other business (taking photographic portraits on location) and did a bit of informal investigating while he was there. Oddly enough, it was during this event rather than on the official investigation that one of the more interesting things happened.

After working on the portraits for a while, he took a break late in the evening and wandered down to the bar and restaurant to rest for a moment. As he approached, he heard the unmistakable sound of piano music coming from the closed restaurant and of course immediately recalled the story of the ghostly quartet of musicians. He quietly snuck in to investigate and found, not a ghost, but none other than famed musician Billy Joel sitting at the piano. He listened for a moment and then wandered back out to complete his business. It may not have been the ghostly encounter we were all looking for, but it is just one of those weird things that can happen when you spend your time in historic and allegedly haunted places.

35   As we've mentioned elsewhere, we always keep our eyes open for any potential hoaxes, and we understand how they're done, but we generally don't find hoaxes to be the explanations for the cases on which we work.

To wrap up, finally, we unfortunately have to consider almost all of the ghostly claims of the Brown Palace to still be open and unsolved investigations. But that's just the nature of the business. Sometimes something weird happens, sometimes you solve a mystery naturally, and sometimes you just have to regroup and try again later.

References & Further Reading

Brick, C. (2018). *Haunted America: Ghost & Legends of Colorado's Front Range*. Charleston, SC: The History Press

Brown Palace (n.d.). Rich Tradition with a Twist at the Historic Brown Palace. *The Brown Palace Hotel*. <https://www.brownpalace.com/our-hotel/history/>

Daily Sentinel (1935). Dr. John Galen Locke is Dead, Former Klan Head and Political Power, Dead. *The Daily Sentinel* (April 2, 1935)

Fink, R. (1970). Brown Palace Hotel: National Register of Historic Places Inventory – Nomination Form. [Register No. 70000157]. *National Register of Historic Places*.<https://npgallery.nps.gov/GetAsset/9b648c8c-45dd-4c67-a9f7-df6a00c05ceb>

Faulkner, D. B. (2010). *Ladies of the Brown: A Women's History of Denver's Most Elegant Hotel*. Charleston, SC: The History Press

Goodstein, P. (1996, 2001). *The Ghosts of Denver: Capitol Hill*. Denver, CO: New Social Publications

Hall, F. (1895). *History of the State of Colorado, Embracing Accounts of the Pre-historic Races and Their Remains*. United States: Blakely Printing Company

Hunt, C. (2003). *The Brown Palace: Denver's Grande Dame*. Denver, CO: The Brown Palace Hotel

Kreck, D. (2003). *Murder at the Brown Palace: A True Story of Seduction & Betrayal*. Golden, CO: Fulcrum Publishing

Lamb, K. (2016). *Ghosthunting Colorado: America's Haunted Road Trip*. Covington, KY: Clerisy Press

McGrath, M. D. (1934). *Real Pioneers of Colorado Volume 1*. Archived by the Denver Library at <https://history.denverlibrary.org/sites/history/files/RealPioneersColorado.pdf>

Noel. T. J. (1997). *Buildings of Colorado*. New York: Oxford University Press

Riley, M. G. (2006). *High Altitude Attitudes: Six Savvy Colorado Women*. Boulder, CO: Johnson Books

Scientific American (1892). Iron and Steel in Large Buildings. *Scientific American, 66*(21) [May 21, 1892]: 325

Simon, S. (2011). Denver's Society Bible Hits the Skids. *The Wall Street Journal*. <https://www.wsj.com/articles/SB10001424052970204450804576623273813179898>

Smith v. People, 120 Colo. 39 (1949)

Williams, B. (2024). Why are streets in downtown Denver crooked? History can explain. *KDVR*. <https://kdvr.com/denver-guide/why-are-streets-in-downtown-denver-crooked-history-can-explain/>

Windsor Beacon (1906). Body Lay in State – Henry C. Brown. *The Windsor Beacon* (March 17, 1906)

6

Our Luck With Weather: Boettcher Mansion

One of the jokes we've been telling for years now is that when Rocky Mountain Paranormal shows up for an investigation, you'd better be prepared for a blizzard. Purely by happenstance, our investigations do seem to disproportionately coincide with cold and blizzardy conditions. This recent investigation was one such example.

There are several mansions bearing the name Boettcher in Colorado. This one, located at 900 Colorow Road on Lookout Mountain in Golden, was Charles Boettcher's private summer home and hunting lodge. Now operated by the county as a wedding and event venue, it's a beautiful old mountain home with great history, amazing views, and a full helping of ghost stories.

Figure 6.1. Boettcher Mansion (photo: Robert Lewis).

The History

Certain names are prominent all around Denver and the surrounding towns. In Chapter 4, we met the Bonfils family, whose various business and philanthropic ventures have put their names on more buildings, foundations, and history books than we know how to count. Another such name and another such family is the Boettcher family.

The patriarch of the family, and the beginning of our story, was Charles Boettcher (1852 – 1948). As is often the case with the successful business pioneers whose histories we've been describing in these books, Mr. Boettcher didn't limit himself to just one enterprise. No, he was one of those people who seemed to have a hand in just about everything.

He was born in 1852 in Cölleda, Kingdom of Prussia (now Kölleda, Germany[1]) but came to the United States at the age of seventeen years to join his brother Herman in Cheyenne, Wyoming, eventually becoming a partner in his brother's hardware store. He relocated to Colorado in 1871 and married Fannie Augusta Cowan (1854 – 1952) in 1874. At the same time, he established a hardware store, the first of his numerous business ventures in Colorado. The couple would have two children together: Claude K Boettcher (1875 – 1957) and Ruth Augusta Boettcher Humphreys (1890 – ?).[2]

Mr. Boettcher's business empire started with a hardware store in Boulder but quickly expanded into Leadville, where he established the first electric light company,[3] and then moved on to influence business throughout the state. By 1880, he expanded his business empire to include mining properties and other investments. A decade later, he built a mansion in Denver's "Millionaire's Row" the same year daughter Ruth was born. Ten more years and he was thinking about retirement, but it wasn't to last and six months later he organized the Great Western Sugar Company and Portland Ideal Cement Company and became president of the Denver and Salt Lake Railroad in 1915. Two years later, he would build the mansion about which we'll discuss more shortly.[4]

Not all was well in the Boettcher family at around this time, though, despite massive successes in business and a growing fortune. Charles and Fannie's relationship was quickly fracturing and the marriage was doomed. Some think the straw that broke the marital camel's back was a sunroom at the Denver mansion. In 1919, Fannie wanted to add such a porch to their home. Charles refused and said he'd leave if she added it. She added it. He left, and they maintained separate residences from then on, though Fannie would remain involved in the Boettcher

1   The change in spelling took place in 1927. The Kingdom of Prussia became the Free State of Prussia following the German Revolution of 1918 – 1919. Prussia itself was abolished in 1947 and Kölleda became part of the German Democratic Republic (aka East Germany) from 1949 until the reunification of Germany.

2   Boettcher Foundation (n.d.). Boettcher Family & Foundation History. *The Boettcher Foundation.* <https://boettcherfoundation.org/who-we-are/history/#event-charles-boettcher-is-born> (accessed August 9, 2025).

3   *Ibid.*

4   Jefferson County (n.d.). Boettcher Mansion: Charles Boettcher. *Jefferson County.* <https://www.jeffco.us/1862/Charles-Boettcher> (accessed August 9, 2025).

business and philanthropic ventures for the rest of her life.[5]

Because once again all of these haunted histories end up being interconnected at some level, Mr. Boettcher responded to the separation by moving into a residence at none other than the Brown Palace hotel (see Chapter 5), of which he became part owner in 1922.[6]

Though separated, both Boettchers remained involved in their various business ventures. Both also shared a deep love of children and their community and wanted to give something back and to establish a legacy for themselves. Charles and Claude established The Boettcher Foundation in December of 1937, and the Boettcher School for Disabled Children in 1940.[7] While the mission of the latter is self-evident from the name, it's worth mentioning that the former is dedicated to more general philanthropic pursuits through grantmaking, leadership development programs, and scholarships.

Charles Boettcher died at the age of ninety-six years in 1948, but his legacy continued to live on through his family and foundations, including the establishment of the Boettcher Foundation Scholarship in 1952, the donation of the Boettcher Mansion (the Denver one, not the one we're going to be discussing) for use as the State Governor's residence in 1959, and establishment of a new funding program for biomedical research as recently as 2008.[8] When Mr. Boettcher died, he left an estate whose value was estimated to be about $5 million, equivalent to about $66.7 million in 2025.

Not all was always well in the Boettcher family, though. Amidst all these business and charitable successes, and in addition to the marital troubles already mentioned, tragedy struck the Boettchers on at least two occasions worthy of mention in this history.

A name once infamous in the United States but now relegated often to historic footnotes was Verne Sankey (1891 – 1934). Back in the day, he was one of the nation's most notorious and most wanted gangsters. A great deal of his then-infamy was owed to his and his accomplices' brief but traumatic intrusion into the lives of the Boettchers. On February 12, 1933, Charles Boettcher II (son of Claude Boettcher, grandson of Charles) and his wife Anna Lou were returning home in Denver unaware Sankey and accomplice Gordon Alcorn, both armed, were waiting for them. The criminals accosted the couple, covered Mr. Boettcher's head, and stuffed him into their car. Before departing, they left Mrs. Boettcher a note demanding a $60,000 ransom (equivalent to just shy of $1.5 million in 2025). The kidnapping immediately captured not only local but national attention. Everyone seemed to be on the lookout and everyone was trying to do something. The newspapers published correspondence (thirteen letters in all) between the kidnappers and Claude Boettcher as they attempted to negotiate Charles'

5   Jefferson County (n.d.). Boettcher Mansion: Fannie Boettcher. *Jefferson County*. <https://www.jeffco.us/1873/Fannie-Boettcher> (accessed August 9, 2025).

6   Jefferson County (n.d.). Boettcher Mansion: Charles Boettcher. *Jefferson County*. <https://www.jeffco.us/1862/Charles-Boettcher> (accessed August 9, 2025).

7   Boettcher Foundation (n.d.). Boettcher Family & Foundation History. *The Boettcher Foundation*. <https://boettcherfoundation.org/who-we-are/history/#event-charles-boettcher-is-born> (accessed August 9, 2025).

8   *Ibid.*

release. The police were of course investigating every possible lead. The FBI got involved, making this one of the high profile cases J. Edgar Hoover would use to build the agency to such new heights. Even the state legislature considered a bill to make ransom kidnapping a capital offense, though the Boettchers asked them to table the bill because it might send the wrong message to the kidnappers if done prior to Charles' release or rescue.[9]

The subsequent investigation makes for fascinating reading which is unfortunately beyond the scope of this history, but we'll hit the important points. The most important point of all is that Charles was eventually (and rather surprisingly) released by his kidnappers on March 1, 1933. Mostly unharmed, he made his way to a local store until police could pick him up. He's reported to have told authorities that his kidnappers had behaved "as gentlemanly as could be expected under the circumstances," displaying a degree of verbal restraint we find quite extraordinary. He was able to provide clues that eventually led to the Sankey ranch, though the criminals had already left by the time police arrived and would not be apprehended until later. Perhaps the most remarkable thing about this entire affair is that even after Charles was released, Claude continued making arrangements to pay the $60,000 ransom! It was paid with police watching but somehow the kidnappers once again evaded capture and would remain on the loose until picked up by authorities one by one over the following months. Verne Sankey himself was apprehended in a sting operation on January 31, 1934 while his face was covered with a barber's towel, but he hanged himself with his tie before he could stand trial.[10]

The other tragedy we must discuss occurred much more recently. Spicer Breeden (1959 – 1996) was one of the heirs of the Boettcher estate who grew up in the very mansion whose own history we're about to discuss. He was son of Charlene[11] Humphreys, who was the daughter of Ruth Boettcher, who herself was daughter of the original Charles Boettcher. On March 17, 1996, Breeden was driving his 1995 BMW 540i at speeds estimated to be approximately one hundred miles per hour when he crashed into the car of *Rocky Mountain News* columnist Greg Lopez, killing the latter in the accident. Overcome with grief and guilt, Breeden took his own life shortly thereafter.[12]

Having met the family, we finally turn our attention to the history of the mansion itself, though its own history may not prove quite as colorful as that of its past inhabitants.

The plan for the mansion goes all the way back to Charles Boettcher himself

9   Lindell, L. (2004). "No Greater Menace": Verne Sankey and the Kidnapping of Charles Boettcher II. *Hilton M. Briggs Library Faculty Publications*, Paper 30. <http://openprairie.sdstate.edu/library_pubs/30> (accessed August 11, 2025).

10   *Ibid.*

11   Sometimes also spelled "Charline." Sources seem divided about equally on the spelling.

12   Shore, S. (1996). When Lives Collide: Denver Hit-and-Run Kills One Man— Then One More. *Los Angeles Times*. <https://www.latimes.com/archives/la-xpm-1996-05-12-me-3295-story.html> (accessed August 11, 2025).
Calhoun, P. (1996). Life in the Fast Lane. *Westword*. <https://www.westword.com/news/life-in-the-fast-lane-5056185> (accessed August 11, 2025).

and to the year 1917. Having become quite successful in all of his various ventures by then, he had the means to construct what would be his dream mountain retreat. As fortune would have it, Mr. Boettcher knew the Lookout Mountain area well. His Ideal Cement Company had recently donated the concrete to pave a road to the summit, so he jumped at the opportunity to buy the choice plot of land when a prior development plan was abandoned. He obtained the services of Fisher & Fisher, a noted Denver-based architectural firm (who also built numerous other still-standing buildings in Denver and the surrounding towns) and arranged for his home away from home to be built with local stone and timber harvested from the construction site. He called it "Lorraine Lodge" and intended to use it as a summer and hunting lodge. However, after his aforementioned separation from Fannie, he increasingly used the home himself (no one knows whether Fannie ever used it at all).[13]

The building was originally called Tudor style in its architecture but has more recently been reclassified as "Arts and Crafts" style. The family quarters consist of a cathedral-beamed living room with large plate glass windows. Deeper into the home, one finds a grand dining room and ballroom complete with an elegant fireplace and a downstairs master bedroom suite. The second floor is equipped with two smaller bedrooms with a shared bath and a sleeping porch. Everything is serviced by a kitchen with a butler's pantry and mudroom, along with a basement storage area. Outside is a large carriage house (now garage) with servants' quarters above, a lovely gazebo, and a well house to supply the property with water.

Though he used it for some thirty years toward the end of his life, the mansion eventually passed to Mr. Boettcher's granddaughter Charlene Humphreys Breeden who raised her family (including the aforementioned Spicer) in the 1960s. However, she was battling cancer toward the end of that decade and arranged to bequeath the entire property including the mansion and some 110 acres of land to Jefferson County, an action which was legally completed upon her death in 1972.[14]

Such an action—donating such a valuable property to the government—probably seems alien to many modern readers. Perhaps it was a different time in the late 1960s and early 1970s, or perhaps such generosity to the public was just in keeping with the Boettcher way of doing things. Either way, the mansion became public property. The county began adding trails around the property for public use. The mansion itself has been used as a wedding and event center since 1975, and still operates in that capacity to this day.[15]

Most elements of the property have been preserved in much the same state as in the original construction, but some changes have been made as recently as 2005 when the current kitchen was added to the property.

In 1984, it was, as has been the case for so many of the remarkable old buildings we've been so honored to visit, added to the National Register of Historic Places.[16]

13   Jefferson County (n.d.). Boettcher Mansion: History. *Jefferson County*. <https://www.jeffco.us/1859/History> (accessed August 11, 2025).

14   *Ibid.*

15   *Ibid.*

16   Becker, S. & Johnston, K. (1983). Lorraine Lodge: National Register of

Paranormal Claims

Numerous ghost stories are connected to this Boettcher Mansion, both in the public literature and which have been privately conveyed to us over the years.

Long before we set out to conduct our investigation (and thus began collecting these stories in earnest), indeed, we received a communication through our website in which an individual claiming to be a member of the events industry (and thus not an employee at the Boettcher Mansion but a regular visitor there) shared some strange occurrences. A lot of strange occurrences, as a matter of fact, consisting of some hearsay and some firsthand experiences.

The first thing this letter told us was of an office manager who'd once worked at the property and claimed to have regularly heard ghostly voices, first in the piano room (there is a wonderful grand piano at the end of the dining hall/ballroom) and then at other locations throughout the house. Our writer claimed to have sometimes mocked this manager for her unusual beliefs and experiences until having experienced some strange things firsthand.

On the first occasion, the very same (or at least assumed to be the very same) voice manifested when our writer was sitting at the reception desk by the front door. This voice, we're told, was a mournful female voice. The caretaker was summoned to investigate but found nothing. This was the first of several experiences the same author reported in this anonymous correspondence.

Other occasions included hearing the sound of footsteps in the kitchen when no one else was about. Guests at the writer's events would report seeing women in 1920s era attire around the house. Sometimes men would be seen standing outside by a car when no one was supposed to be outside. Our personal favorite of this person's stories occurred while the narrator was sitting alone on a cold and foggy October evening (just the sort of night for a ghost story, we think). Suddenly, the piano began to play by itself! It is not, we hasten to mention, a player piano. We can skip ahead to our own investigation report just enough to point out that we checked on that.

The same author also mentioned that several psychics had visited the property but were always hesitant to reveal what they'd sensed, felt, or detected. Elsewhere, we have explained in detail why psychic reports are troublesome things. On the one hand, we're committed to neutrality regarding the existence of psychic phenomena—it is, after all, a class of the sort of paranormal claims we'd want to investigate. So we don't rule out that such claims are truthful and accurate. On the other hand, it hardly actually proves anything. Because psychics themselves are unproven, their reports can't be used reliably to justify a claimed haunting. Instead, they should be collected, studied, and reported along with the ghost stories both because they represent part of the location's lore and because it might be possible in some cases to independently verify their findings depending on how detailed they are.

One final note from this letter we received before we move on to all the

Historic Places Inventory – Nomination Form. [Register No. 84000858]. *National Register of Historic Places*. <https://s3.amazonaws.com/NARAprodstorage/lz/electronic-records/rg-079/NPS_CO/84000858.pdf> (accessed August 11, 2025).

other claims. The author claimed that the voices the manager heard occurred within a week or two of "the death [by suicide] of a Boettcher heir." Though not specifically named in the letter, it was obvious to us the person was talking about Spicer. That's relevant because other witnesses have claimed one of the most haunted locations in the mansion is what's now called the Columbine Room on the second floor. That had been Spicer Breeden's room when he was growing up at the mansion, and though his death occurred elsewhere, some people think perhaps his spirit has returned to his childhood home.

Care is necessary when talking about events like this. We're reporting ghost stories connected to a tragedy that happened during living memory of many of our readers and many of the players involved. Therefore, we emphasize that we're only reporting the speculations we've heard in connection to the ghost stories and not passing any judgment on the individuals involved nor on the paranormal claims associated with them in the absence of some compelling evidence.

Other stories were relayed to us by staff at the Mansion and were either those employees' own firsthand experiences or one degree separated and reported to us as they were reported to our contacts by the witnesses.

One told to us directly by a witness was of a phantom whistling sound in the front lobby. It was like an old tea kettle whose steam was up but happened when the witness was alone (and importantly, not making tea). Furthermore, rather than from the kitchen, she said it seemed more like it came from a closet (or a corner near the closet) and perhaps somewhere between the first and second floor.

Another witness saw a woman standing outside the Mansion, looking into the main hall through one of the large windows. When staff went to investigate, no one was there.

A similar occurrence happened in the hall between the kitchen and the basement stairs and dumbwaiter. This witness saw a ghostly woman walk (or more accurately, float) past the kitchen door, crying. A crying woman has also been reported in the dining area (located at the same end of the great hall as the kitchen).

A final visual manifestation, and the most chill-inducing one if you ask us, was that a witness reported seeing a woman dressed in historic garb literally floating in the main hall in front of the fireplace.

No set of ghost stories is complete without a trip to the restroom. Often, for whatever reason, the women's restroom is the culprit.[17] Here, it was actually the men's room at the heart of one of the ghost stories. The men's room is located between bedrooms on the second floor and witnesses report hearing unusual sounds from this room. Sometimes, they're ordinary sounds—toilets flushing, faucets running—even if they happened at extraordinary times and when no one was in the room. But other collections of odd sounds ranging from footsteps to whispers have also been reported from time to time.

Two related stories happened under similar circumstances. First, one of the staff members was preparing to close up and leave for the evening. The very moment she stepped outside and moved to lock the front door, all the lights in the entire Mansion came on all at once. She went back in, checked for anyone

---

17   We could write a doctoral thesis on why this might be the case, but for now we just observe that it does seem to be the trend.

who might be lurking inside—of course, there was no one—turned off all the lights, and left for the night, a bit shaken by the experience. On another occasion, a different staff member was just arriving to open up in the morning, but every time he turned the key to unlock the front door, the door just kept locking itself.

Figure 6.2. The piano that played itself (photo: Robert Lewis).

Returning to the letter we mentioned above and the piano playing itself, we have to add that other witnesses have had similar experiences. Our correspondent came up with radio interference causing a radio to tune into a jazz station as a potential rational explanation (there are, after all, a lot of radio towers on Lookout Mountain). Other witnesses have not made these kinds of concessions and insist they've heard *that* specific piano playing itself in an otherwise empty room on multiple occasions.

Finally, we're heard reports of a phantom car showing up at least once. The way it was told to us, an antique model black car showed up at the front of the mansion as if dropping off passengers for a party or an event. Sure enough, a load of people got out and the car drove off. But no one recognized who any of the people were and none of them were ever seen again.

Our Investigation

We were contacted by Jefferson County staff working at the Boettcher Mansion to see if we'd be willing to give a public lecture at and about the Mansion to tell ghost stories for some of their patrons. Obviously, we're always happy to do that sort of thing, but we suggested it would be even better if we'd also be able to conduct a full and proper paranormal investigation first so we could speak on the subject more intelligently and share our firsthand results. They agreed and we arranged two dates, about a week apart, for our investigation. First, we'd do one daytime investigation, both because some of the paranormal claims have happened during the day and so we could have an opportunity to get the lay of the land in full light. And then the following week we'd return for an overnight investigation.

On both days, the building was closed to the public, but on the first day a few staff members were still in their on-site offices getting some work done. They agreed to be as silent as possible while we collected data, but that did limit our ability to monitor certain locations. However, on the overnight investigation we were joined only by one staff member to help us make sure everything got locked up at the end of the night, and she maintained her place at the front desk so we could work in silence as we usually do.

For ease of discussion, we'll start by describing our overall setup, which was identical on the two days except that the second one included a few extra cameras. Then we'll walk through our experiences in a more linear fashion.

Because this is a large property, we used eight surveillance cameras (in addition to our various handheld cameras) to get a view of as much of the property as we could. Two cameras offered a full view of the lobby and the hallways as well as the second-floor balcony. Another camera was placed between the kitchen and the basement stairs to monitor the location where someone saw a ghostly figure (on the second night, we mounted the camera in the dining room but with a view of the same space). We placed two cameras near the fireplace in the great hall pointing in opposite directions to afford us a view of the entire room (including the dining room and the piano room at either end). Upstairs, one camera had a view from the hallway into both the Columbine Room (Spicer's old bedroom) and the men's room through their open doors. Yet another camera was in the Columbine room with a view of the sun porch. Finally, back on the first floor, three additional cameras were used on the second day of the investigation with a view of the basement stairs, the first floor bedroom, and the first floor sun porch. We placed thermometers in the kitchen hallway and the upstairs hallway. Ordinarily, we'd want to monitor audio on numerous tracks in such a large building. However, because the Mansion's design is so open, we were able to use just three microphones: one in the front lobby, one in the kitchen hallway, and one in the upstairs hallway. Between those three, we knew we'd be able to record anything that happened in the house, and because we had three microphones spread out in different locations, we also knew we'd be able to use volume differentials on each microphone to triangulate any sound's actual point of origin if it should become necessary.

As always, in addition to the stationary monitoring, we thoroughly documented the site with still photography and spherical video prior to each investigation and conducted hourly walk-throughs to measure electromagnetic radiation on both AC and DC EMF meters. We also used air quality monitors and seismometers on the hourly walkthroughs. As is also typically the case, we loaded in cases full of all our other gizmos to have on hand in case we should need them, but we don't always take every possible kind of measurement if nothing in the lore suggests they should be necessary.

The first (daytime) investigation was mostly uneventful for most of the time. Because it was our first time setting up there, we took a little extra time to arrange and document the site and then began our typical process of silently monitoring for about an hour and then doing a walkthrough to take all of our measurements.

Because there were some staff still on hand, we ended up with fairly noisy audio recordings throughout the entire day even though they were mostly doing a good job of staying as quiet as possible and remaining out of our cameras' sight. The one exception was when someone had to walk through to take a delivery at the front door, but no harm done because we knew exactly who was doing what and when so we could cross that off our list of potential anomalies while listening and watching back all the footage. We did, however, get a Grammy-award-worthy audio recording of one of the most powerful sneezes we've ever heard from one of the staffers who we couldn't identify. You never know what's going to show up in paranormal investigation records.

Figure 6.3. The kitchen hallway where the ghost was seen (photo: Robert Lewis).

We did obtain two unusual audio recordings during this first daytime investigation. About thirty-three minutes into our monitoring, we heard a door slam followed by an odd whistling sound we could not account for. Obviously this caught our attention because it somewhat matched with the ghost story of the odd whistling sound we'd been told in advance. Their story, though, described something more like a whistling teapot, whereas ours was what we can only call a quieter sort of nondescript whistling sound. It didn't last for long and didn't recur. The door slam, we think must have been one of the staff members closing a door, but based both on the quality of the sound itself and the relative volume picked up by our microphones, it sounded like the front door. However, video monitoring of the front door revealed no one in the room (much less at the door) at that time. Therefore, an employee closing a door is probably still the most parsimonious explanation but it was strange enough to merit inclusion in our report.

Then, about an hour in, we heard another—entirely different—whistling sound. This one was a sort of two-tone whistle. Once again, quite distinct from the way the earlier witness described her own anomalous whistle sounding but intriguing enough to catch our attention.

On that topic, we need to offer a couple words more generally. Eyewitness testimony is far from the most reliable kind of testimony. Often, researchers have to go with what they've got, and that means giving witness testimony all due credence. However, the human perceptual systems are not particularly good at processing anomalous information in the first place, and our memory systems are even worse at accurately storing, retrieving, and describing what we've perceived. People can witness the exact same event and report drastically different interpretations thereof. Because most of our investigations do start with some kind of eyewitness testimony, we'd be foolish to discount it. And we're certainly not accusing anyone of lying or even of stretching the truth. But we also have to be careful to realize that what they actually saw or heard could be somewhat different from what they eventually report to us, especially if the report comes months or years (sometimes even decades) later.

In the case of the whistle, we have a certain suspicion that at least one of the whistling sounds we recorded might have been the same or similar to what our witness reported to us. No, it didn't sound like a teapot, but we can easily imagine that might have just been the first thing to come to her mind when she heard a strange whistle. We're not saying that *is* the case, but it's one quite plausible interpretation of the evidence available to us. Unfortunately, we weren't able to track down the true source of either of the whistles we heard—natural or paranormal—so conjecture is all we have for now.

Regarding our other types of monitoring, we don't have anything anomalous to report. Nothing of interest showed up on any of our video. The seismometers were remarkably steady. And EMF readings were largely consistent with baseline readings taken at the beginning of the investigation. Levels were elevated slightly above what one might expect from the property itself, but we also had to account for numerous broadcast antennae on Lookout Mountain which would account for those elevated readings.

Before moving on to our second evening, we should mention one completely mundane occurrence one of us experienced which any would-be paranormal investigators should take into account. About halfway through the investigation one of us who'd been staring silently at video monitors all day thought he saw movement on Camera 7 near the great hall fireplace, but it was brief and out of the corner of his eye, so he couldn't be sure. He noted the timestamp and after we returned home, we checked his note against the video recording. Nothing was there.

The important lesson is that when you're staring at a largely static image for long periods of time, your eyes and your mind can begin to play tricks on you. In this case, it was nothing but camera fuzz combined with psychological tricks that made him think he saw something. We've seen other paranormal groups, though, who take any slight perception as proof positive that a ghost is about. Jumping to conclusions is not the best way to gather quality evidence. Note everything you see, by all means, however insignificant it might seem. But don't forget to do the all important follow-up step of reviewing the footage with fresh eyes and a sober mind to determine whether you really saw something or whether your mind was just playing the kinds of tricks minds are wont to do when one puts oneself in unusual situations for long periods of time.

Our second investigation was to take place overnight, and it was far more eventful than the first, but not necessarily in the way you might think.

We've often joked, in the introduction to this chapter and elsewhere, that when Rocky Mountain Paranormal is investigating, there's likely to be a blizzard. Well, in this case, it wasn't a blizzard *per se* but we still had unusual luck with the weather. On our drive up the mountain, we noticed how ghastly the winds were. And they were truly terrible. If we could put video in this book, we would show you footage of giant trees swaying. Not the tops, mind you; the trunks. Other footage shows structures attached to the Mansion similarly swaying in the wind. Indeed, the winds were blowing so cold that the entire walkway—the entire *heated* walkway, no less—leading from the parking lot to the front door had completely iced over. Make no mistake: it is *not* fun to carry crateloads of delicate investigative and surveillance equipment up an icy slope. None of us wanted to fall, but we were honestly less concerned for our personal safety than for the tens of thousands of dollars' worth of fragile equipment!

When we arrived, we could hear the wind howling all through the building, so we knew for one thing we'd have to be extra careful with our interpretations of our audio recordings. We also checked reports from the National Weather Service for that evening and found warnings of wind gusts up to seventy-five miles per hour! For reference, this is considered "Storm Force" winds, and had speeds actually exceeded those seventy-five miles per hour, the winds could be considered "Hurricane Force."[18] This was going to be an interesting night, but we noted the weather conditions, set up as before, and began our investigation.

Sure enough, it was a noisy night, but for the most part we were able to attribute all the sounds we were hearing to the wind outside blowing through cracks and rattling the windows and doors.

At one moment partway through the investigation, we thought we really got something interesting. While we sat watching the video monitors, we all suddenly gasped as a window in the piano room slowly crept open and then slammed shut!

We immediately abandoned our silent monitoring protocol and rushed over to the piano room to investigate. Obviously we can't show you the video of it happening in print, but Figure 6.4 shows that exact window in its open condition. Just after that photograph was taken, the window again slammed shut. This was all a bit exciting for a moment because we knew all the windows had been closed and latched prior to our investigation.

It didn't take us long to figure out exactly what had happened. The windows had only simple latches rather than true locks, and the one on this particular window was just a little bit loose in its setting. The winds that night were so strong that over the course of a couple of hours, the window rattling in its frame provided sufficient force to open the latch. Then it was a simple matter for the wind to blow the window open and for the air pressure differential between the inside and outside to cause it to close again. Problem solved and case closed. Despite the simple solution, and even knowing the explanation was completely natural rather

18   National Weather Service (n.d.). Estimating Wind Speed. *National Weather Service.* <https://www.weather.gov/pqr/wind> (accessed August 12, 2025).

than paranormal, there's something magical about sitting in a supposedly haunted old mansion and watching a window open and close itself. Ghostly explanations might be preferable, but that was fun anyway.

Figure 6.4. This window opened by itself (photo: Bryan Bonner).

Before returning to our work, we affixed a small piece of gaffer's tape to the latch so it wouldn't continue to interfere with our investigation and, even more importantly, so it wouldn't blow open again once we'd all left and no one would be there to close the window. We also alerted our host to the issue so they could see to any necessary repairs or modifications later on.

As the evening wore on, we continued our monitoring and measurements as normal. Just like on the previous daytime investigation, we didn't detect any usual seismic, video, or EMF anomalies. Temperatures dropped considerably throughout the night but that was entirely consistent with the wind blowing windows open and letting the cold air in.

Audio was another matter entirely, though. This time we recorded several strange audio anomalies. The first was another one of those odd whistling sounds. This one, we might have dismissed as likely due to the wind except that it was consistent with the whistling we'd heard on the prior investigation, so we're not ruling out any potential explanations for that one. For what it's worth, we asked the staff member who was hosting us for this second evening and she couldn't think of anything in the building that would make such noises.

A bit later in the recording, we heard what sounded to all the world like a bird squawking. This seems unlikely to be the true source of the sound. It was late at night so most diurnal birds wouldn't have been about. And any nocturnal birds surely would have taken shelter from the wind long before the time of the recording. Chalk that one up to another unsolved mystery.

Our recordings are also full of various clicks and thumps. Most likely, these are due to the wind, but a few of them were strange enough that we're not yet prepared to rule anything out. We'd like to repeat our investigation some day when the winds are calmer and determine if the same sounds manifest again. For now, just know that we have already ruled out all the clicks, taps, creaks, whistles, and thumps we know to be certainly (or at least almost certainly) caused by the wind. For example, we were able to match one set of thumps to an outdoor structure swaying in the wind and knocking against the building. Only a few are left unexplained, and those, of course, are the interesting ones.

A final unexplained sound happened to us live while we were listening but did not show up on any of our recordings. Twice throughout the evening a couple of us looked at each other with puzzled expressions. We'd heard music. Not like a full band, and certainly not the piano playing itself (we were hoping for the latter but were disappointed). It was a very faint sound and hard to pin down as anything in particular except that it sounded like sweet music, perhaps of someone humming or some unidentifiable instrument being played in the distance.

Could this also have been caused by wind howling through particular pipes or cracks? We can't rule it out. After all, entire classes of instruments operate by pushing air through certain shapes of pipes and holes. But something about what we heard sounded more "intentional" than mere wind sound. Beyond that, we're not making any claims about its source. We do know that it wasn't a figment of the imagination because multiple investigators confirmed hearing it independently. Most parsimoniously, it probably was the wind. But there was something just eerie enough about it, coupled with the fact that we couldn't hear it on our recording, that gives us cause for thought.

When we packed up in the wee hours of the morning, we thought our adventure was over. Little did we know, it was only just beginning.

Recall the iced-over walkways from the beginning of our discussion? They'd only gotten worse. It took several trips to get all the equipment down the sidewalk even on a good day, and this time took even more because we were taking small loads to be extra careful. Yes, a few people took a tumble. No serious injuries, but there were many bruises and sore joints the next morning!

To get back down the mountain, we all piled into one car (we often carpool to locations to save on gas money or to minimize the number of parking spaces needed). Our host from the Mansion staff had her own car. She drove in the lead and we followed. It's important to realize that it wasn't snowing that day, so other than wind gusts, the roads should not have been particularly hazardous. Unfortunately, it had snowed earlier in the season. Though the roads started the day clear, the winds were strong enough to create a ground blizzard and to blow fallen snow back onto the road.

Sure enough, our host's car got stuck in a drift on the road ahead of us. Sure enough also, the road was too narrow to pass. None of our attempts to push met with any success. And so we spent the next several hours gently moving our car forward and backward a couple feet so it wouldn't also get stuck while we waited for Colorado State Patrol to eventually show up and tow the other car out of the snow.

Figure 6.5. The after-investigation adventure (photo: Bryan Bonner).

Never let it be said paranormal investigation is dull business, even when we don't find the conclusive proof of ghosts we're searching for. When you put yourself in unusual situations, unusual things tend to happen.

At the end of the day, this remains another open investigation and as is often the case, we'd love to return one day to follow-up on some of the unusual recordings we captured. But we were thrilled with what we were able to learn about the location. As for the lecture we gave after the investigation, it was also a great success. Though it was also a frigid and snowy day (this time actual snow instead of a ground blizzard), people still came out to hear about the history and the ghosts, and that's what this work is really all about.

References & Further Reading

Bean, G. R. (1976). *Charles Boettcher: A Study in Pioneer Western Enterprise.* Boulder, CO: Westview Press

Becker, S. & Johnston, K. (1983). Lorraine Lodge: National Register of Historic Places Inventory – Nomination Form. [Register No. 84000858]. *National Register of Historic Places.* <https://s3.amazonaws.com/NARAprodstorage/lz/electronic-records/rg-079/NPS_CO/84000858.pdf>

Boettcher Foundation (n.d.). Boettcher Family & Foundation History. *The Boettcher Foundation.* <https://boettcherfoundation.org/who-we-are/history/#event-charles-boettcher-is-born>

Calhoun, P. (1996). Life in the Fast Lane. *Westword.* <https://www.westword.com/news/life-in-the-fast-lane-5056185>

Goodstein, P. (1996, 2001). *The Ghosts of Denver: Capitol Hill.* Denver, CO: New Social Publications

Jefferson County (n.d.). Boettcher Mansion: History. *Jefferson County*. <https://www.jeffco.us/1859/History>

Jefferson County (n.d.). Boettcher Mansion: Charles Boettcher. *Jefferson County*. <https://www.jeffco.us/1862/Charles-Boettcher>

Jefferson County (n.d.). Boettcher Mansion: Fannie Boettcher. *Jefferson County*. <https://www.jeffco.us/1873/Fannie-Boettcher>

Lindell, L. (2004). "No Greater Menace": Verne Sankey and the Kidnapping of Charles Boettcher II. *Hilton M. Briggs Library Faculty Publications*, Paper 30. <http://openprairie.sdstate.edu/library_pubs/30>

National Weather Service (n.d.). Estimating Wind Speed. *National Weather Service*. <https://www.weather.gov/pqr/wind>

Shore, S. (1996). When Lives Collide: Denver Hit-and-Run Kills One Man—Then One More. *Los Angeles Times*. <https://www.latimes.com/archives/la-xpm-1996-05-12-me-3295-story.html>

7

The Pillar of Fire: Westminster Castle

Some buildings just *look* like they ought to be haunted. Such is the case with Westminster Castle, also known to locals as the Pillar of Fire, located at 3455 West 83rd Avenue in Westminster, Colorado. Built to truly look like a castle from the pages of history, it's the kind of place that people would just think surely must have a ghost or two. But it's not just appearances that lend it a haunted reputation. It has plenty of specific stories attached to it as well.

Unfortunately, our own access has been limited so far so this will also be one of our shorter stories despite the location's rich and colorful history.

Figure 7.1. The Pillar of Fire (photo: Bryan Bonner).

The History

Westminster is of course a London borough and Westminster Abbey is the Anglican church that's been the site of dozens upon dozens of coronations and royal weddings. Why such a name should also be the name of a municipality in Colorado is an interesting story which is at the heart of our history in this chapter.

A common thread throughout the histories described in this series is gold and silver mining, and the same is true of Westminster. What eventually became the City of Westminster began as a settlement of individuals chasing their fortune after gold was discovered in the South Platte River in the middle of the nineteenth century.

Our own story, though, begins a couple decades later when, in 1882, the Presbyterian Synod of Colorado began to contemplate opening a religious college in Colorado associated with their denomination. Originally, they thought to build their new institution in Denver, but Henry J. Mayham had some different ideas. He owned a substantial amount of property in what would later become Westminster and realized that if the Presbyterians built there instead of in Denver proper, it would be close enough to the growing "big" city to still be a desirable location but would substantially increase the value of his own property. As such, he got the Reverend T. H. Hopkins to back his plan and thus convinced the Denver Presbytery to build instead on 640 acres of land he donated, of which forty acres would be used for the university proper and the rest for support purposes or to be sold for additional funding.[1]

The new college was to be called Westminster University, presumably in honor of the English Westminster, and its founders wanted it to become the "Princeton of the West." Architect E. B. Gregory was commissioned to design the building, and what a job he did! The Westminster Castle truly looks like one of the grand historic castles from across the pond, including a six-story tower. Gregory's original design was a castle in the Richardsonian Romanesque style constructed of gray stones, but construction delays convinced Mayham and the other individuals involved to recruit another architect, Stanford White, to revise the design and commence construction. The biggest change to the design was the substitution of red sandstone from the Manitou and Red Rocks region for the gray stones.[2]

For our part, we think this was probably a wise substitution. Gray stone may be more classical in appearance, at least in the popular imagination, but the red stone is certainly striking—the building is impossible to miss. That red sandstone from those formations is also sometimes implicated in paranormal lore (see Volume 2, Chapter 13) is completely beside the point. Likely the reason a disproportionate number of Colorado's famous haunted locations were constructed of this signature red sandstone is simply that it's a visually striking stone that appealed to the people who created the grand structures whose histories and physical appearances caught the public imagination both for its looks and because it was readily

1   Colorado Encyclopedia (n.d.) Westminster University. *Colorado Encyclopedia.* <https://coloradoencyclopedia.org/article/westminster-university> (accessed August 12, 2025).

2   *Ibid.*

available from a nearby supply when these buildings were under construction.

The cornerstone for the remarkable building was laid down in 1892 and construction completed in 1893, but it sat vacant until 1907.[3] Once again, we can't help but notice a coincidence between this structure and another building made from the same stone. The Croke-Patterson Mansion in Denver (see the aforementioned Volume 2, Chapter 13) was also constructed of the same material from the Manitou formation and also went unoccupied by its original owner for some time, at least according to legend. It would be easy for those prone to pattern-detection and inclined toward the paranormal to leap to conclusions. However, at least in the case of Westminster Castle, we know the reason the University didn't move in.

Recall from Chapter 4 that there was an economic depression at this time. The so-called Panic of 1893 was the most severe economic depression in the United States until the Great Depression. Causes of the depression have been widely studied and hotly debated ever since, but Westminster University was another victim of this financial downturn and lacked the funds to open their doors.

Del Norte College of the Southwest closed its own doors a few years later in 1902, prompting the Presbyterians to revisit their plans to open Westminster University, which they finally did in 1907 (ironically, the year of another infamous economic panic). Despite best plans, business was always somewhat shaky for the budding college. They were eventually able to retire their debt and open a law school by 1912, and things started to look up. In 1915, they switched from a coeducational institution into an all-male school. People argue even to this day about the relative merits of sex segregation in higher education and this book isn't the place to settle that debate. We can say with certainty, though, that at least in this case, the decision probably spelled the end of Westminster College. In 1917, the United States entered World War I and suddenly almost the entire student body had either voluntarily enlisted or been drafted into military service. They shut their doors permanently that year, with only their evening law program surviving until it was absorbed by the larger University of Denver's Sturm College of Law in 1957.[4]

Despite never quite managing to get off the ground as an institution of higher learning, the school's name still lives on. The City of Westminster was named after Westminster University, not (as most would probably guess) the other way around.

After sitting empty for another approximately three years, the property was purchased by Bishop Alma Bridwell White of the Pillar of Fire church (about which more in a moment) for only $40,000 (equivalent to just over $640,000 in 2025[5]). To get such a price, they had to stipulate in the contract that the building would continue to be used for private religious education, which the new church owners were glad to do after investing an additional $75,000 into remodels and

---

3   *Ibid.*

4   *Ibid.*

5   That's still quite the bargain! While it may sound like a lot of money—and it is, especially to mere mortals like your friendly local paranormal investigators—it's only about $100,000 more (in the 2025 market) than the average single-family home in the same region.

repairs and then reopening as a coeducational Christian school called Westminster College and Academy.[6]

Some locals who know that Westminster Castle is commonly referred to as the Pillar of Fire don't realize that it got its name from a religious movement. When we asked around in a highly-unscientific and informal poll of locals, those who'd heard the name and knew what building it referred to mostly thought its name originated with the building's tallest tower and red sandstone construction. We're happy to clear up the confusion.

Pillar of Fire, as a church, has quite a colorful history of its own. More formally called Pillar of Fire International, it is a church founded by the aforementioned Alma Bridwell White in Denver in 1901. She originally founded it as the Pentecostal Union, but married a Methodist minister, though she would eventually withdraw from the formal Methodist Church when their leadership opposed some of her messages. The church was renamed Pillar of Fire in 1917 with White serving as its bishop.[7]

Though to all appearances from the outside today it looks like just about any other Methodist-aligning Christian denomination, its history is far more controversial. On the one hand, White was the first female bishop of a Christian denomination in the United States and led the church a reputation as a staunch defender and champion of women's rights.[8] On the other hand, White—and thus her Pillar of Fire church—were also strongly aligned with the Ku Klux Klan and her philosophy became known as a sort of witch's brew of feminism, antisemitism, racism, anti-Catholicism, anti-Pentecostalism, white supremacy, temperance, and anti-immigration or nativism.[9]

The close affiliation between the KKK and Colorado history has been mentioned elsewhere in this volume and is well documented in general.[10] However, the Klan connections with Alma White and the Pillar of Fire exceeded expectations even for those days in Colorado history; it was the only Christian church that formally and openly associated with the Klan even then.[11] The Pillar of Fire church was well known to host Ku Klux Klan meetings and cross burnings at

---

6   *Ibid.*

7   Encyclopaedia Britannica (n.d.). Pillar of Fire (American religion). *Britannica.* <https://www.britannica.com/topic/Pillar-of-Fire-American-religion> (accessed August 12, 2025).

8   Kurian, G. T. & Lamport, M. A. (2017). *Encyclopedia of Christianity in the United States.* Lanham, MD: Rowman & Littlefield.

9   Kandt, K. E. (2000). Historical Essay: In the Name of God; An American Story of Feminism, Racism, and Religious Intolerance: The story of Alma Bridwell White. *Journal of Gender, Social Policy & the Law, 8*(3): 753-794.

10   Goldberg, R. A. (1981). *Hooded Empire: The Ku Klux Klan in Colorado.* Urbana, IL: University of Illinois Press.
Goodstein, P. (2006). *Denver from the Bottom Up Vol. 3: In the Shadow of the Klan: When the KKK Ruled Denver, 1920-1926.* Denver, CO: New Social Publications.

11   Kandt, K. E. (2000). Historical Essay: In the Name of God; An American Story of Feminism, Racism, and Religious Intolerance: The story of Alma Bridwell White. *Journal of Gender, Social Policy & the Law, 8*(3): 753-794.

numerous of its properties including the Westminster Castle.[12] The building was even host to the 1927 statewide KKK convention.[13] There are even rumors—completely unsubstantiated but arguably plausible given what we've just read—that the organization may have hosted or participated in events at which black people were tortured or even murdered.

Pillar of Fire International still exists to this day, with churches, schools, and missionary operations around the world. Fortunately, it has publicly repudiated its Klan-affiliated past and issued multiple statements of apology and clarification of their current (more welcoming) message.[14]

Returning to the more mundane history of Westminster Castle itself, the Pillar of Fire organization renamed Westminster College as the Belleview College and secured formal accreditation in 1926. It continues to operate to this day, now as the Belleview Christian School, offering programming for children in kindergarten through grade twelve. The building also hosts Pillar of Fire's *KPOF* radio station which has been continually operating since 1928.[15]

In 1979, it was added to the National Register of Historic Places under the name of Westminster University, presumably honoring its original history as the "Princeton of the West," rather than Pillar of Fire's own early history; indeed, the nomination form does mention Pillar of Fire's ownership of the property and its use for religious education, but omits the sketchier parts of the organization's past.[16]

Paranormal Claims

Our introduction to this chapter mentioned that Westminster Castle just *looks* like it ought to be haunted. With a history like the one you've just read, it should come as no surprise to anyone that it does indeed have quite the haunted reputation. We don't know what—if anything—might actually cause a building to become haunted, but lore does seem to suggest that places with more colorful pasts are the most likely to have haunted reputations.

12   Dunn, M. (2019). Colorado Women of the Ku Klux Klan: Part 5 in a Series. *Northern Colorado History*. <https://www.northerncoloradohistory.com/colorado-women-of-the-ku-klux-klan/> (accessed August 12, 2025).

13   Colorado Encyclopedia (n.d.) Westminster University. *Colorado Encyclopedia*. <https://coloradoencyclopedia.org/article/westminster-university> (accessed August 12, 2025).

14   Otterman, S. (2017). A Booming Church and Its Complicated, Ugly Past. *The New York Times*. <https://www.nytimes.com/2017/09/15/nyregion/zarephath-christian-church-new-jersey-pillar-of-fire.html> (accessed August 12, 2025).

15   Colorado Encyclopedia (n.d.) Westminster University. *Colorado Encyclopedia*. <https://coloradoencyclopedia.org/article/westminster-university> (accessed August 12, 2025).

16   Dallenbach, R. B. (1979). Westminster University: National Register of Historic Places Inventory – Nomination Form. [Register No. 79000572]. *National Register of Historic Places*. <https://s3.amazonaws.com/NARAprodstorage/lz/electronic-records/rg-079/NPS_CO/79000572.pdf> (accessed August 12, 2025).

Most of the experiences people have reported have come to us second- or third-hand at best, and tales have been passed around on local interest and paranormal blogs for many years, so it's difficult to separate credible reports of anomalous experience from mere rumor and speculation, but the paranormal stories tend to cluster around a few distinct ideas.

First, several witnesses over the years have reported seeing ghostly faces in the windows late at night. Those who have gained access to the property without permission in the past have reported seeing similar faces, hearing footsteps, and experiencing strange feelings of uneasiness in and around the property. One blog poster described an incident in which several teens snuck into the property after hours with a spirit board and experienced similar phenomena, and goes on to speculate that the property might have been used by the Klan during its dark history because it sits on a Ley Line.[17] We'll have some thoughts regarding the later claim in particular but will hold them until we discuss our investigative activities below.

Most of the other claims are more nebulous. A lot of people believe the Castle is haunted by spirits connected to past murders. Others think it's haunted by spirits connected to a cemetery on a plot of land adjacent to the Castle. Still others think at least one spirit belongs to a young girl who died of a head fracture on the entrance steps to the University in 1910, though whether the injury was due to accident, suicide, or foul play is not specified.[18]

With stories like those and a history like the one we discussed, we knew we had to look into this place.

Our Investigation

As we said, this is he kind of place we knew we had to investigate. It had been on our radar for years, but we were never able to arrange access. It's still owned by the Pillar of Fire church, operates as a school and radio station, and because it's still in active use, public tours and after hours access are difficult if not impossible to arrange most of the time. It could also be that the Pillar of Fire people want to move on from their past rather than to dredge it up with a bunch of literal or metaphorical ghost stories. Or perhaps they might even have religious objections to paranormal investigations or are concerned because they've been broken into in the past by wannabe ghost hunters. Whatever the reason, we never gained access to the property.

Then one day opportunity struck. The Castle was opened to the public for an open house event. It's not the kind of conditions under which we would ordinarily like to investigate, but we'll take what we can get, so we dispatched a small crew equipped with cameras and a few measuring devices to see what they could discover.

Unfortunately, other than a remarkable old Castle, the answer is: not very

---

17   Morera, A. (2014). Westminster Castle of Strange? *Haunted Northside Tales (Haunted Denver)*. <https://ascarymorera.blogspot.com/2014/09/westminster-castle-of-strange.html> (accessed August 13, 2025).

18   *Ibid.*

much. We didn't detect any photographic or electromagnetic anomalies and no one had any strange feelings or bizarre experiences while we were there, but that's about all we can say about our on-site miniature "investigation."

However, we have done some armchair sleuthing to see if we could learn anything about some of the more specific claims people have made about incidents in the past that might have led to the hauntings. Proving those wouldn't prove the ghost stories, of course, but they would at least be an important piece of the puzzle. Similarly, disproving them wouldn't mean there are no ghosts, but would likewise be useful information to know.

The first and easiest thing to check on was the existence of a cemetery adjacent to the campus. That part is true. Belleview Cemetery (sometimes misspelled on various maps and documents as "Belleview Cemetary") is a small, fenced-in private graveyard owned and maintained by Pillar of Fire and is located at the northeast corner of the Westminster Castle campus. It is the final resting place of a variety of church members and leaders as well as some historic figures with connections to the site.

No records, reports, or even rumors we could find, though, explicitly connect any of the individuals buried there to any of the paranormal lore. Cemeteries and graveyards are commonly reputed to be haunted in general, and the presence of such a property on or near the Castle grounds surely contributes to its haunted reputation, but it seems to contribute only in the abstract sense rather than as a result of any particular claim.

We also wanted to consider the idea that the site might be on a Ley Line and that this could be why there is paranormal activity there and why the KKK might have chosen the site for some of their rituals.

Ley Lines themselves are hypothesized straight lines that connect ancient monuments and locations of spiritual significance. The idea seems to be that the flow of spiritual energy is stronger along these lines and that's why people have built important structures on or near them. We don't want to go too far afield into a discussion of Ley Lines themselves because that would require a complete case file of its own (one of these days, we might do so), but it's worth pointing out at least that sources don't even agree on where the lines are, and that most researchers consider them speculative at best and completely discredited at worst. However, even if they are largely discredited, it would be possible to consider that the site might have been chosen because individuals in charge *believed* in them, so we wanted to dig a little bit deeper.

Regarding whether the Klan conducted their rituals at Westminster Castle because they knew it to be on a Ley Line, we consider the claim highly dubious. Ley Lines as a concept were first described (or invented, if you're skeptical) by businessman and antiquarian Alfred Watkins first in his 1922 book *Early British Trackways* and in more detail in his 1925 *The Old Straight Track*. Given that Pillar of Fire purchased the property prior to these publications, they would have had no way of knowing about the lore unless they were somehow mystically drawn to the property, and that would be an impossible claim for us to test.

We also conducted a search of various maps of "known" Ley Lines (in this case we mean the term "known" to indicate locations where people who believe in Ley Lines think they are or might be) in Colorado. Though the Centennial State is home to some suspected lines, none of them pass through the Westminster Castle property, nor even the City of Westminster more generally. Conclusive disproof? Certainly not! But at the very least we can say there is nothing in the paranormal literature to suggest Ley Lines have anything to do with Westminster Castle or the Pillar of Fire organization.

Outside of some fictional interpretations, further, we're unaware of any sources in either the historical or the paranormal literature suggesting a more general connection between KKK activities and Ley Lines. Though the Klan did perform some strange rituals from time to time, they're not known for their allegiance to the kind of New Age thinking of which Ley Lines are typically a component.

Rather, it seems the most parsimonious explanation here is simply that the Pillar of Fire took a good real estate bargain to establish their school back in the 1920s.

Next, we turned our attention to the 1910 incident in which a girl fractured her skull on the property. We didn't have a way to look for the ghost directly, but we tried to find any evidence or reports we could locate describing the death itself. So we sent our research monkeys out to crawl through all the archives we could access, conducted extensive web searches, browsed old newspapers, and we simply couldn't find anything except the one source describing the incident in the context of paranormal lore. Failing to find anything that way, we expanded our search to years other than 1910, just in case the report we found simply got some details wrong. Still nothing. Indeed, we can find no records of any kind, from any year, that match any of the details of the alleged incident. Records from more than a century ago are often incomplete, and even if they were complete, we can't absolutely guarantee we searched every conceivable nook and cranny. But we did conduct as thorough a search as we could and that nothing turned up suggests the story is probably apocryphal.

Finally, there was the claim that there had been a murder or multiple murders on the site over the years. This claim is so general it was even tougher to search for records, but search we did. Once again, we found no credible reports in official public documents, newspaper articles, history books, or honestly anyplace other than the paranormal literature alluding to any specific murder cases in Westminster Castle or on its grounds.

There are rumors going back to the Pillar of Fire's past KKK connections alleging some tortures and/or murders may have taken place in those days. We certainly can't rule that out. But there are no specific incidents we could find in any credible records or reports. Cross burnings and other KKK meetings, yes. But stories of specific acts of violence are found only in rumor and innuendo.

Alas, that was all we could manage on this one. We'd love some day to be invited in for a full on-site investigation, but for now we have to leave this as an open file in our records.

References & Further Reading

Colorado Encyclopedia (n.d.) Westminster University. *Colorado Encyclopedia.* <https://coloradoencyclopedia.org/article/westminster-university>

Dallenbach, R. B. (1979). Westminster University: National Register of Historic Places Inventory – Nomination Form. [Register No. 79000572]. *National Register of Historic Places.* <https://s3.amazonaws.com/NARAprodstorage/lz/electronic-records/rg-079/NPS_CO/79000572.pdf>

Dunn, M. (2019). Colorado Women of the Ku Klux Klan: Part 5 in a Series. *Northern Colorado History.* <https://www.northerncoloradohistory.com/colorado-women-of-the-ku-klux-klan/>

Encyclopaedia Britannica (n.d.). Pillar of Fire (American religion). *Britannica.* <https://www.britannica.com/topic/Pillar-of-Fire-American-religion>

Goldberg, R. A. (1981). *Hooded Empire: The Ku Klux Klan in Colorado.* Urbana, IL: University of Illinois Press

Goodstein, P. (2006). *Denver from the Bottom Up Vol. 3: In the Shadow of the Klan: When the KKK Ruled Denver, 1920-1926.* Denver, CO: New Social Publications

Kandt, K. E. (2000). Historical Essay: In the Name of God; An American Story of Feminism, Racism, and Religious Intolerance: The story of Alma Bridwell White. *Journal of Gender, Social Policy & the Law, 8*(3): 753-794

Kurian, G. T. & Lamport, M. A. (2017). *Encyclopedia of Christianity in the United States.* Lanham, MD: Rowman & Littlefield

Morera, A. (2014). Westminster Castle of Strange? *Haunted Northside Tales (Haunted Denver).* <https://ascarymorera.blogspot.com/2014/09/westminster-castle-of-strange.html>

Otterman, S. (2017). A Booming Church and Its Complicated, Ugly Past. *The New York Times.* <https://www.nytimes.com/2017/09/15/nyregion/zarephath-christian-church-new-jersey-pillar-of-fire.html>

Uncover Colorado (n.d.). Westminster Castle – Westminster. *Uncover Colorado.* <https://www.uncovercolorado.com/activities/westminster-castle/>

8
Night Court: The Weld County Courthouse

Sometimes even we get surprised by the unusual situations we find ourselves in. In this case, we found ourselves given greater access to government offices than we ever could have hoped for, all because there were some interesting ghost stories to investigate.

The Weld County Courthouse, located at 901 9th Avenue in Greeley, Colorado, is remarkable for numerous reasons. The first thing you'll notice if you visit is the stunning architecture, of course. But if you dig a bit deeper, you'll find both some fascinating history and some chill-inducing ghostly lore.

Figure 8.1. The Weld County Courthouse (photo: Bryan Bonner).

The History

Some of our members have a certain affinity for law and legal proceedings. Indeed, you'll get a small taste of that for yourself when you read our essay at the beginning of the third section of this volume. As such, we have a certain temptation to go into a real deep dive into the entire history of Weld County jurisprudence and Colorado's Nineteenth Judicial District. We are determined to resist this temptation, fascinating though it may be, and will try to keep our history focused on the courthouse itself. Be that as it may, we do need to go back a little further in history to explain how the courthouse came to be.

Borders have shifted a little bit as history has progressed, but what is now Weld county, for the most part, used to be St. Vrain County, originally part of the Nebraska Territory. We've described across various chapters in our previous two volumes at least a rough sketch of how Colorado grew into a state from its origins as a collection of small mining settlements. President James Buchanan made the first major move toward Colorado's statehood on February 28, 1861 when he signed a bill establishing the Territory of Colorado.[1] Colorado's General Assembly then followed by officially creating 17 counties, including Weld County, named for Lewis Ledyard Weld (1833 – 1865), a lawyer, Secretary of State of Colorado, and acting Governor of Colorado who died of exposure while serving the Union in the American Civil War.

The newly founded county needed a courthouse, and the magnificent structure whose story we're getting to was still well into the future. Indeed, the first courthouse the County officially recognized was a one-room log cabin on the farm of one Andrew Lumry, who offered it for this use from 1861 to 1869 (as an interesting aside, it has been reconstructed and can be visited as part of the Centennial Village Museum in Greeley).[2]

Other courthouses for Weld County preceding the current one were: Fort Latham, from 1869 to 1870; a building in a region called Block 36, from 1870 to 1877; a single-story frame building on 7th Street from 1877 to 1883; and a brick building built at the current courthouse's location, from 1883 to 1914.[3]

However, in that final year, the powers that be realized that the current building was simply going to be insufficient in size for the operations of the growing Weld County's government. Therefore, the building was demolished and construction on a new building, the one which still stands, began in 1915, though it would take an additional two years to complete. It was designed by architect William N. Bowman (another of those architects responsible for a substantial proportion of Colorado's historic landmarks) in the Neo-Classical style and with an exterior constructed of Indiana limestone. Even today, it stands out as one of

1   "An Act to provide a temporary Government for the Territory of Colorado," of the 36th Congress of the United States. Quoted in: Sanger, G. P. (1863), ed. *The Statutes at Large, Treaties and Proclamations of the United States of America from December 5, 1859 to March 3, 1863*, volume 12: 172-177.
2   Weld County (n.d.). The History of Weld County's Courthouses. *Weld County, Colorado.* <https://history.weld.gov/Courthouse-100/The-History-of-Weld-Countys-Courthouses> (accessed August 13, 2025).
3   *Ibid.*

the most visually striking buildings in the area.[4] It came at a steep cost, though. The county had to shell out over $400,000 in cash for the construction (equivalent to nearly $13 million in 2025) over the objections of many locals.[5]

Some interesting features make its interior just as fascinating as its exterior. Marble used in its construction was sourced from the quarry in Marble, Colorado, and represents the last of the building grade stone removed from that site. It also features a unique clock system. Eight clocks throughout the courthouse are actually "slave" units connected to a central "master" grandfather clock which pneumatically controls the displays on all the others. There are only perhaps half a dozen such systems left in the state. Unfortunately, over the years, the system has stopped working and while the master clock has been repaired, most of the other clocks are now battery-powered.[6]

Figure 8.2. The Courthouse clock (photo: Bryan Bonner).

4   Too often, modern government buildings are built in a soulless Brutalist style. Back in the day, buildings belonging to the public were designed to convey a sense not only of beauty but of majesty. We think it's important, both aesthetically and politically, to return to the architectural philosophy of the past.

5   Kavalec, C. (1976). Weld County Courthouse: National Register of Historic Places Inventory – Nomination Form. [Register No. 78000886]. *National Register of Historic Places.* <https://s3.amazonaws.com/NARAprodstorage/lz/electronic-records/rg-079/NPS_CO/78000886.pdf> (accessed August 12, 2025)

6   Ehnert, R. (n.d.). The Courthouse Clocks. *Weld County, Colorado.* <https://history.weld.gov/Courthouse-100/The-Courthouse-Clocks> (accessed August 13, 2025).

The Courthouse was officially dedicated on Independence Day itself, July 4, 1917. In the 1930s, work crews repainting the building unfortunately destroyed many of the decorations, friezes, symbols, and the original choice of paint color. Restoration to its original design was completed on November 7, 1973 with the dedication of new Courthouse Chimes. Originally, it held all of the County government offices but eventually became overcrowded, so on April 17, 1974 the Chief Judge of the Nineteenth Judicial District ordered that the building should from then on be used only to house the County and District Courts.[7]

Other changes throughout the years have been relatively minor and many of them have involved restoring the original design. Also in the 1970s, the heating system was upgraded to a more modern design, the manual elevator was replaced with an automated one,[8] and the ceiling tiles and original jury seating were replaced. Stained glass windows were restored in the 1990s.

That about does it for the history of the building itself. The most fascinating parts of the rest of its history are all the cases and trials its seen over the years, and here is where we have to stifle our desire to overwhelm the reader with historic and legal minutiae. There is one story we have to tell, though: the tale of the lynching of Wilbur D. French (1844 – 1888), the only known lynching in Greeley history.

Relatively little is known about the principal players in this drama except that W. D. French was a wealthy but brash and disliked (not to mention often drunk) businessman known for his temper and rumored (though never proven) to have murdered his own wife.[9] He had a brief but fraught business relationship with a local shoemaker named Harry Woodbury. After their business relationship failed, they parted ways with a settlement including 850 pounds of flour, 250 pounds of which were delivered and the other 600 pounds of which caused another disagreement between the former partners. Woodbury filed a lawsuit against French and won. On December 14, 1888, Woodbury was at home in the back portion of the house he leased from French when the latter arrived with a posse of several other men and, furious over the legal ruling, attempted to gain entry. When he was denied, he went around the back, unlocked the door, pushed Woodbury's wife into another room, and shot and killed Woodbury. He, along with John Hogan and John Samples (members of his party) were arrested that night, but jurors at the Coroner's Inquisition couldn't be sure which member of the group actually fired the fatal shot and so rumors began to circulate among the community that French might be allowed to walk free.[10]

7   Kavalec, C. (1976). Weld County Courthouse: National Register of Historic Places Inventory – Nomination Form. [Register No. 78000886]. *National Register of Historic Places.* <https://s3.amazonaws.com/NARAprodstorage/lz/electronic-records/rg-079/NPS_CO/78000886.pdf> (accessed August 12, 2025)

8   We almost wish they hadn't made this change. Obviously an automated elevator system is more consistent with a busy government building in modern times, but there's something undeniably charming about an old fashioned manual elevator operated by an attendant.

9   For whatever our opinion is worth: we don't think he did.

10   Whitman, S. (n.d.) Two men died. Two lives were lost – all over 600 pounds of flour. *19th Judicial District.* <https://www.courts.state.co.us/userfiles/file/

On December 28 of the same year a man whose identity has been lost to history dressed as a woman, put a veil over his face, and entered a Greeley hardware store asking to purchase fifty feet of rope. You might expect the store owner to throw such a person out, especially in 1888. But he knew what was about to happen. "Take this and forget it," he said, referring to the rope. "If you need any more, come back."[11]

Later that night, actually about 1:00 a.m. on December 29, a group of some fifteen masked men appeared at the Sheriff's house and demanded the key to the jail, their spokesman speaking in a false Irish accent. Knowing what was about to happen, the Sheriff refused but the mob bound and gagged him and proceeded to the jail where they used hammers and crowbars instead of the key to break into French's cell. They grabbed French, dragged him to a large tree which stood between the jail and the Courthouse (this would have been the old courthouse at the current location), and hanged him on the spot, exacting the "justice" they feared the Court would not supply. According to legend, the "hanging branch" died shortly thereafter and French had no mourners at his funeral. The Court would find that his death was the result of a felony but that it was committed by unknown persons and no charges were ever filed.[12]

Paranormal Claims

This place has got more paranormal claims and ghost stories than we could shake an EMF meter at! Unfortunately most are just of the "I heard something odd" variety and are hard to track down or to spin a fascinating yarn about. But there are at least some reports worthy of mention.

We can begin where we left off in the previous section: with Wilbur D. French. Actually relatively few of the claims we've heard in modern times specifically identify French as the ghost in any of the ghost stories, but there certainly are rumors that his spirit still haunts the courthouse. If the lore is correct that violent ends tend to prompt a haunting, it would make a certain degree of sense.

There are persistent stories about a "shadow man" who haunts the fourth floor of the Courthouse, particularly the women's restroom.[13] Some authors have specifically connected that spirit to French, suggesting perhaps he still wanders the courthouse waiting for the verdict in his case that never came.[14] Whether or not they're all connected to French's story, numerous witnesses say they've seen figures on the fourth floor, which seems to be the hotspot of paranormal activity in the Courthouse.

Court_Probation/19th_Judicial_District/Court_House_History/WD%20French.pdf> (accessed August 13, 2025).

11   Peters, M. (1999). The Story of W. D. French. *Weld County, Colorado.* <https://history.weld.gov/County-150/Justice/WD-French> (accessed August 13, 2025).

12   *Ibid.*

13   As we've said before, we have no idea why the ghosts seem drawn to women's washrooms, but that does seem to be a noticeable trend.

14   Brick, C. (2018). *Haunted America: Ghost & Legends of Colorado's Front Range.* Charleston, SC: The History Press.

Also related to the French lynching was the story that the hanging branch withered and died after the deed was done. Because that occurred in 1888, if at all, it's well beyond our ability to directly investigate, but it has become part of the lore. And of course if the story is true, it would be a paranormal claim, so it bears repeating in any inventory of the paranormal stories at the Courthouse.

Throughout the years, other stories have involved witnesses seeing people who weren't really there, hearing footsteps or other unusual sounds, having strange feelings, and all the usual kinds of things we would expect. Multiple witnesses have reported seeing "demonic" faces reflected in the clock faces. One janitor working late one night to clean those clocks said he saw a such a frightening image and never returned to work again.

Shortly before our investigation, one employee said she was setting up a courtroom for the day's proceedings. Before anyone else entered the room, she heard a loud breathing sound coming from the benches in the gallery. No one was there. She called for help and some other employees also checked but found no person and no sign of anything that could have made such a sound.

Audio recordings are often taken as a part of court proceedings. During one hearing, a tape recorder was running. The hearing went as expected and no one in the room at the time noticed anything strange. When Court personnel listened back to the recording, though, there is a clip about thirty seconds in length during which the voices are drowned out by sounds of banging, footsteps, and unintelligible voices belonging to people who weren't present in the hearing. After that, the recording of the hearing continues as normal. No one in the room stopped talking because the disturbance occurred only on the recording.

There's a back staircase used by jurors moving between the third and fourth floors of the Courthouse. It's supposed to be haunted by the spirit of a young boy. No one knows what his real name would have been (assuming, of course, the story is true), but Court staff have taken to calling him Jonathan, and they say he often likes to move things around while he's at play.[15] Employees have even taken to leaving toys out specifically to either give him something to do or to test whether he might move them.

Once, an overhead storage compartment closed when nobody was near it, and some think this might also have been "Jonathan's" work.

Our Investigation

Every once in a while, we're given even more access to a location than we might expect, and this was such an occasion. We were given access to any locations we wanted and allowed to return multiple times. The notion of giving paranormal investigators free run not just of *a* building but of the County Courthouse, where serious government business is ongoing, flabbers our ghasts, but we weren't about to look a gift horse in the mouth.

Because this was to be a large operation with tons of ground to cover (the Courthouse is huge), we started with a preliminary investigation during business hours. On this visit, we interviewed several Court staff members who shared the various ghost stories they'd either heard or experienced. They were also kind

15   *Ibid.*

enough to give us a complete tour of the building, during which we were able to photographically document most of the site, take initial baseline EMF and other measurements, and determine the best areas to set up our monitoring equipment.

Electromagnetic fields in the building were mostly within normal expected range for an office space, with one exception. There was one location where EMF levels were uncharacteristically high (both by general expectation and by comparison with the rest of the building). We were quickly able to identify its source as a particular wall behind which was a master power line for the building. One mystery solved, but that still left plenty of work to do.

On the day of our first overnight investigation, we arrived mid-afternoon to begin hauling our mountains of equipment from the parking garage to the third floor, where we'd chosen to establish our base of operations in a staff office and corridor. A staff member was kind enough to show a few of our team members around who'd been absent on the first preliminary meeting.

We established video and audio monitoring throughout the courthouse, focused primarily on the third and fourth floors. Cameras were placed in the courtrooms we and our staff hosts had determined to have the most interesting histories and most prominent ghost stories, including the fourth-floor women's room.[16]

For the back staircase between the third and fourth floors, we established video monitoring and placed some toys for Jonathan the ghost to play with. Specifically, we set out a stuffed owl and a stuffed dragon. Whenever we place such "control objects," we always carefully document their location and orientation at the beginning and end of the investigation to see if they've moved, and also keep them under constant video surveillance.

Before we get into more of a linear narrative of the evening, we should mention that temperature readings were consistent all night. The building does have an HVAC system and of course outdoor temperatures fell as the night wore on. These easily explain slight decreases in our measurements throughout the evening. Similarly, EMF readings were consistent throughout the night and consistent with our preliminary readings, with the sole exception that the "hot" wall we'd detected earlier didn't seem hot anymore.

As night fell, we had to acclimatize ourselves to the sounds of the building. Even when empty, it was remarkably noisy. Any old building, even one as sturdy and well-built as this one, is going to have some pops and creaks from time to time, and it's the paranormal investigator's responsibility to get to know these sounds intimately so they can be differentiated from any anomalous noises detected later on. In addition to the normal "building sounds" common to most locations, we noticed time stamp machines would click regularly (this was normal but caught us off guard the first time we heard it), compressors inside vending machines would run a cycle, and toilets would automatically flush.

Besides the normal or easily explainable sounds, though, we heard some we can't so easily explain. Throughout the entire evening, we heard the sounds of footsteps or shoe heels scraping against the floors. As always, we knew exactly

16   We repeat here our warning that when we're investigating, it's a good idea to check with one of us before using the restroom because there's a good chance we've set up video monitoring.

where all our people were and what they were doing, so we cannot account for those sounds in an otherwise empty building. These occurred mostly on the second and fourth floors.

At one point, we heard a strange persistent clicking sound on the second floor, where no one was stationed at the time. Two of our people went down there and sat in silence for about twenty minutes, hearing nothing (including the clicking sound). As soon as they left, the clicking resumed. This, too, is unexplained.

Several times, we heard doors slam in distant parts of the Courthouse where (of course) none of our people were nearby to investigate. None of these occurred on video, but we know that we were the only ones in the building and locked in so we're not sure how the doors came to slam. Perhaps odd air currents could have done it. Arguably that's the most parsimonious explanation. But on the other hand, it happened often enough that we still consider it at least a little strange.

Much of this investigation was plagued by technical difficulties, though it seemed to only affect our fourth-floor team. Equipment we use regularly and which we had tested prior to the investigation failed to work. The wireless LAN system we were using to connect some of our cameras to some of our computers simply didn't want to connect. These things can happen, especially with wireless systems.[17] Lore does suggest that equipment failures are a common sign of paranormal activity. While we can't rule that out, it has not been our experience, and we're often more likely to assume amateur ghost hunters simply misused their tech or forgot to charge batteries. When it happens to ourselves, we pay a little more attention since we know that we obsessively check our equipment prior to investigations. Even still, and even recognizing that this was an unusual circumstance fully in line with what others have often taken as evidence for the paranormal, we're inclined to just assume that sometimes electronics don't behave as they're meant to.

But then there was the camera. One of our investigators had placed a video camera (a stand-alone one, not one connected to our surveillance system) in the fourth-floor jury room near the staircase. He left it there about midnight to record for about fifty minutes and then returned to retrieve it and see if he'd recorded anything. It had been turned off and had recorded nothing. Now, when we say it had been turned off, we don't mean it lost battery power or something. No, the physical switch on the camera had been turned off. That's something for which we have no explanation.

Later, our employee host had a strange feeling about Courtroom 3 and thought it would be the most active throughout the rest of the evening. While we don't take psychic or intuitive feelings as proof positive of anything, we're more than happy to follow up on them from time to time and did so here, just to see if anything would happen. One of our team members sat in the room in darkness and in silence for about half an hour. When he returned, he reported that nothing had happened. Still, the Court employee grew agitated. Something malevolent was in Courtroom 3, she said, and no one should go back in there alone.

---

17   These days, we're more likely to run several miles of cable so we don't have to rely on wireless technologies, but every investigation has different requirements.

Most likely, we think that was just a good old-fashioned case of the hee-bie-jeebies, but we were about ready to pack up for the night anyway, so we heeded her advice for the time being and started collecting our gear at about five in the morning.

When we went to collect the stuffed animals, we were in for quite the shock!

Figure 8.3. The stuffed dragon moved (photo: Bryan Bonner).

The stuffed dragon we'd placed near the staircase—along with a plush owl as a control object to see if Jonathan the ghost (or anyone else for that matter) might take the opportunity to play—had moved slightly. It wasn't much. Originally, we placed them propped up one against the other, and the dragon had fallen so it was lying on the floor but still in the same place. Could it just have fallen because its original position was slightly more precarious than we thought? That's the logical explanation. That's the probably true explanation. But we can't prove it.

When we went to put away some of our camera equipment, we got another shock. We were going to store them in a black bag we'd newly purchased for this investigation. It had been sitting empty on the floor and no one paid it any attention. After all, it was just a storage bag. But now, when we went to put the cameras back in, its inside was coated with chips of white paint and plaster. Nothing we carried with us could leave such a residue, so we started looking around to see if we could figure out what had landed in our pack.

It didn't take long to find the culprit. When we looked up, we found a child-sized handprint, consistent with a child of about six or eight years, on the ceiling right above the bag! This was about ten feet away from the back staircase where the control objects had also moved, and the ceiling was of such a height that no child would be able to reach it under ordinary circumstances. Importantly, we can't prove with 100% certainty that handprint showed up during our investigation. We hadn't thought to photograph the ceiling upon setting everything up, so we don't have a "before" image.[18] Still, we're as certain as we can be that the handprint showed up at the time of our investigation because where else might the plaster and paint chips have come from if not from the handprint scraping off some of the textured ceiling into our bag? But if that's the case, who left the handprint and why was it on the ceiling?

An interesting addendum to that story came to our attention when we discovered a subsequently-published collection of Colorado ghost stories now includes the handprint on the ceiling as one of the tales.[19] Rocky Mountain Paranormal was not cited in the book. *C'est la vie.*

Figure 8.4. The ghostly child's handprint (photo: Bryan Bonner).

Yet another handprint, or more accurately, a finger streak, seemed to manifest itself on a full-length mirror in the jury room. In this case, we can't entirely rule out the possibility that one of our team members might have accidentally left a fingerprint while absentmindedly wandering through, but it seems unlikely. Regardless of the source, we do know it showed up during our investigation because it was not present on the mirror at the start of the evening.

Many people have claimed Jonathan is a playful spirit who likes to play jokes and pranks on people in the way many children do, even in life. We're not necessarily saying any of our experiences should be attributed to Jonathan—or to any

18   These days, we have added 360-degree or spherical cameras to our kits and we use those to document the entire location at the start of an investigation. That way, if something does happen later on, we know we can find a "before" image because the spherical camera doesn't leave anything outside of its vision.

19   Brick, C. (2018). *Haunted America: Ghost & Legends of Colorado's Front Range.* Charleston, SC: The History Press.

ghost for that matter. They're strange, and we have no explanations, but we still lack the level of proof required to definitively declare the undisputable discovery of the afterlife. However, we have to admit, the various anomalies we experienced that night—equipment malfunctions, objects moving, the handprint on the ceiling, strange noises, the fingerprint on the mirror, etc.—are all consistent with prankster behavior.

As we finished packing up, the ghost(s)—if indeed that's who or what was at work that night—had one more playful surprise in store for us. When one of our people went to put away a camera we hadn't even used or turned on all night, we found it (but not any of its surroundings) was hot to the touch.

But we weren't quite finished yet. Though we'd finished for the day, and already managed to collect more anomalous experiences than at most other locations, we arranged for a second overnight investigation to follow up. We hoped we'd be able to get some answers to some of the questions posed by our first night. We arrived with an even larger crew, ready to chase down whatever we'd been experiencing the previous time, and moved our base of operations to the fourth floor to be closer to where we'd experienced the most unusual happenings before.

The staff member who greeted us filled us in that the "shadow man" had been seen again as recently as the week before this investigation, again in and around the ladies' restroom on the fourth floor and also peering down from the fourth-floor atrium balcony.

For control objects this time, we stationed our noble stuffed dragon on a judge's desk. He seemed to have earned the place of honor during the prior investigation. Additionally, one of our members placed a crucifix on a piece of paper on a table in the judge's chambers. On the paper, we'd drawn an ink outline of the crucifix to mark its exact orientation. We also added a stuffed tiger doll at the foot of the (allegedly) haunted staircase under constant video monitoring.

Alas, despite all the weird things during the previous investigation, this evening was a lot quieter. Both literally and figuratively. Other than the normal building sounds, we didn't notice the consistent footsteps, clicking, or slamming doors throughout the evening as we had before. But that doesn't mean we didn't still add a couple new strange experiences. Instead of repeating the same things that happened before, thus giving us a chance to investigate more thoroughly, the Courthouse treated us to new anomalies.

First, several times throughout the night, we heard a telephone ring. Though it was late at night and no one would be answering phones in the Courthouse, it's certainly possible someone could have been calling. However, we noticed it sounded like an old mechanical bell phone rather than a modern one. To check, we had someone look up the telephone directory for the Courthouse and dial each and every number for every phone and even every fax machine in the entire building. None of them matched the sound we heard.

At another point, just as one of our team members announced he was going to head downstairs to meet with some of the other investigators, the elevator rumbled to life and opened on the lower floor with no one in it. Everyone confirmed that no one had pressed any button to summon it. None of our team were anywhere close to the elevator at the time. Electrical glitch or mechanical failure?

It certainly sounds like the likeliest explanation. After all, elevators do sometimes just do funny things if their electronics have a hiccup. Staff members confirmed for us after the fact, though, that this elevator had never been known to malfunction or to move without being summoned.

Same as before, we decided to call it a night at about five in the morning and started to pack up. We were pretty excited to check on all of our control objects to see if anything had moved this time. No such luck. However, the door to the personal restroom in the judge's chambers had been opened. We knew it was securely shut when we started, and it was a large, heavy door with a stiff handle, so it's unlikely to have been blown by an air current or something of that nature. No one had been in the room all night and we have no idea how it came to be open.

All in all, this was one of the most exciting investigations in terms of the sheer number of strange experiences we got to document. None of them rise to the level of conclusive proof, but a lot of them are impressive enough to leave us all scratching our heads. As is most often true, the case of the Weld County Courthouse remains open and unsolved. But certainly one of the most intriguing ones in our quarter-century of work. Also as is often the case, we would love to be invited back to follow up on the strange experiences we had when we were there several years ago.

References & Further Reading

Brick, C. (2018). *Haunted America: Ghost & Legends of Colorado's Front Range*. Charleston, SC: The History Press

Ehnert, R. (n.d.). The Courthouse Clocks. *Weld County, Colorado*. <https://history.weld. gov/Courthouse-100/The-Courthouse-Clocks>

Geffs, M. L. (1938). *Under Ten Flags: A History of Weld County, Colorado*. Greeley, CO: McVey Printery

Kavalec, C. (1976). Weld County Courthouse: National Register of Historic Places Inventory – Nomination Form. [Register No. 78000886]. *National Register of Historic Places*. <https://s3.amazonaws.com/NARAprodstorage/lz/electronic-records/rg-079/NPS_CO/78000886.pdf>

Peters, M. (1999). The Story of W. D. French. *Weld County, Colorado*. <https://history. weld.gov/County-150/Justice/WD-French>

Sanger, G. P. (1863), ed. *The Statutes at Large, Treaties and Proclamations of the United States of America from December 5, 1859 to March 3, 1863*, volume 12: 172-177

Weld County (n.d.). The History of Weld County's Courthouses. *Weld County, Colorado*. <https://history.weld.gov/Courthouse-100/The-History-of-Weld-Countys-Courthouses>

Whitman, S. (n.d.) Two men died. Two lives were lost – all over 600 pounds of flour. *19th Judicial District*. <https://www.courts.state.co.us/userfiles/file/Court_Probation/19th_Judicial_District/Court_House_History/WD%20French.pdf>

9
An Elusive Legend: Third Bridge

Every city, state, or region has its collection of famous (or infamous) haunts. The kinds of places all the paranormal enthusiasts and thrill seekers know about, but which are mostly discussed only in whispers and innuendo. When we give our lectures, one of those locations that people often ask about is the so-called Third Bridge. People want to know if we've ever investigated there. And the answer is: yes, to a certain extent, but the first thing one needs to figure out is which bridge is actually supposed to even be the haunted one in the first place!

Figure 9.1. Third Bridge (photo: Robert Lewis).

Paranormal Claims

Though Third Bridge is a part of a publicly accessible road and thus a part of this first section of public-facing venues and locations, we've chosen to omit our usual history section in this chapter simply because there's not a lot of credible information about the location's history. Indeed, there are even disagreements about which Colorado bridge is supposed to be the Third Bridge. Without a definitive location, it's hard to write much of a history. To the extent that we think we've figured it out and established some of the location's genuine history, we'll confine those findings to our investigative report below and instead jump right in with the paranormal claims and urban legends.

Even the paranormal stories are shrouded in mystery. This is one of the more elusive urban legends in Colorado because it seems like just about everyone has heard of it. Maybe they even have a second- or third-hand connection to someone who experienced something there. But the stories often disagree on their particulars, and when you dig a little deeper you find that in a majority of cases, people telling the story are telling you what they've heard rather than what they've seen. Obviously that doesn't mean there's nothing to be found there, but it does mean we have our work cut out for us in terms of even starting to figure out what the more credible paranormal claims are.

The basis of the urban legend seems to be that of land that's either haunted or cursed, which results in higher than usual levels of a variety of different paranormal phenomena. The trouble is, if you ask different people which land is the cursed land, or which bridge is the haunted one, you'll get multiple different answers. Over the years, we've heard dozens of bridges claimed to be "the" Third Bridge and other than the haunted reputation, their characteristics are quite different. A few of the most popular ones we've heard people talk about include:

- A small overpass on County Line Road near Bennett, Colorado,
- On Quincy Road just east of Parker Road in Aurora, Colorado,
- A bridge on Smoky Hill Road in Aurora, Colorado near the Kiowa – Bennet line,
- County Line Road just past Powhaton Road (an intersection in southern Aurora), or
- An unspecified pedestrian bridge in Aurora, Colorado.

When we discuss our investigative activities, we'll do our best to narrow the field to what we think is the correct location, but for now just recognize that there are several candidate locations to which the supernatural lore has been attributed.

In most variations of the story, the land and/or the bridge is/are cursed because it was the site of a Native American massacre in the distant past. Some sources explicitly connect the curse to the infamous Sand Creek massacre. That chapter in the American Indian Wars was an attack on a Cheyenne and Arapahoe village by the Third Colorado Cavalry under the command of Colonel John Chivington in which an estimated 150[1] Native American people, including women

---

1   Numbers vary from 70 people to over 600 people. Chivington himself claimed 500 – 600 fatalities, but modern scholars have reached a tenuous and still-debated consensus that the true number was closer to 150.

and children, were killed.[2] Other reports lack the specificity but instead connect the story to a more generic massacre of Native Americans somewhere on the relevant plot of land.

Most of the stories do bear some similarity to the details of the Sand Creek Massacre, though. It was meant to be a massacre by American settlers or soldiers of Native Americans that claimed the lives mostly of women and children (some even specifying the attack was carried out while the men were away on other business). When the men returned, legends say, they started beating their war drums and went to seek vengeance. According to the paranormal lore, sometimes those war drums, now in ghostly form, can still be heard today.[3]

Among the paranormal claims most explicitly related to this origin story are tales that people visiting Third Bridge have heard the aforementioned war drums, sometimes nearby and sometimes in the distance. Others have claimed to have even seen ghostly manifestations of the Natives thought to haunt the Bridge. A few reports take a turn toward the even more frightening and claim witnesses have heard disembodied screams, perhaps belonging to the massacre victims themselves.

Other paranormal claims aren't specifically related to the lore about the massacre but take on a more generic ghostly nature. Several people have reported car trouble on the bridge, including cars that simply refused to run while at the site. Drivers say they've had to push their car off the bridge but that once they got away from the haunted location, their car started running just fine from then on and they could never figure out any mechanical source of the failure.

Shadow figures seen at the site are sometimes and sometimes not associated with the Native American connections. Similarly, witnesses who've heard the otherworldly sounds of either screams or children playing late at night are divided on whether they've specifically mentioned Native American spirits or just described the sounds more generally.

Throughout the years, a lot of ghost hunting groups have investigated the location. Actually, they've investigated several different locations claimed to be the Third Bridge, and several have claimed to have had paranormal experiences even at the disparate locations, so we have to assume at least some of the reports are either false, mistaken natural phenomena, or, in the best case scenario, genuine paranormal experiences from an entirely different haunting that's only been misapplied to Third Bridge. The trouble is, only some reports have pointed out which bridge they've visited, so we have to report the wide variety of claims and then do the difficult-to-impossible task of trying to sort out which ones are more or less credible.

Among the most noteworthy of reports from these various paranormal enthusiasts are a (now-defunct) group of paranormal enthusiasts who reported hearing the ghostly drumming, saw shadow people running all over, experienced strange feelings, and captured EVPs[4] of a woman screaming. They also claimed

2   Michno, G. F. (2004). *Battle at Sand Creek*. El Segundo, CA: Upton and Sons. Hoig, S. (1977). *The Sand Creek Massacre*. Norman, OK: University of Oklahoma Press.

3   Soto, J. (2016). Third Bridge – A Classic Aurora Haunt. *Out Front Magazine*. <https://www.outfrontmagazine.com/third-bridge/> (accessed August 17, 2025).

4   Electronic Voice Phenomena. This is a common practice in which paranormal

to experience a complete failure of one of their cameras which subsequently resumed working normally the next day.

Another Third Bridge enthusiast who claimed to have been visiting the site (which he also connects with the Sand Creek Massacre) for over a decade claimed to have heard the ghostly drumming, heard a "gurgling snarl," and even saw a ghostly silhouette on the bridge itself.[5]

Other reports include people who saw a tall, dark, featureless figure on the bridge that seemed to stare at the witnesses, spirit orbs showing up in photographs taken at the site (for more on which, see Volume 1, Chapter 24), and even being touched by unseen entities.

A common thread through many of the reports, both those offered first-hand and those only repeating what they'd heard, is that the site is known for a high incidence of car crashes, which some attribute either to the Bridge being cursed or to ghosts manipulating people's automobiles. Speaking of cars, some claim that if you park your car on Third Bridge, small handprints will appear in the dust or dirt on your car.

Finally, in addition to the Native American massacre, numerous reports have connected Third Bridge to a murder or murders which occurred on or near the Bridge.

All in all, it's one of those places that just seems to collect ghost stories and urban legends. In that regard, it's a lot like another famous Colorado haunt: Riverdale Road, which will be the subject of Chapter 13. Both locations are described as exhibiting so much paranormal activity one might wonder how anyone could live an ordinary life anywhere in their vicinity. Both also collect so much disparate paranormal lore that entire volumes could be written just to collect all the ghost stories.

But because Third Bridge has become such a prominent Colorado haunt, we wanted to see what we could do to trace the development of the urban legend and to separate fact from fiction. And, of course, we were also hoping as always to experience something paranormal for ourselves.

Our Investigation

Call us crazy, but we had a certain reluctance to spend all night on a poorly-lit bridge famous for car accidents. We're not saying we wouldn't do it, but if we

---

investigators or ghost hunters will use any of a broad range of devices available on the market in an attempt to audio record the spirits' voices. In some cases, the device rapidly scans radio signals and the theory is that the spirits can manipulate the signals to convey their own messages. In other cases, the device records only static or white noise and any vocal message is thought to be the spirits' attempt at communication. Both variations are widely used. For our part, we're not opposed to such things in theory but have concerns about the reliability of their results and would insist upon strict scientific protocols which are often not obeyed by other ghost hunting groups.

5   James, C. (n.d.). Haunted Places USA – Third Bridge Aurora, Colorado Ghost Hunters Drums. *Vocal Media*. <https://vocal.media/horror/haunted-places-usa-third-bridge-aurora-colorado-ghost-hunters-drums> (accessed August 17, 2025).

were to put ourselves in that kind of situation, we'd want to make sure we knew exactly what we were doing. Add to that the fact that one of the most important questions about Third Bridge is: which bridge is it in the first place? Between those factors, we figured the first part of our investigation probably ought to be more about digging through archives and trying to figure out where the stories came from.

The first thing we can put to rest is any direct connection to the Sand Creek Massacre. That took place in what is now Eads, Colorado, roughly 150 miles east of what is now Colorado Springs and thus some 150 miles (give or take) southeast of *any* of the locations claimed to be Third Bridge. That was absolutely a real historic event, and it certainly shaped some of Colorado's history, but we just can't fathom why it should be connected to any location so far away.

The idea that there could have been a different Native American massacre closer to the relevant locations makes some degree of sense. Just about every location in the entire country was once occupied by one tribe or another, and if you go far enough back in history, it's reasonable to suspect that there have been massacres or violence just about everywhere, either between Native tribes and settlers or between different Native tribes. Of course, many such battles, massacres, or acts of violence have long since been lost to history. So we're perfectly comfortable with the idea that Third Bridge might be haunted by Native American spirits, but not that they're from Sand Creek.

However, while we were doing some digging to determine if there might have been a different such massacre that could have contributed to the urban legend, we did come across the 1864 Hungate Massacre, which occurred at a homestead along Running Creek near modern-day Elizabeth, Colorado. While none of the proposed locations for Third Bridge are in or particularly close to Elizabeth, it's a lot closer to the mark than the site of the Sand Creek Massacre. Interestingly enough, this massacre, in which the entire Hungate family (Nathan, twenty-nine; Ellen, twenty-five; Laura, two; and Florence, under six months) were "feloniously killed by some person or persons…unknown, but supposed to be Indians," has been seen as one of the precipitating events that led directly to the Sand Creek Massacre.[6]

No other noteworthy Native American massacres in the immediate vicinity of any of the reported Third Bridge locations have come to our attention. Therefore we suspect that Sand Creek connection probably came from someone initially connecting Third Bridge to Hungate Massacre (even though it didn't take place at precisely the same location). Subsequent retellings of the tale, then, might have dropped the Hungate part and tried to connect the story more directly to Sand Creek.

We then tried to figure out which bridge is the true Third Bridge and, for that matter, where the name "Third Bridge" came from in the first place.

Lets start by process of elimination.

The first proposed location is a small bridge on County Line Road near Ben-

6   Campbell, J. (2014). 1864's Turning Point: Hungate Family Murdered. *National Park Service*. Archived from the original at <https://web.archive.org/ web/20200806215051/https://www.nps.gov/sand/learn/news/hungate-family- murdered.htm> (accessed August 17, 2025).

nett, Colorado over the dry bed of what was once Kiowa Creek (the creek still exists but doesn't flow under the bridge). It's by far the most commonly-reported location and we think for historic reasons we'll explain in a moment that it's the correct one. But just to cover our bases, we should address the other contenders.

Some have said it's located on Quincy Road just east of Parker Road in Aurora. Well, there is no bridge of any kind near that intersection. Indeed, if you start from that intersection and travel east on Quincy Road, the first bridge of any kind you'll come to is the overpass where E-470 crosses over Quincy. That's just shy of six miles from the intersection in question, so this proposed location doesn't even exist.

Another proposed location is a bridge on Smoky Hill Road in Aurora. This is more plausible as there are some rather remote bridges on that road. However, relatively few sources cite this location compared to our first contender, and those that do have mentioned more than one potential bridge on Smoky Hill as "the" right bridge. Most likely, this branch of the lore stems from someone mistaking the correct location of the bridge sometime in the past, with subsequent reports stemming from the original error.

Yet another contender is on County Line Road just past Powhaton Road. Admittedly, we haven't visited this site in person, but we did check it out using satellite imagery and there is no bridge anywhere near that intersection. However, if you travel along County Line Road about twelve and a half miles east from that location, you'll come to the very spot we think is the real one. How did the lore migrate? Our best guess is that someone tried to describe how to get to the real location from the Powhaton intersection and someone else misunderstood the directions and assumed the bridge was *at* or *near* Powhaton. That's just a guess on our part, but is consistent with how folklore develops.

Finally, some have considered it to be an unspecified pedestrian bridge in Aurora, Colorado. That seems unlikely for numerous reasons. First of all, that no one has ever specified which of several pedestrian bridges in Aurora is the culprit suggests that none of the people repeating the story have ever been there. This is at odds with the number of reports we've read of people visiting the site. Plus, most of the stories involve car crashes or automobile malfunctions, and you can't take cars on pedestrian bridges. We're not sure how the lore could have been so corrupted as to turn a remote automobile bridge into a suburban pedestrian one, but we're convinced it's not the true location regardless.

Now let's do some digging into the history of the folklore and see if we can identify how what we believe to be the true bridge (whose GPS coordinates are approximately 39.563734, -104.442795) on County Line Road came to be associated with such supernatural rumors.

The earliest online citation we could find was from a 2003 repository of paranormal lore which identified Third Bridge as being on Smoky Hill Road but also mentioned a separate bridge over Kiowa Creek near Bennett as a *separate* haunted location.[7] Though it mentions two distinct locations, it's worthy of note that one of the locations is another commonly cited location for Third Bridge, so

---

7   Shadowlands.net (2003). Haunted Places in Colorado. *Shadowlands.net.*
Archived from the original at <https://web.archive.org/web/20031209110422/http://www.theshadowlands.net/places/colorado.htm> (accessed August 17, 2025).

it seems likely that these two mentions might have originally been the same location but that at some point the lore bifurcated. This might account for some of the confusion regarding the true Third Bridge location. It's also worthy of note that, though this was the earliest online citation we can find, it clearly references back (albeit without direct citations) to existing lore.

An online folklorist traced the interest in Third Bridge further back to the mid-1990s and suggests that for reasons unknown the bridge started to develop a haunted reputation among youths and that the urban legends really took off when some teenagers hoping to see something spooky on the bridge suffered a fatal car crash in 1997.[8] That part, at least, is absolutely true. On June 21, 1997, a group of youths (some of whom had been drinking despite being underage) went looking for the reputedly haunted bridge; when they found it, they hit the guardrails at some seventy miles per hour (the posted speed limit was twenty-five miles per hour) and crashed into the dry creek bed, leaving two dead and one paralyzed.[9]

If the modern Third Bridge lore largely stems from that tragic accident, what was the original lore that caused the youths to look for ghosts there in the first place? That remains something of an open mystery for us. We haven't yet been able to locate the first citation of ghost stories at that location. The earliest ones we've found have all referred back to pre-existing lore. However, we did manage to identify a *potential* source of the rumors, though we can't confirm it absolutely.

On May 21, 1878, a Kansas Pacific train was traveling east through the region when it came to cross a bridge over Kiowa Creek. Unfortunately, the region had recently flooded and the train crashed into the water and was washed away. Depending on who you ask, it was either recovered and rebuilt shortly thereafter or was never seen again and still remains buried somewhere on the banks of Kiowa Creek (the former being the far more likely scenario).[10]

Here's where things get interesting. Though that train crash isn't a commonly-repeated bit of lore among Colorado locals today, it was a famous enough case to even catch the attention of bestselling author and adventurer Clive Cussler, who spent some time investigating to see whether he could discover the truth

8 White, M. (2016). The Ghosts of Third Bridge. *Colorado Urban Legends*. <http://www.coloradourbanlegends.com/third-bridge/> (accessed August 17, 2025).

9 Perez-Giese, T. (1998). A Bridge Too Close. *Westword*. <https://www.westword.com/news/a-bridge-too-close-5058218> (accessed August 17, 2025).

10 Cussler, C. (n.d.). Lost Locomotive of Kiowa Creek. *National Underwater and Marine Agency*. <https://numa.net/expeditions/lost-locomotive-of-kiowa-creek/> (accessed August 17, 2025).
White, M. (2016). The Legend of the Lost Locomotive of Kiowa Creek. *Colorado Urban Legends*. <http://www.coloradourbanlegends.com/lost-locomotive-kiowa-creek/> (accessed August 17, 2025).
Bangs, R. (1997). Kiowa Creek Myth Solved. *Douglas County News-Press, 105*(33). Archived from the original at < https://www.coloradohistoricnewspapers.org/?a=d&d=DNP19970514-01.2.102&e=-------en-20--1--img-txIN%7ctxCO%7ctxTA--------0------> (accessed August 17, 2025).

behind the missing locomotive.[11] That's precisely the kind of lore that can easily lead to reputations of hauntings. Even minor tragedies can often lead to ghost stories, and this one had some real heft to it.

And then we discovered something even more intriguing. An article claiming the mystery of the lost locomotive had been solved was published in the *Douglas County News-Press* just a little more than a month before the 1997 car accident![12]

Now, this is all speculation on our part. We have no way of knowing. But it seems possible that word had spread among the young people about the 1878 train crash and that it might have been a topic of conversation in the summer of 1997. Perhaps ghost stories from the original train crash already existed in the zeitgeist or maybe some young people just read about the crash and decided it *must* be a haunted location. Either way, that interest may have led to the crash, and the crash may have created or reinvigorated the story that Third Bridge is cursed or haunted.

Perhaps counting against that theory is the fact that the train bridge and the modern bridge aren't the same. They're not even in quite the same location. But they're close to each other. It's entirely possible that lore originating from the train crash just got slightly mangled and misunderstood by young people looking for ghosts (or even adults looking for ghosts) and that subsequent crashes have only further reinforced the lore.

That also makes sense in light of where we think the name "Third Bridge" comes from. One folklore enthusiast told us that if you start in Aurora and head east on County Line Road, beginning as soon as that stretch of road takes on the name "County Line Road," at—wait for it…the intersection at Powhaton—the bridge in question will be (or at least once was; the landscape has changed a lot over the years) the third one you encounter. This can also bring some of the other reputed locations into some greater focus. For instance, though the bridge is not on Smoky Hill Road, one way of getting there is to head east on Smoky Hill, follow the curve as the road changes its name, then turn left on County Line Road.

What about the rumors that there was a murder at or near Third Bridge? That one turns out to be partially true. There was a murder, but it actually took place not on or in the immediate vicinity of Third Bridge, but about a mile away. That's close enough to have had the murder dragged into the folklore but we think distant enough to discount it as being in any way related to the paranormal claims.

The murder in question was that of local high school teacher Randy Wilson. On June 13, 2010, Wilson was seen at a gas station. The next morning, his body was discovered in an empty field with a bag over his head, a belt tied around his

11   Cussler, C. (n.d.). Lost Locomotive of Kiowa Creek. *National Underwater and Marine Agency.* <https://numa.net/expeditions/lost-locomotive-of-kiowa-creek/> (accessed August 17, 2025).

12   Bangs, R. (1997). Kiowa Creek Myth Solved. *Douglas County News-Press,* *105*(33). Archived from the original at < https://www.coloradohistoricnewspapers. org/?a=d&d=DNP19970514-01.2.102&e=-------en-20--1--img-txIN%7ctxCO%7ctx TA--------0------> (accessed August 17, 2025).

Perez-Giese, T. (1998). A Bridge Too Close. *Westword.* <https://www.westword.com/news/a-bridge-too-close-5058218> (accessed August 17, 2025).

neck, and his hands bound behind his back. No clear motive for the crime was found, and though officers were able to collect DNA samples from the scene, they couldn't match them to anybody. The case went unsolved for a long time and though there have been some subsequent developments, it remains deeply mysterious to this day. The developments? In 2017 one Daniel Pesch was arrested and charged with the murder—to which he even confessed—but investigators were dissatisfied with contradictions between the confession and the facts of the case and with their failure to match Pesch's DNA to samples collected from the crime scene. Eventually the murder charge was dropped in an arrangement in which Pesch pleaded guilty to the lesser charge of attempted escape, for which he was sentenced to probation. It's unclear what the truth of the matter is and the case is still listed as an unsolved cold case by the Colorado Bureau of Investigation.[13]

Once again, this is the sort of situation where we can understand how and why this murder case got mixed in with the paranormal lore. Unsolved cases tend to breed speculation in and of themselves and its proximity to Third Bridge makes it an obvious candidate for discussion. But as we said, the actual crime scene was far enough removed from the bridge that we don't see much connection. Sure, it's entirely possible (even probable) both the victim and perpetrator (whoever the latter may have been) crossed over Third Bridge on their travels that fateful day, but that alone is insufficient to justify paranormal claims. Lots of people travel that bridge without incident, so we'd need something more concrete before leaping to any conclusions.

One interesting thing about these kinds of locations that become famous in folklore and urban legend is that they can become self-perpetuating, whether or not there's anything supernatural afoot. That's certainly true in the case of Third Bridge, where we know of multiple car crashes, at least some of which involve drunk youngsters driving unsafely while trying to catch a glimpse of the ghost or ghosts. It makes sense. Most ghost hunting expeditions are done at night and this is a dark bit of road. There's also a nearby hill and if cars in either direction aren't paying attention because they're looking for ghosts and perhaps driving a bit fast-

13   CBS Colorado (2011). Investigators Stumped Over High School Teacher's Murder. *CBS News.* <https://www.cbsnews.com/colorado/news/investigators-are-stumped-over-high-school-teacher-murder/> (accessed August 17, 2025).

Queen, J. (2017). Former Summit County man charged with 2010 Elbert County murder of schoolteacher. *Summit Daily.* <https://www.summitdaily.com/news/crime/former-summit-county-man-charged-with-2010-elbert-county-murder-of-schoolteacher/> (accessed August 17, 2025).

Elbert County News (2018). Judge clears way for trial in death of Kiowa teacher. *Elbert County News.* <https://coloradocommunitymedia.com/2018/06/04/judge-clears-way-for-trial-in-death/> (accessed August 17, 2025).

CBS Colorado (2021). Daniel Pesch Sentenced To Prison After Tri-County Health Department Threats & Vandalism. *CBS News.* <https://www.cbsnews.com/colorado/news/daniel-pesch-sentenced-prison-health-department-threats-vandalism/> (accessed August 17, 2025).

Colorado Bureau of Investigation (n.d.). Case Detail: Randall Wilson. *Colorado Bureau of Investigation Cold Case Files.* <https://apps.colorado.gov/apps/coldcase/casedetail.html?id=259000> (accessed August 17, 2025).

er than they ought, collisions are almost guaranteed from time to time. Increased traffic in what is otherwise a pretty remote area due to the ghostly lore can also make conditions less safe, as can individuals who stop on the road to get a better look over the sides of the bridge.

In addition to the 1997 crash we already described, there was a much-discussed and widely-reported accident in October of 2016 in which five teenagers aged fifteen to nineteen years were out thrill-seeking, trying to get a look at Third Bridge. The teens were speeding along the road when the driver lost control of the car, causing it to flip and killing all five occupants.[14]

We've also been told law enforcement have stepped up traffic monitoring around that area for exactly this reason. Off the record, we've been informed of at least one other case—thankfully not involving fatalities—in which officers issued traffic citations to people looking for ghosts on or near Third Bridge. And that's just the one our contact happened to know about off hand. We're sure there are plenty more where it came from.

Though the bridge itself is part of a public roadway, the land to the sides is privately owned, and the landowner has also become concerned about the dangers should ghost hunters keep visiting the property; still, even though police and landowners alike have tried to warn people away, thrill seekers (especially of the teenage variety) seem to keep going back.[15]

So that about covers all the history and folklore. What about the paranormal claims themselves? What about the sounds of Native American drums? The shadow figures? The sounds of people screaming? Well, we found ourselves at a bit of an impasse. On the one hand, the only surefire way to know what's going on with regard to paranormal claims is to conduct a proper investigation. On the other hand, we've read all those stories of car crashes and really didn't want to get added to their number. So we settled on a mixed approach. We have indeed visited Third Bridge numerous times, but we've done our investigation piecewise, just to see what we could see while also following all due safety protocols.

Throughout those several visits, we didn't see any of the visual manifestations anyone has reported. That doesn't necessarily mean none of them ever occur, but unless and until we witness them for ourselves, all we can really do is shrug our shoulders and say "we didn't see it."

14   Philips, N. (2016). Five teenagers killed in fiery rollover crash identified by Arapahoe County coroner. *The Denver Post.* <https://www.denverpost. com/2016/10/05/arapahoe-county-rollover-crash/> (accessed August 17, 2025).
Sallinger, R. (2016). Grandmother: 5 Victims Killed Were Thrill Seeking. *CBS News.* <https://www.cbsnews.com/colorado/news/grandmother-5-victims-killed-were-thrill-seeking/> (accessed August 17, 2025).
Kovaleski, J. L. (2016). Other teens crashed at 'Third Bridge' years before deadly accident that killed five. *Denver 7.* <https://www.denver7.com/news/local-news/other-teens-crashed-at-third-bridge-years-before-deadly-accident-that-killed-five> (accessed August 17, 2015).

15   Boyle, M. (2016). Reported haunted bridge in Arapahoe County drawing teens, concerns from landowner. *Denver 7.* <https://www.denver7.com/news/local-news/reported-haunted-bridge-in-arapahoe-county-drawing-teens-concern-from-landowner> (accessed August 17, 2025).

Similarly, we didn't hear any of the sounds of screeching or children playing or crying or any of the other ghostly sounds people have reported. The bridge is located in the middle of a group of trees lining both sides of the road, and we observed that there seems to be a small but thriving wildlife population there. It's possible, though purely speculative at this point, that some of the sounds at least some witnesses have heard might have been wildlife noises like the call of a screech owl (which are present in the region).

Screech owls, if indeed they can be blamed for some of the sounds being misinterpreted as either crying, screaming, or children at play, are one thing. What about the sounds of Native American war drums heard around Third Bridge? Well, another researcher did an excellent piece of investigation we wish we'd first discovered for ourselves but we have to give credit where it's due and this discovery was first published on the *Colorado Urban Legends* website. There are at least two oil pump jacks less than a mile from Third Bridge, and they make a kind of drumming sound that could easily be mistaken, especially if one is pre-conditioned to think there are Native American spirits in the area, as ghostly war drums.[16] They're distant enough that the sound is faint from the Bridge, but when conditions are right, they absolutely can be heard that far away. Up close, you'd never mistake their sound for drum music. But from half a mile or a mile away, they sound eerily like someone beating drums in the distance.

During the rest of our time out there, we didn't see or hear anything note-worthy except a small group of cows on adjacent land, one of which seemed rather annoyed or confused as to why another damned group of paranormal investigators would be foolish enough to explore that bit of road.

Figure 9.2. We confused a local cow (photo: Bryan Bonner).

---

16   White, M. (2016). The Ghosts of Third Bridge. *Colorado Urban Legends.*
<http://www.coloradourbanlegends.com/third-bridge/> (accessed August 17, 2025).

There we have it. That's what we've been able to learn so far about Colorado's infamous Third Bridge. We'll still call it an open and unsolved case in our files because we haven't been able to definitively rule out all of the paranormal claims. Still, we don't hold out too much hope for this one being genuinely haunted. It has all the hallmarks of pure folklore or urban legend rather than a "legitimate" haunting. Under normal circumstances, we advise people to check out all the locations we write about (at least the publicly accessible ones), whether we think they're haunted or not, because they have such wonderful history and fascinating stories connected to them. In the case of Third Bridge, though, we're going to advise would-be ghost hunters to stay away unless you have a legitimate need to travel that stretch of road. No reason to add to all those automobile accident statistics.

References & Further Reading

Bangs, R. (1997). Kiowa Creek Myth Solved. *Douglas County News-Press, 105*(33). Archived from the original at < https://www.coloradohistoricnewspapers.org/?a=d&d=DNP19970514-01.2.102&e=-------en-20--1--img-txIN%7ctxCO%7ctxTA--------0------>

Boyle, M. (2016). Reported haunted bridge in Arapahoe County drawing teens, concerns from landowner. *Denver 7.* <https://www.denver7.com/news/local-news/reported-haunted-bridge-in-arapahoe-county-drawing-teens-concern-from-landowner>

Campbell, J. (2014). 1864's Turning Point: Hungate Family Murdered. *National Park Service.* Archived from the original at <https://web.archive.org/web/20200806215051/https://www.nps.gov/sand/learn/news/hungate-family-murdered.htm>

CBS Colorado (2011). Investigators Stumped Over High School Teacher's Murder. *CBS News.* <https://www.cbsnews.com/colorado/news/investigators-are-stumped-over-high-school-teacher-murder/>

CBS Colorado (2021). Daniel Pesch Sentenced To Prison After Tri-County Health Department Threats & Vandalism. *CBS News.* <https://www.cbsnews.com/colorado/news/daniel-pesch-sentenced-prison-health-department-threats-vandalism/>

Colorado Bureau of Investigation (n.d.). Case Detail: Randall Wilson. *Colorado Bureau of Investigation Cold Case Files.* <https://apps.colorado.gov/apps/coldcase/casedetail.html?id=259000>

Colorado Haunted Houses (n.d.). Third Bridge – Bennet CO Real Haunts. *Colorado Haunted Houses.* <https://www.cohauntedhouses.com/real-haunt/third-bridge.html>

CULZParanormal (2017). Expedition Culz: Ghost Bridge (Third Bridge). *YouTube.* <https://www.youtube.com/watch?v=i0VeCOamIXA>

Cussler, C. (n.d.). Lost Locomotive of Kiowa Creek. *National Underwater and Marine Agency.* <https://numa.net/expeditions/lost-locomotive-of-kiowa-creek/>

Elbert County News (2018). Judge clears way for trial in death of Kiowa teacher. *Elbert County News.* <https://coloradocommunitymedia.com/2018/06/04/judge-clears-way-for-trial-in-death/>

Hoig, S. (1977). *The Sand Creek Massacre.* Norman, OK: University of Oklahoma Press

James, C. (n.d.). Haunted Places USA – Third Bridge Aurora, Colorado Ghost Hunters Drums. *Vocal Media.* <https://vocal.media/horror/haunted-places-usa-third-bridge-aurora-colorado-ghost-hunters-drums>

Kovaleski, J. L. (2016). Other teens crashed at 'Third Bridge' years before deadly accident that killed five. *Denver 7*. <https://www.denver7.com/news/local-news/other-teens-crashed-at-third-bridge-years-before-deadly-accident-that-killed-five>

Michno, G. F. (2004). *Battle at Sand Creek*. El Segundo, CA: Upton and Sons

Perez-Giese, T. (1998). A Bridge Too Close. *Westword*. <https://www.westword.com/news/a-bridge-too-close-5058218>

Philips, N. (2016). Five teenagers killed in fiery rollover crash identified by Arapahoe County coroner. *The Denver Post*. <https://www.denverpost.com/2016/10/05/arapahoe-county-rollover-crash/>

Queen, J. (2017). Former Summit County man charged with 2010 Elbert County murder of schoolteacher. *Summit Daily*. <https://www.summitdaily.com/news/crime/former-summit-county-man-charged-with-2010-elbert-county-murder-of-schoolteacher/>

Sallinger, R. (2016). Grandmother: 5 Victims Killed Were Thrill Seeking. *CBS News*. <https://www.cbsnews.com/colorado/news/grandmother-5-victims-killed-were-thrill-seeking/>

Shadowlands.net (2003). Haunted Places in Colorado. *Shadowlands.net*. Archived from the original at <https://web.archive.org/web/20031209110422/http://www.theshadowlands.net/places/colorado.htm>

Soto, J. (2016). Third Bridge – A Classic Aurora Haunt. *Out Front Magazine*. <https://www.outfrontmagazine.com/third-bridge/>

White, M. (2016). The Ghosts of Third Bridge. *Colorado Urban Legends*. <http://www.coloradourbanlegends.com/third-bridge/>

White, M. (2016). The Legend of the Lost Locomotive of Kiowa Creek. *Colorado Urban Legends*. <http://www.coloradourbanlegends.com/lost-locomotive-kiowa-creek/>

10
The Weeping Piano: Hudson Gardens

If you're looking for a place to enjoy the beauty of nature as well as comfortable indoor accommodations, you could do a lot worse than The Hudson Gardens and Event Center (which we'll simply call Hudson Gardens in the interest of brevity), located at 6115 South Santa Fe Drive in Littleton, Colorado. It may not be Colorado's most famous botanical gardens, but its beauty is something to behold, making it the perfect place for a stroll through nature or to watch some bees under the care of local beekeepers.

Supernatural entities like ghosts may not be the first thing to come to one's mind in the presence of such natural wonders, but the property nevertheless has its share of ghost stories, including one that manifests in a different way from most of the claims we've heard through our decades of research.

Figure 10.1. The Inn at Hudson Gardens (photo: Bryan Bonner).

The History

What would eventually become Hudson Gardens began as the dream of Evelyn Hudson (1905 – 1988), who built the estate with her husband Colonel King C. Hudson (1893 – 1984) over the course of several decades.

Sometimes when we write these histories, we find ourselves describing the major movers and shakers who founded major institutions and helped to direct the course of entire cities or states. This is not such a story. This is the story of a couple who were equally impressive but on a smaller scale. They weren't deeply involved in politics, they weren't multi-billionaires, but they nevertheless built something that has lasted through the decades and offered countless moments of peace and joy to their local community. These are the histories most in danger of being lost. Too often, they're omitted from the history books. Nevertheless, these are the histories that matter most to the individuals involved and are arguably most in need of preservation.

Evelyn Leigh was born on February 13, 1905 in Aurora, Illinois, a suburb of Chicago (not to be confused with the Aurora in Colorado). She studied restaurant operation at the Lewis Institute in Chicago and long harbored dreams of opening her own tea house. During her early adult years, she worked hard to achieve that dream by managing no fewer than four Chicago department store restaurants.[1] Though largely now a relic of the past, department store restaurants were all the rage back in the day and were a popular way for stores to attract shoppers in big cities including Chicago.[2] A few such relics of the dining and shopping past still exist in the Denver area and elsewhere and we recommend revisiting that tradition during your own shopping once in a while, though that's neither here nor there.

King C. Hudson, for his part, was born on June 4, 1893 in Topeka, Kansas, earned a degree in dentistry and then his law degree from the Chicago Kent College of Law. During World War I, he served in the United States Army's dental corps and continued his service as a reserve officer while practicing dentistry. He earned the rank of Colonel.[3]

King Hudson married Evelyn Leigh on June 17, 1940 in Chicago. Readers paying close attention to the dates might realize they were married just as World War II was brewing and so Colonel Hudson was recalled to active service and relocated with his new wife to Colorado where he served at both Fitzsimmons Army Hospital and at Fort Logan.[4] Thus did the Hudsons bring their dream of opening a restaurant to Colorado.

1   Hulse, D. F. & Larison, P. (2021). Hudson Family and Hudson Gardens. *Littleton Museum.* <https://www.museum.littletonco.gov/Research/Littleton-History/Biographies/Hudson> (accessed August 18, 2025).

2   Ewbank, A. (2022). The Lost Glamour of the Department-Store Restaurant. *Atlas Obscura.* <https://www.atlasobscura.com/articles/department-store-restaurant> (accessed August 18, 2025).

3   Hulse, D. F. & Larison, P. (2021). Hudson Family and Hudson Gardens. *Littleton Museum.* <https://www.museum.littletonco.gov/Research/Littleton-History/Biographies/Hudson> (accessed August 18, 2025).

4   *Ibid.*

Reports have varied a bit concerning when the restaurant actually opened, with some claiming they opened their Country Kitchen restaurant in 1940 and others saying it was 1942. The latter is far closer to the truth. Indeed, it would be difficult to imagine the Hudsons managing to have a new business up and running as early as 1940 considering they were still in Chicago when they married in June of that year. The real answer is that they purchased the property which would eventually expand into what is now Hudson Gardens in December of 1941 when Colonel Hudson retired from the Army, that they built a log cabin structure on said property which became their Country Kitchen restaurant, and that they opened for business in May of 1942. The Hudsons lived in an apartment in their new restaurant.[5]

Locals didn't think it would be successful. The property was too far from Main Street. It was inconvenient. People wouldn't want to go so far out of their way just to have dinner. By God, those people were wrong! They offered a menu of some seventy delicious dishes, most made with fresh produce grown right on the grounds, for a grand price of $1.25 each (which, inflation-adjusted, is roughly equal to about $25 in 2025—still a bargain as far as we're concerned for what everyone seemed to consider some of the finest food around).[6] They even published their own series of cookbooks between 1948 and 1960 and were featured in *Life* magazine's *Famous Roadside Inns for Travelers in America* and an issue of *Ford Times*.[7] The word was out and the restaurant was a hit!

In fact, it was so successful that it wasn't long before they were able to survive on only half a year's income. They ran the restaurant during the summers and managed to generate enough profit that the Hudsons could spend the winters traveling the world and gathering stories and recipes for their *Tummy Travels* series of booklets.[8]

Throughout the couple of decades following their grand opening, the Hudsons continued to expand the estate to approximately forty acres, with the restaurant as the centerpiece and the rest dedicated to preserving the natural beauty of the land and the carefully cultivated gardens. But by the 1960s, they were starting to think of retirement. Colonel Hudson built an additional home on the property, one with a magnificent view, where they could enjoy their golden years and then leased the restaurant building to Fred and Pat Maten in 1962. The latter would continue to operate the restaurant, but now under the name of North Woods Inn.[9] That's when the ghost stories started, but we'll get to those after we finish

5   *Ibid.*

6   Waring, H. (1988). Hudson estate lives on in area. *Littleton Sentinel Independent* (July 13, 1988).

7   Hudson, E. L. & Hudson, Col. K. C. (1955). *South of the Border (Tummy Travels)*. Littleton, CO: Country Kitchen.
Hulse, D. F. & Larison, P. (2021). Hudson Family and Hudson Gardens. *Littleton Museum.* <https://www.museum.littletonco.gov/Research/Littleton-History/Biographies/Hudson> (accessed August 18, 2025).

8   Waring, H. (1988). Hudson estate lives on in area. *Littleton Sentinel Independent* (July 13, 1988).

9   Hulse, D. F. & Larison, P. (2021). Hudson Family and Hudson Gardens. *Littleton Museum.* <https://www.museum.littletonco.gov/Research/Littleton-History/

our brief history.

Colonel Hudson died on March 2, 1984 at the age of ninety years.[10] Though they had some relatives, the Hudsons had no children of their own, but Evelyn Hudson seemed to have a certain affinity for organizations supporting children and the broader community, often encouraging Boy Scouts and other organizations to plant flowers on the Hudson Gardens grounds and serving as a member of the Littleton League of Women Voters and on the board of the Littleton YMCA. Following her husband's death, she dedicated her final years and the bulk of her estate to the organization of the Colonel King C. and Evelyn Leigh Hudson Foundation, which incorporated in 1986, marshalling the nearly $6 million Hudson estate to keep the beautiful Gardens open for the public to enjoy.[11]

Over the following years, some of the land was sold, leaving approximately thirty acres of useable property for the North Woods Inn restaurant and the Hudson Foundation's gardens.

Eveyln Hudson died in July of 1988, leaving the Foundation in charge of her estate and of the Hudson Gardens, which they continued to manage in accordance with her and her late husband's wishes.[12]

The North Woods Inn came to an end in 1997. When their lease expired, the Hudson Foundation resumed control of the restaurant building and began operating it as a part of the Hudson Gardens open space and event center, in cooperation with the South Suburban Park and Recreation District, which purchased the land but left the Foundation in control of management. The same year, the original log cabin underwent extensive restorations, maintaining its original charm while making it fit for continued use, now bearing the name of The Inn at Hudson Gardens.[13]

To this day, members of the public can visit Hudson Gardens free of charge except on days when the property is closed to host private events such as wedding receptions.

Biographies/Hudson> (accessed August 18, 2025).

Smith, J. (2002). Professionals investigate the haunting of Hudson Gardens. *Littleton Independent* (October 31, 2002).

10   Hulse, D. F. & Larison, P. (2021). Hudson Family and Hudson Gardens. *Littleton Museum.* <https://www.museum.littletonco.gov/Research/Littleton-History/Biographies/Hudson> (accessed August 18, 2025).

11   *Ibid.*

Waring, H. (1988). Hudson estate lives on in area. *Littleton Sentinel Independent* (July 13, 1988).

12   Waring, H. (1988). Hudson estate lives on in area. *Littleton Sentinel Independent* (July 13, 1988).

13   Hulse, D. F. & Larison, P. (2021). Hudson Family and Hudson Gardens. *Littleton Museum.* <https://www.museum.littletonco.gov/Research/Littleton-History/Biographies/Hudson> (accessed August 18, 2025).

Paranormal Claims

Though Fred and Pat Maten, owners of the North Woods Inn which occupied the main building at Hudson Gardens between 1962 and 1997, were far
from the only people to have reported paranormal happenings in the building,
it was during their tenure as restauranteurs there that most of the ghost stories
found their origin. As far as we know, the Hudsons themselves never experienced
anything unusual, and Colonel Hudson himself publicly denied any such experience.[14]

But during the days of the North Woods Inn, it seemed like there was plenty
of ghostly activity. Public notoriety for the alleged hauntings can be traced at least
to 1977 (with reports of experiences going back even further than that). According to Fred Maten, the very first paranormal phenomenon he experienced there
was that after all the candles had been snuffed out, they would suddenly relight
themselves. But thereafter, he and his wife seemed to almost always experience
weird things. On one occasion, a woman's necklace seemed to float straight out in
front of her. Windows and doors would open. Chandeliers would swing. Curtains
would open and close. Utensils would vanish and reappear in the incorrect locations. Trays would fall in the kitchen even when no one was around. Supposedly
there was one spot just outside the north end of the building where no plants
would ever grow.

Whatever ghost or ghosts haunted the location never seemed threatening.
Maten called it or them "ornery." But diners during the day or at the evening
dinner service never needed to worry about anything. The events only seemed to
happen after the restaurant was closed.[15]

Even the police got involved on at least two occasions. First in 1969, police
making their ordinary rounds outside the restaurant heard what they described
as running footsteps and the clatter of tons of pots and pans crashing in the
kitchen. They were let inside and searched for nearly an hour but found no sign
of any intruders. Some years later, police again entered the building to investigate
a disturbance and found nobody there. But when they went upstairs to search the
attic, they found quite the surprise—a footprint which disappeared into a wall, as
if someone left the print while stepping through the solid material.[16]

Later on, Maten would start calling the resident ghost (or at least one of
them—some reports allude to multiple spirits) "The Bastard," for all the prank-
ster-ish and poltergeist-style activity.[17] Apparently this spirit had taken up residence in the storage area above his office and he would often converse with the
ghost.

In 2002, a group of ten paranormal "professionals" were invited to explore the restaurant and reported numerous positive findings of paranormal
activity ranging from communications through dowsing rods and spirit boards

14   Reynolds, J. (1977). Popular Littleton haunt provides haunting tales. *Littleton
Independent* (January 28, 1977).
15   *Ibid.*
16   *Ibid.*
17   Smith, J. (2002). Professionals investigate the haunting of Hudson Gardens.
Littleton Independent (October 31, 2002).

to the faint smell of pipe or cigar smoke even though the venue was (and is) a non-smoking one and neither of the Hudsons ever smoked to the best of any-one's knowledge.[18]

One thing they paid particular attention to on that outing was the piano. Evelyn Hudson, who was still alive when the first paranormal phenomena began to be reported during the Matens' time running the restaurant but who had died prior to this group investigation, was known to have a particular fondness for the antique piano, and numerous people over the years have seen shadows or figures they thought were spirits moving about near the same instrument. On this inves-tigation, they didn't hear or see anything there.[19]

However, that piano is the subject of one of the more unusual claims we've come across in our years of investigations. According to reports originating with Fred Maten himself, the piano sometimes wept. That is, he believed there was a ghost (he thought probably a female one due to the crying behavior) which caused tears to stream down the front of the piano.

The same piano was relocated to a storage area in the outbuilding which had been used as the Hudsons' residence and later became administrative offic-es. Papers stacked on the piano were reported by several witnesses to blow off or scatter themselves. This occurred even with doors and windows shut and no breeze in the room.

We were told of more recent events in the 2010s in which witnesses saw a little girl running through the Gardens but when people went to investigate, she was nowhere to be found. Around the same time, a contractor working on renovations would enter a room with all of his tools, step out for a moment, and when he returned, some or all of the tools had been moved to a different room. Spooked by these experiences, he stopped working in the evenings.

Our Investigation

Alas, this is going to be a short one, despite all the great ghost stories asso-ciated with Hudson Gardens (and particularly with the Inn at Hudson Gardens, formerly the North Woods Inn and, before that, the Country Kitchen).

We were invited to conduct a single evening overnight investigation in preparation for a public lecture we were to give at Hudson Gardens to tell guests about the history of the property, about the ghost stories, and about our findings, whatever they may be. These are always great arrangements for us because they give us the opportunity to investigate another location and we think they're also great for the attendees because they get to hear all of the ghost stories in the very place where they took place. It's one thing to read a book like this one—in fact, we think that's one of the best ways one can spend one's time—but it's quite an-other to hear the same tales *on location*.

Before the investigation, we tried to think of some hypotheses about some of the claims. Objects moving, doors opening and closing, curtains blowing, and those kinds of things are pretty common paranormal claims so we already knew exactly what things to look for. If we experienced any of those, we'd want to

18   *Ibid.*
19   *Ibid.*

check for air currents and vibrations because those are simple and straightforward naturalistic explanations that need to be ruled out before considering the paranormal cause.

Objects disappearing and reappearing elsewhere are much more difficult to explain unless they're simply the result of forgetfulness. To preclude that possibility, because we knew so many of the claims were of displaced items, we made sure to document exactly who left which of our equipment in which location throughout the evening.

Smells of pipe or cigar smoke are another common ghostly manifestation in the lore. We're not sure why that should be except perhaps that many ghosts are thought to have lived during times when such things were more common. Given the location is a non-smoking venue, such odors would be quite unusual. However, this is another place we have to be careful. If someone is psychologically primed to expect tobacco smoke, they tend to interpret *any* smoky aroma as that of tobacco. We ran into that directly in our first volume (chapter 14) when some witnesses thought they smelled the ghostly leaf burning but it turned out to just be a hotel guest burning some bacon on a hot plate. In this case, because the main building under investigation was a restaurant, we wanted to make sure we accounted for any cooking smells so we could rule those out in case anyone smelled the paranormal pipe smoker.

Candles that relight themselves, unless you've purchased a novelty "trick" candle designed to do so as a joke for a birthday cake, are a bit tougher to explain. On rare occasions, if the flame has been extinguished but a small ember is still burning on the wick, this can occur naturally. Also, if a spark or open flame is exposed to the smoke emanating from a recently-extinguished candle, the flame can actually travel down the smoke and relight the candle. But though we were aware of such possibilities, we didn't think that it sounded like a plausible explanation for the re-igniting candles as reported in the ghost stories. Another thing just to keep our eye out for.

But the one that really caught our attention was that weeping piano. Rarely do we hear of those kinds of ghostly manifestations. Indeed, tear-like emanations from inanimate objects are much more commonly associated with claims of religious miracles like weeping or bleeding statues of Jesus or Mary than with ghost stories. When we arrived for our investigation, we hoped to start by checking it out to make sure it hadn't been rigged with hoses or something as part of a hoax. Assuming not, we were hoping we'd get to see the piano weeping so we could monitor the activity closely and perhaps take some samples of the "tears" for later laboratory analysis (either with our own equipment or, if necessary, by third party forensic scientists). Unfortunately, the piano was no longer there by the time of our investigation, so we have to leave that as just one more massive question mark in our ledgers.

The other one that's tougher to think of a natural explanation for was the footprints disappearing into the wall. That sort of thing would be quite easy to do as a hoax. But if we rule out a deliberate hoax—and we have no reason to suspect there was any deliberate hoax here[20]—then we have a much harder time explain-

---

20   In our experience, intentional hoaxes are rare. To the extent we've gotten involved in hoax cases in the past, they've been situations in which we introduced

ing what could have happened. Perhaps a trick of the light if the officers who reported seeing the footprint only caught a quick glance at it? That's plausible, but police tend to be trained observers so we don't really like that as an explanation. Once again, we were hoping to find such footprints for ourselves so we could really dig in and investigate properly.

When we arrived, we were given access to whatever locations we wanted, but chose to focus on the Inn at Hudson Gardens (the main building) and the former residence. Almost all of the paranormal claims centered on those two buildings, with only a few odd stories here and there branching out to other parts of the Gardens. We went through the usual process of documenting the entire location, taking baseline readings for vibrations and EMF, and setting up our surveillance systems to give us a clear view of the property.

Unfortunately, the entire evening passed without incident. Despite best efforts, we didn't see any of the paranormal manifestations while we were there, so we can't definitively supply any explanations—natural or supernatural.

Sometimes that's just the way of things. People often ask us how often we experience something seemingly paranormal and how often we debunk the claim. Our rough estimate is that our cases are approximately divided into thirds. About a third of the time, we see something genuinely weird that we can't (yet) explain. About a third of the time, we see something weird but we *are* able to explain it naturally. And about a third of the time, as was the case with our Hudson Gardens investigation, the thing simply doesn't happen when we're watching.

Therefore, we still consider the Hudson Gardens case to be open.

References & Further Reading

Ewbank, A. (2022). The Lost Glamour of the Department-Store Restaurant. *Atlas Obscura*. <https://www.atlasobscura.com/articles/department-store-restaurant>

Hudson, E. L. & Hudson, Col. K. C. (1955). *South of the Border (Tummy Travels)*. Littleton, CO: Country Kitchen

Hulse, D. F. & Larison, P. (2021). Hudson Family and Hudson Gardens. *Littleton Museum*. <https://www.museum.littletonco.gov/Research/Littleton-History/Biographies/Hudson>

McCormack, K. (2000). *The Garden Lover's Guide to the West*. Hudson, NY: Princeton Architectural Press

Reynolds, J. (1977). Popular Littleton haunt provides haunting tales. *Littleton Independent* (January 28, 1977)

Smith, J. (2002). Professionals investigate the haunting of Hudson Gardens. *Littleton Independent* (October 31, 2002)

Waring, H. (1988). Hudson estate lives on in area. *Littleton Sentinel Independent* (July 13, 1988)

---

ourselves into a story that was getting some media attention. Clients who reach out to us for investigative services tend not to be trying to pull a fast one on us.

11

The Phantom Rocking Chair: Francisco Fort Museum

The Francisco Fort Museum, also known as Francisco Plaza, located at 306 South Main Street in La Veta, Colorado, is the oldest structure in that town. Its history goes all the way back to the Colorado Territory (before Colorado became a state) and is currently operated as a museum dedicated to telling the story of that local history. The museum actually consists not only of one building but an entire campus with one central location and several outbuildings.

As one might expect, any place that's seen so much history probably has its share of ghost stories, and indeed this one does, ranging from flickering lights to a floating lady to a rocking chair that rocks itself.

Figure 11.1. The Francisco Fort Museum (photo: Bryan Bonner).

The History

This is one of those locations that's seen so much history, it's difficult to know what to include and what to omit. Entire volumes could easily be written about all of the things that have occurred at the Francisco Fort Museum, so we'll have to do our best to just provide a brief (but we think interesting) overview of some of the highlights.

Our introduction at the beginning of this chapter mentioned that the original building now serving as the centerpiece of the broader Franciso Fort Museum campus predates the State of Colorado. That's true. As we've mentioned elsewhere both in this and other volumes in this series, Colorado statehood was granted in 1876 (hence Colorado's nickname: "the Centennial State"). Before that, the Colorado Territory was established in 1861 in a bill signed by President Buchanan.[1] Prior to that, Huerfano County, which contains the town of La Veta, was part of Mora County in the New Mexico Territory.

Though not strictly related to the Fort itself, the history of the land on which it rests is interesting enough to merit some brief discussion. It begins not with American history, but with Mexican history. Beginning in the sixteenth century, the Mexican government (first as a Spanish territory and after 1821 as an independent nation) issued a variety of land grants throughout much of what is now the southwestern United States including parts of what is now Colorado. The largest such land grant in Colorado is known as the Vigil and St. Vrain grant, spanning 4.1 *million* acres. In 1843, Ceran St. Vrain and Cornelio Vigil petitioned for the grant which was approved only one day later and they took possession of the land on January 2, 1844. Students of history may recall that the United States annexed Texas (which then included what is now southern Colorado) in 1845. This led to the Mexican-American War in 1846 which was ultimately resolved by the Treaty of Guadalupe Hidalgo in 1848 when the Americans won the war and took possession of most of the land which became several states (including Colorado). The Treaty's terms, though, required the American government to honor the contractual terms of the Mexican land grants, including the Vigil – St. Vrain Grant. Thus Congress approved the grant in 1860 but reduced its acreage in 1869 in accordance with acreage limits under Mexican law, thus partitioning the 4.1 million acres into plots, some of which were retained under the terms of the grant and the rest of which was repossessed by the United States government.[2]

Our story now takes a slight detour to introduce the key players in what was to follow. Colonel (honorary) John M. Francisco (c. 1820 – 1902) first saw the Cucharas Valley (home of what later became La Veta) on a prospecting trip in 1839 and knew from then that he eventually wanted to settle the area. While selling provisions to the army at Fort Garland, he met Henry Daigre (1832 –

1   "An Act to provide a temporary Government for the Territory of Colorado," of the 36th Congress of the United States. Quoted in: Sanger, G. P. (1863), ed. *The Statutes at Large, Treaties and Proclamations of the United States of America from December 5, 1859 to March 3, 1863,* volume 12: 172-177.

2   Simmonds, R. (n.d.). Mexican Land Grants in Colorado. *Colorado Encyclopedia.* <https://coloradoencyclopedia.org/article/mexican-land-grants-colorado> (accessed August 18, 2025).

1902) and the two formed a business partnership.[3] In one of the relatively few transactions from the Vigil – St. Vrain land grant approved by Congress, the duo purchased the land that would become Francisco Plaza and the surrounding settlement in 1861.[4]

The U-shaped Fort itself was built to protect the newfound settlement from attacks by the Ute tribe, prevalent in the region and hostile at that time. Twenty men were hired to build the fort with two-foot thick adobe walls and construction was finished in 1862. Their fears were not unfounded. In 1863 the settlement was attacked by a band of Ute. The settlers rallied in the Fort, manning gun ports and lying atop the roofs. A horseman was dispatched to Fort Lyon (not a short ride in the days before automobiles) to recruit assistance, but by the time the cavalry returned the Ute had given up the attack. The Fort held.[5]

From then on, the settlement grew, with Francisco Plaza at its center. It became an important focal point for Hispanic and Anglo settlers in the 1870s. The first post office was established in 1871, then under the name of Spanish Peak.[6]

Following the Civil War, there was an influx of settlers who had been Southern/Confederate sympathizers, including Green Russell, who founded Russell Gulch, a Gilpin County mining settlement.[7]

Ute attacks continued occasionally throughout the 1860s but increasing settlement and treaties with the Ute tribe eventually brought a period of peace. Peace and settlement brought further development. In 1876, the famous Denver and Rio Grande Railroad reached the settlement, ushering in a period of accelerated development. At the same time, Spanish Peak changed its name to La Veta, a name chosen by Railroad founder William Jackson Palmer (1836 – 1909) and former territorial governor Alexander Cameron Hunt (1825 – 1894), who filed the town's articles of incorporation themselves. A portion of the Fort was converted to a railroad depot and also hosted various other services and amenities railroaders might need including a general store, guest accommodations, and a freight company.[8]

Sometime in the 1870s (the exact date is unclear), Francisco and Daigre

3   Colorado Encyclopedia (n.d.). Francisco Plaza. *Colorado Encyclopedia.* <https://coloradoencyclopedia.org/article/francisco-plaza> (accessed August 18, 2025).

4   Christofferson, N. H. & Pearce, S. (1985). Francisco Plaza (Francisco Fort Museum): National Register of Historic Places Inventory – Nomination Form. [Register No. 86002950]. *National Register of Historic Places.* <https://npgallery.nps.gov/NRHP/GetAsset/NRHP/86002950_text> (accessed August 18, 2025).

5   Jessen, K. (2014). Francisco Fort became the center of La Veta. *Loveland Reporter-Herald.* <https://www.reporterherald.com/2014/09/25/francisco-fort-became-the-center-of-la-veta/> (accessed August 18, 2025).

6   *Ibid.*
Christofferson, N. H. & Pearce, S. (1985). Francisco Plaza (Francisco Fort Museum): National Register of Historic Places Inventory – Nomination Form. [Register No. 86002950]. *National Register of Historic Places.* <https://npgallery.nps.gov/NRHP/GetAsset/NRHP/86002950_text> (accessed August 18, 2025).

7   *Ibid.*

8   Colorado Encyclopedia (n.d.). Francisco Plaza. *Colorado Encyclopedia.* <https://coloradoencyclopedia.org/article/francisco-plaza> (accessed August 18, 2025).

parted ways, leaving Fracisco in charge of the Plaza. In the 1890s, the rail depot moved several blocks north and the Plaza resumed its function as residential housing. Fransico died in 1902 and the property subsequently passed through a variety of different owners and uses. In 1918, the southwest corner was replaced with an adjacent (but internally disconnected) adobe house.[9]

Among the different faces of the Francisco Plaza over the years have been: the original fort, a hotel, a general store, a telegraph office, an outfitter's shop, a granary, a town hall, private residences, a practice room for a brass band, and a physician's office.

Ultimately, the town of La Veta acquired the property and converted it into the Francisco Fort Museum, which opened for business in 1958. Throughout the 1960s, several other historic buildings from the area were moved to the Fort Museum campus and incorporated as part of the historic exhibition.[10] Among those are:

- The Ritter Schoolhouse, a one-room schoolhouse originally located about three miles east of its present location which served until it was sold as storage for the Ritter Ranch in 1897 and replaced by a larger school,
- A recreation of an 1800s saloon,
- The old Town Hall,
- The Mining Museum, exhibiting mining and medical artifacts,
- The 1899 Inn, and
- A recreation classic blacksmith's shop exhibiting vintage tools.[11]

Under both the historic name of the Francisco Plaza and the modern name of the Francisco Fort Museum, the property was added to the National Register of Historic Places in 1986.[12] Two grants from the State Historical Fund in the 1990s provided for restoration and renovation.[13] One interesting side note from the restoration: to maintain the historic character of the site, new adobe bricks were required; while they were drying overnight, a bear walked across and left a paw print, which is now on display in one of the bricks.

The Francisco Fort Museum (or, we should say, the Fort itself) is the only complete adobe fort in Colorado.

Because the Fort was such a center for the town, the entire history of La Veta touched it at some point or other. A complete history of all of those happenings, though, is far beyond the scope of this book. For such a small town (with a population of only 862 as of the 2020 census), it's quite a rich and colorful history and we recommend interested readers look it up.

---

9   *Ibid.*

10   *Ibid.*

11   Christofferson, N. H. & Pearce, S. (1985). Francisco Plaza (Francisco Fort Museum): National Register of Historic Places Inventory – Nomination Form. [Register No. 86002950]. *National Register of Historic Places.* <https://npgallery.nps.gov/NRHP/GetAsset/NRHP/86002950_text> (accessed August 18, 2025).

12   *Ibid.*

13   Colorado Encyclopedia (n.d.). Francisco Plaza. *Colorado Encyclopedia.* <https://coloradoencyclopedia.org/article/francisco-plaza> (accessed August 18, 2025).

Paranormal Claims

The first of the paranormal claims might very well have been placed in the history section instead of here because it's one of those that bridges the gaps between real history, supernatural belief, and folklore. It has to do with the town's location and some early mining operations.

The town of La Veta, and thus the Francisco Fort Museum, is nestled in the shadows of twin mountains called the Spanish Peaks. Prior to Spanish exploration (followed by Mexican and then, ultimately, American), the land in the area was known to the Commanche people who called the mountains "Huajatolla" (sometimes spelled "Wahatoya"[14]), which sources alternately translate as either "double mountains" or "breasts of the world."[15] The former is the more accurate translation, but we suppose the proliferation of innuendo and double entendre even in such popular restaurant names as "Twin Peaks" renders the difference somewhat academic in informal language.

Though the word is Comanche, it was primarily the Ute who inhabited the region prior to European settlement, and the mountains were of spiritual significance to the Native people of the area. Though religious claims are supernatural, it's not our practice to investigate or criticize those directly unless and until they manifest in the physical world in some (at least potentially) measurable or observable way. One such claim was that gold in the mountains was (or perhaps still is, if one is inclined toward such belief) guarded by demons. Alone, that urban legend might be insufficient to earn mention as a specific paranormal claim, but it touched the real world in a now locally-famous urban legend involving the Spanish explorer Juan de la Cruz in or around 1541. Not one to let demons scare him away from a good haul of gold, he hired Native crews to mine the mountains (in some versions of the story, he hired them so as to not anger the demons personally and in other versions he didn't believe in the demons and just hired a crew to help with the mining operation). They found the gold, sure enough. But as the Native crews were packing the gold, they were killed. Later, de la Cruz and his party were themselves killed by the demons and the gold rained over the mountains. Gold found in 1811 has been rumored to be the very same gold, but of course this cannot be proven.[16]

Obviously, we can't investigate claims originating in the mid-1500s. Neither

14   Obviously the Comanche language did not use the Roman alphabet. Linguistic anthropologist Alice Anderton developed a writing system which was adopted by the Comanche Nation in 1994 as part of an effort to revitalize the severely endangered language.

15   Smith, S. (2005). Magical Mountains. *The Pueblo Chieftain*. <https://www.chieftain.com/story/opinion/columns/2005/10/06/magical-mountains/8582946007/> (accessed August 19, 2025).
Christofferson, N. (2015). The Spanish Peaks: Legends. *The Huerfano World Journal*. <https://worldjournalnewspaper.com/the-spanish-peaks-legends/> (accessed August 19, 2025).

16   Pemberton, P. (1997). Highway of Legends. *The Pueblo Chieftain*. <https://www.chieftain.com/story/special/1997/05/29/highway-legends/8502199007/> (accessed August 19, 2025).

do we have much of a chance of checking out the claim from 1811. But we include this story just to set the stage for the kinds of paranormal beliefs common in the region.

Moving on to the Francisco Fort Museum itself, most of the paranormal phenomena reported take on a more ghostly characteristic and have been reported by museum staff, visitors, and prior ghost hunting expeditions.

The most common story we've seen reported around the paranormal discussions about the museum are reports of a floating lady in the west wing of the main building. A few witnesses have described seeing this phantasm on multiple occasions, though we also need to point out that a lot of the published reports are repetitions of other people's prior experiences.

Coming right on the heels of the haunted piano in the previous chapter is... another haunted piano, also located in the west wing. This one doesn't weep, as far as we know, but multiple witnesses have said they've heard a ghost pressing keys on the piano as they walked by. Never fully playing music, mind you. Just pressing some keys and producing one or two notes.

Others have seen ghosts carrying candles throughout the entire main Plaza building. But the ghosts don't seem to discriminate about their lighting choices. Electric lights are also said to flicker throughout the building at unpredictable times.

Most notably for us, there's an antique rocking chair in the Museum which has been seen to gently rock back and forth as if someone were resting in it. But of course no one was (nor had been recently).

Figure 11.2. The phantom rocking chair (left; photo: Bryan Bonner).

Prior to our own investigation, we were shown notes and video from an earlier group of ghost hunters who'd explored the property, and they reported some unusual findings of their own.

First, they said there were electromagnetic anomalies in the specific location where the floating lady had been reported by other witnesses. On another occa-

Christofferson, N. (2024). Vignettes of the Spanish Peaks: Part 2 of 3. *The Huerfano World Journal.* <https://worldjournalnewspaper.com/vignettes-of-the-spanish-peaks/> (accessed August 19, 2025).

sion, while exploring the saloon, they saw a light flicker and determined it didn't come from any reflective surface in the room. Throughout their entire experience, they reported strange feelings, unusual sounds, and all the kinds of things people typically report experiencing in haunted locations.

Armed with all the stories from past witnesses, we were ready for our own investigation.

Our Investigation

Our investigation was the kind of setup we really like. Most of our investigators live in the Denver area. That's quite a drive to La Veta and back, especially if we want to keep the ridiculous nocturnal hours often required of paranormal researchers. All the more true if we wanted to investigate for more than a single day. And we did want to investigate for more than a single day. The Francisco Fort Museum is a large compound with numerous buildings and we wanted to make sure we gave it all due effort. Things worked out perfectly in our favor this time, and we were put up in local lodgings for several days and given the run of the Museum during the evenings.

Before we got there, we did take a closer look at the prior ghost hunters' footage and found we could explain at least some of it. The unusual feelings and sounds they reported can't easily be conclusively debunked, but neither are they strong evidence. Believers tend to view such things as paranormal experiences while skeptics tend to dismiss them as ordinary building sounds coupled with exhaustion and confirmation bias. So for those, we're not necessarily saying they were wrong, but just that we're not convinced by those claims to our own evidentiary standards.

More intriguing were their EMF anomalies and the flickering light. We examined their use of the EMF meters closely and found that we could not explain away those findings as simply the result of improper use of the equipment.[17] But since we weren't present when the measurements were taken, we simply noted what they'd found and resolved to check it out with our own meters when we arrived.

When we looked at their flickering light, we determined that its source was a reflection off a piece of glass from their own infrared camera. So we ruled out at least one claim before the investigation started, but we still had a lot of work to do.

Upon arrival, we went through our usual steps of documenting the entire place and figuring out where to monitor. We paid particular attention to the rocking chair that sometimes rocks itself (see Figure 11.2) and to the haunted piano that sometimes plays itself (see Figure 11.3) . With regard to the latter, we (in our twisted and horror-obsessed minds) were delighted to discover not only is the piano rumored to be haunted, but lots of antique dolls (of the sort many modern people would find creepy, probably due in large part to horror movies like the *Annabelle* series) surround it.

---

17   You'd be amazed how often EMF anomalies *are* explainable simply as misused technology.

Figure 11.3. The haunted piano, with creepy dolls (photo: Bryan Bonner).

We chose to focus our investigation on the main Plaza building because that's where most (if not all) of the ghost stories have been reported). We did document the entire property but thought a more focused investigation would be a better use of our limited time. In particular, we wanted to make sure our surveillance system had a view of as much of the west wing as possible, and we managed to monitor essentially the entire thing.

While some of our people were setting up the cameras and microphones, some others started collecting baseline measurements on all of our assorted gizmos. Much to our surprise, we immediately detected the same EMF anomaly the prior team had found. Throughout the building, EMF readings were consistent at about one milligauss, perfectly within the range expected in such a building. Near a religious display in the west wing where the floating lady has been reported to manifest in the past, we picked up electromagnetic fields that periodically spiked to as high as four milligauss.

Now, four milligauss is not a frighteningly high reading. It's about the strength of an electromagnetic field one would expect to be produced by a (running) toaster at about one foot of distance. Nothing to be afraid of. But it was uncharacteristically high *for the location*, so we felt the need to investigate more closely and see if we could determine its source.

After half an hour or so of searching, we found it. Unfortunately, the source was not ghostly in origin. Rather, it came from a Wi-Fi router in the library on the other side of the wall. Alas, not every anomaly turns out to be the paranormal evidence we're always hoping for. But we're nevertheless glad to have not only found the cause of our own seemingly anomalous measurement but that of some prior investigators.

One mystery solved, we went about our usual practice of alternating between silent monitoring and hourly walk-throughs to take new measurements. Instead of providing a chronological account of the rest of the investigation, we'll proceed according to topic.

Video and audio monitoring didn't yield any interesting results. No floating ladies or anything else we couldn't explain. We did pick up a couple flickering lights, but nothing we can't easily attribute to bad wiring in an old building which has gone through numerous renovations. Similarly, there were some sounds on our recordings, but they were just the pops, creaks, and groans we expect from any old building. Beeps from a security system periodically showed up in the recording, and occasionally caught us by surprise when we were listening back to all of our data, but their source is known to us.

Interior temperatures dropped from 79 degrees (Fahrenheit) during the day to 76 degrees at night, consistent with falling outdoor temperatures as the evening wore on. There were no anomalies in any of those measurements.

That brings us to the ghost rocking the rocking chair. We actually saw it! At one point during our walkthroughs, one of our investigators noticed the chair was gently rocking. No one had gone anywhere near it. These are the moments that get us really excited. Still hoping to find strong evidence of a ghost, we set about looking for any natural explanations we could think of, because ruling out the natural is the only way to be sure of any result.

Loose floorboards seemed like the most likely natural culprit. Perhaps if someone stepped on one end of a floorboard, it might move the opposite end, on which the chair was resting. To test that theory, we set seismometers all over the floor and monitored both them and the chair itself as we had one of our team members stomp around the floor to see if anything would happen.

Sure enough, it did. It wasn't so much a single loose floorboard, however, as it was general vibration of the entire floor if someone stepped heavily enough. That provided just enough motion for the chair (which rocks at the slightest provocation) to start gently swaying.

But that wasn't all. While we were stomping around to see what would happen with the rocking chair, we heard the piano produce a couple of notes. As it turns out, the same natural explanation applied to both of the alleged paranormal claims. When we looked more closely at the piano, we found it was an antique instrument in poor repair, and the vibrations in the floor were sufficient to vibrate the strings, producing a couple of faint notes.

Solutions like these are bittersweet. We're always thrilled to solve any mystery. But we're also always disappointed when the explanation doesn't turn out to be "a ghost did it." At the end of the day, we did manage to debunk a couple of the specific claims, but we always have to remind people that just because we found a solution to some of the mysteries, that doesn't necessarily prove the Museum is not haunted. There are other experiences witnesses have reported which we simply didn't see. Therefore, as with the majority of our cases, we still consider this one open, even if we have closed the books on a couple of its subsidiary claims.

References & Further Reading

Christofferson, N. H. & Pearce, S. (1985). Francisco Plaza (Francisco Fort Museum): National Register of Historic Places Inventory – Nomination Form. [Register No. 86002950]. *National Register of Historic Places.* <https://npgallery.nps.gov/NRHP/GetAsset/NRHP/86002950_text>

Christofferson, N. (2015). The Spanish Peaks: Legends. *The Huerfano World Journal.* <https://worldjournalnewspaper.com/the-spanish-peaks-legends/>

Christofferson, N. (2024). Vignettes of the Spanish Peaks: Part 2 of 3. *The Huerfano World Journal.* <https://worldjournalnewspaper.com/vignettes-of-the-spanish-peaks/>

Colorado Encyclopedia (n.d.). Francisco Plaza. *Colorado Encyclopedia.* <https://coloradoencyclopedia.org/article/francisco-plaza>

Jessen, K. (2014). Francisco Fort became the center of La Veta. *Loveland Reporter-Herald.* <https://www.reporterherald.com/2014/09/25/francisco-fort-became-the-center-of-la-veta/>

Pemberton, P. (1997). Highway of Legends. *The Pueblo Chieftain.* <https://www.chieftain.com/story/special/1997/05/29/highway-legends/8502199007/>

Sanger, G. P. (1863), ed. *The Statutes at Large, Treaties and Proclamations of the United States of America from December 5, 1859 to March 3, 1863,* volume 12: 172-177.

Simmonds, R. (n.d.). Mexican Land Grants in Colorado. *Colorado Encyclopedia.* <https://coloradoencyclopedia.org/article/mexican-land-grants-colorado>

Smith, S. (2005). Magical Mountains. *The Pueblo Chieftain.* <https://www.chieftain.com/story/opinion/columns/2005/10/06/magical-mountains/8582946007/>

12
Fondue (Or Don't): The Melting Pot

The Melting Pot, located at 2707 West Main Street in Littleton, Colorado, is one of the state's most famously haunted locations. Diners and staff alike have reported a wide variety of paranormal or ghostly occurrences. Whereas some restaurants might shy away from such a reputation, the Melting Pot seems to appreciate its own cadre of ghosts.

Actually one location of a chain of Melting Pot restaurants, the Littleton location's building certainly has a colorful enough history of its own to merit the stories.

Figure 12.1. The Melting Pot in Littleton (photo: Bryan Bonner).

The History

When people in Colorado (or at least in the Denver area) think about "the" Melting Pot restaurant, they're probably thinking of the Littleton location, which is the primary subject of this chapter. Doubly so if they're talking about the Melting Pot's ghost stories. However, it's not the only Melting Pot location. Before we get to "the" Melting Pot location in question, though, it makes sense to briefly mention the restaurant chain's history as a whole.

The chain began with a single location in Maitland, Florida in 1975 offering just Swiss cheese, beef, and chocolate fondue options. But as its popularity grew, so did its menu, and eventually, its number of locations. Waiter Mark Johnston opened his own location in Tallahassee, Florida in 1979, followed by a second in 1981. Then, along with his brothers Mike and Bob, he purchased the entire brand in 1985 and began franchising locations to others the following year. Since then, they've grown the chain to include more than ninety locations across thirty-two states and one in Canada.[1]

Four of those locations are in Colorado: in Colorado Springs, Fort Collins, Louisville, and of course the location in Littleton we're discussing now.

But that location wasn't always the Melting Pot. Indeed, it wasn't always a restaurant at all, and its storied history may be a part of why it has attracted so many ghost stories over the decades. Its first incarnation was a public library.

Prior to taking up residence in the building in question, the first public library in Littleton was a small operation run out of a drugstore located elsewhere on Main Street which opened in 1897 and then moved from place to place over the next several years as demand grew. By the mid-1910s, Littleton decided they would reach out to the Carnegie Foundation for support to build a new and more permanent location.[2]

During his later years, businessman and philanthropist Andrew Carnegie (1835 – 1919) engaged in a program of funding seemingly as many public libraries as he could. After his death, his Carnegie Foundation continued this work, ultimately funding or helping to fund more than 2,500 libraries, some thirty-six of which were (or are) in the State of Colorado.[3] For more about Mr. Carnegie's support of other Colorado libraries with haunted reputations, see Volume 2, Chapter 7 of this series. For our purposes here, the important thing to realize is that the Littleton library was one of the ones that gained Carnegie Foundation support.

Financial support for the project didn't come without some strings. The Carnegie Foundation wanted to support the library projects but they also wanted the local municipalities to be able to sustain the libraries into the future, so they

1   Melting Pot (n.d.). History of Melting Pot. *The Melting Pot*. <https://www. meltingpot.com/history.aspx> (accessed August 19, 2025).

2   Hulse, D. F., Christensen, K., & Larison, P. (2021). Carnegie Library: Local Landmark—1973. *Littleton, Colorado*. <https://www.littletonco.gov/Building-Development/Historic-Preservation/Historic-Littleton-Buildings/Carnegie-Library> (accessed August 19, 2025).

3   Denver Public Library Special Collections and Archives (n.d.). The History of the Denver Public Library. *The Denver Public Library*. <https://history.denverlibrary.org/exhibit/history-denver-public-library> (accessed June 18, 2024).

conditioned their support on Littleton imposing a local tax to maintain the library. This was approved in April of 1915, the Carnegie funding was secured, and the land on Main Street was purchased for the princely sum of $500 (equivalent to just about $16,000 in 2025).[4]

Locals were concerned about the amount of control the Carnegie Foundation had over the project, but it proceeded anyway, with Mayor J. E. Maloney (1863 – 1948) selecting Jules Jacques Benois Benedict (1879 – 1948) as the architect to lead the project, which would ultimately be built by contractors V. W. Robbins and the Watts Brothers. However, the local concern about Carnegie Foundation oversight was perhaps not without cause. It took until July of 1916 for Benedict and the Carnegie Foundation's secretary to agree to plans and for the latter to release $8,000 (about $237,000 in 2025) in funding to complete the project. But complete it they did, and the Library, designed in an Italian Renaissance style with Beaux-Arts influences, opened in 1917 under the name of The Woman's Club.[5]

Why The Woman's Club? In addition to being a public library much like any other, it was also part of a Colorado Works project providing employment to local women who were skilled in bookbinding.[6] Additionally, the library became a community center, often offering services to those in need during difficult times. During the Second World War, it was home to a Red Cross "Bundles for Britain" program offering support to British troops during the war.[7]

It was relocated to a larger building on South Datura Street in 1965 and the Main Street building became the Littleton Police Department and jail, in which capacity it served until 1977.[8] This was a period of modernization for the Police Department. They'd finally moved from their small offices at Town Hall into a proper Police Station and took the opportunity to improve radios, vehicles, and other equipment, as well as to hire new officers, dispatchers, and other employees. When they outgrew the Main Street location, they relocated in 1977 to their new (and current, as of this writing) quarters on West Berry Avenue.[9]

4   Hulse, D. F., Christensen, K., & Larison, P. (2021). Carnegie Library: Local Landmark—1973. *Littleton, Colorado.* <https://www.littletonco.gov/Building-Development/Historic-Preservation/Historic-Littleton-Buildings/Carnegie-Library> (accessed August 19, 2025).

5   *Ibid.*

6   A dying trade we're desperate to preserve. On that note, if you happen to have an antiquarian books lying around your home (antique family Bibles are a common one), you might consider taking them to your local skilled bookbinder for repair and preservation, thus simultaneously preserving your own artifact and keeping a skilled tradesperson employed.

7   Hulse, D. F., Christensen, K., & Larison, P. (2021). Carnegie Library: Local Landmark—1973. *Littleton, Colorado.* <https://www.littletonco.gov/Building-Development/Historic-Preservation/Historic-Littleton-Buildings/Carnegie-Library> (accessed August 19, 2025).

8   *Ibid.*

9   Littleton Police Department (n.d.). Littleton Police History: A Brief History of the Littleton Police Department. *Littleton, Colorado.* <https://www.police.littletonco.gov/About-LPD/Littleton-Police-History> (accessed August 19, 2025).

Likely part of the haunted history of the building (though we'll get to those details in a moment) stems from its time as a police station and a jail. Though we're unaware of any specific major incidents at the facility (riots, murders, escapes, etc.), nor any particularly infamous criminals housed there during the 1965 – 1977 period, the fact of the matter is by the very nature of police work, jails and police stations are likely to see people at the lowest moments of their lives. Low moments tend to be associated with paranormal lore. Plus, jails themselves are just inherently creepy places, if you ask us.[10]

After the police moved on to their new offices, the building lay vacant for about two years until it became "Pistachio's Bar & Dance Club" in 1979. During this dance club's ownership, the interior became much more "restaurant-like" and additions were made to the south and west sides of the building. Since then, it has been used exclusively as a restaurant—by Café Kandahar, offering European cuisine and an Alpine atmosphere; by Alpine Café, offering family dining; by The Old Library (an appropriate name, we think), a beer garden; by Scribbles Cafe, owned by two football players for the Denver Broncos; and finally, since 1997, by the Melting Pot, which currently occupies it and offers their popular fondue dining.[11]

Not much has changed with the building since then. With only a single company owning it and using it for the same purpose for the last few decades, its history seems to have largely stabilized.

There are two rumored but unconfirmed events that took place at the building, but because they're unconfirmed and related directly to the paranormal lore, we'll get to those in our next section.

Paranormal Claims

Oddly enough, the Littleton Melting Pot isn't the only Colorado Melting Pot location reputed to be haunted. The Louisville location is also rumored to have been built above a mine shaft that collapsed and to be haunted by the miners killed in the disaster.[12] But it's the Littleton location that's supposed to be even *more* haunted and which was the result of our work, so we'll confine our remarks to it.

Two specific ghosts (albeit unnamed ones) meant to haunt the building are connected to the rumors we mentioned at the end of the History section.

When the building was a police station and jail, one rumor goes, an inmate was killed after murdering one of his jailers during a botched escape. One or both

---

10   For more about a jail that *does* have a long history of infamous events and characters, see our discussion of the Denver County Jail in Volume 1, Chapter 10 of this series.

11   Hulse, D. F., Christensen, K., & Larison, P. (2021). Carnegie Library: Local Landmark—1973. *Littleton, Colorado.* <https://www.littletonco.gov/Building-Development/Historic-Preservation/Historic-Littleton-Buildings/Carnegie-Library> (accessed August 19, 2025).

12   Brick, C. (2018). *Haunted America: Ghost & Legends of Colorado's Front Range.* Charleston, SC: The History Press.

of those individuals—the prisoner and the jailer—are meant to still haunt the premises.[13] Specifically, his favorite haunt is supposed to be a small hallway behind the bar on the south side of the building. Multiple witnesses have reported hearing disembodied footsteps in that hallway and around the nearby bar.

Figure 12.2. Location of the former jail cells (photo: Bryan Bonner).

And then there's the fountain. On the south side of the building is a non-functional fountain said to be the site of a young girl's drowning.[14] Further, her spirit is supposed to still haunt the women's restroom on the lower floor.[15] This particular story was also the subject of "The Fountain of Death," an episode in the third season of the Biography Channel's *My Ghost Story*.[16]

Beyond those two rumors connected to specific alleged spirits are more ghost stories than we know how to count. Some follow predictable patterns. Many seem to repeat each other across multiple incidents and with multiple witnesses. And a few stand out as unique or nearly so.

Earlier, we mentioned people often heard the mysterious sounds of ghostly

13   Reedy, A. (2019). Are these Colorado restaurants haunted? That depends on who you ask. *The Denver Post.* <https://www.denverpost.com/2019/08/23/haunted-denver-restaurants/> (accessed August 19, 2025).

Yoe, S. (2023). Haunted Dining in Denver. *Denver Center for the Performing Arts.* <https://www.denvercenter.org/news-center/haunted-dining-in-denver/> (accessed August 19, 2025).

14   Xu, H. (2024). Eat and Drink Your Way Around Historic Littleton. *Westword.* <https://www.westword.com/restaurants/eat-and-drink-in-historic-downtown-littleton-20846115> (accessed August 19, 2025).

15   What have we told you? It's almost always the women's washroom!

16   Harris, S. C. (Executive Producer). (2011). Lizzie Borden Took an Axe (Season 3, Episode 6) [TV series episode]. In Phillips, M. & Ayalon, H. (Executive Producer), *My Ghost Story.* Mark Phillips Philms & Telephision; Biography Channel.

footsteps in the hallway supposed to be haunted by the would-be jailbreaker. But those aren't the only footsteps heard in this restaurant. No, all throughout the building, staff and diners alike report having heard similar sounds. One additional hotspot for them is the lower level of the building, near where the jail's holding cells used to be.

Employees locking up some nights say they've heard people still inside the building. Of course it was really empty and the voices and other sounds of inhabitants can't be explained.

By far the most common set of ghostly occurrences there are objects that move of their own accord. To hear some employees describe it, one would think glasses, bottles, plates, and other objects are just constantly flying around the place in a scene reminiscent of the "paranormal investigators" scene in the original *Poltergeist* film. Others paint a milder portrait of the phenomenon but still insist that glasses and bottles move across the bar without anyone pushing them and even that tables and chairs have rearranged themselves in the restaurant. Another variant of the moving objects claim involves a storage room just outside the kitchen, where employees say sometimes the heavy storage shelves shake themselves so hard objects fall from them.

One moving object seemed to outshine all the others. We were told that one day, the large commercial coffee and espresso machine literally "walked" itself across the room. There was even video of it, we were told. Glasses moving across a bar are one thing. Commercial equipment doing the same is quite another— those things are heavy! And if there were video evidence, we couldn't be more excited to check it out.

Children seem to also haunt the place. The one in the women's restroom who drowned in the fountain is one, but others have reported seeing multiple children enter the building on the south side, never to be found inside or seen again.

We already alluded to the women's restroom being haunted potentially by that one child. Other claims in that room include guests hearing someone breathing in the center stall of the women's room even when they're alone in the restroom. One witness even said she felt something grab her ankle while she was in that restroom. Multiple reports also claim the faucets in the same restroom will turn themselves on and off by themselves. And of course, the phantom footsteps are heard there, too, following guests even into the most private of rooms in the restaurant.

And then there's one final ghostly manifestation that could only occur in a fondue restaurant. Apparently sometimes the table heaters guests use to melt their cheeses or chocolates find themselves reconnected and turned back on even after the staff had disconnected them and cleaned them to prepare to turn over the table to a new party of guests.

That's a lot of potential ghosts to investigate!

Our Investigation

We were fortunate enough to investigate at the Melting Pot (and yes, to enjoy some of their fondue cheeses) several times over the years, as well as to conduct some off-site research in historic archives. Each on-site investigation followed our usual practice which by now readers are certainly familiar with, so instead of describing each investigation individually in its entirety (as that would be horrifyingly redundant), we'll focus on the results we obtained over the full course of our investigations.

Let's start with what we've been able to learn about the two specific incidents meant to have precipitated two of the hauntings. Because both of these have come to our attention as unsubstantiated rumors, we've struggled a bit to try to put names or dates to the incidents. In both cases, we have completely unable to find any public records matching the events as described.

As far as we know, there were no murders in or near the building when it was a jail, whether or not during a botched escape attempt. That's the sort of event that should be written up not only in police records but in all the local (perhaps even national) newspapers, so the lack of any corroborating reports suggests this was either a complete fabrication by somebody at some point in the literature or perhaps it was a real event at a *different* jail and was just misapplied to the one that became the Melting Pot.

We've tried to find botched jail or prison escapes whose details match the description of the one associated with the Melting Pot's location. It's difficult work because there have been numerous such cases all throughout the country and internationally. For example, we described in one of our own prior cases multiple incidents in which prisoners attacked and sometimes killed guards during attempted escapes from the Denver County Jail (see Volume 1, Chapter 10 for the full story), but in all of those cases, the prisoners were not killed in the attempt so it doesn't quite match the description. Perhaps one could argue that the relatively close proximity of the Littleton facility to the Denver County Jail simply allowed some of those stories to get mangled in the retelling over the years, and that is a distinct possibility, but we don't like it as an explanation. Too many loose ends and too many discrepancies.

Other events, like the infamous Battle of Alcatraz in 1946, match different elements of the story in that both guards and inmates were killed in the attempt. But that event was a much larger-scale one and no one could ever confuse the little jail in Littleton for Alcatraz Federal Penitentiary.

With that in mind, we're not saying with absolute perfect certainty that no such escape attempt occurred, but it seems highly unlikely. More likely, though we still don't have any documentary evidence, would be that some other violent event took place near the jail and that the story has just morphed over subsequent retellings into the version that exists today.

Drowning in a fountain, on the other hand, is an entirely different kind of story. We started from the same point of ignorance. We had no idea whether the story was true or not so we went searching for any records we could find. Similarly, we came up empty. A case like this is less likely to have been widely reported so it's more plausible that it could be a true story that's simply not identifiable in public records. Still, it's poor logic to say something is true when there's no direct evidence for it.

Unlike the jail escape story, though, we had pretty much no chance of connecting this tale to any particular true (and documented) event from any other location and we knew it from the start. Children sometimes drown in fountains, ponds, or pools. It's a tragic reality. Without other details, there's no way we can connect this drowning to any of the numerous such deaths we have found (albeit at different locations) in the public literature.

Figure 12.3. The kitchen where the coffee machine walked away (photo: Bryan Bonner).

Once we got to the location to start looking into things in person, one of the stories we were most interested to start with was the haunted coffee or espresso machine that had walked away. We'd been told there was even video of it and we wanted to see it in action. Our first request was to examine the machine itself. Maybe we could weigh it, see if it had wheels or was on an uneven surface, anything like that. Our first disappointment, therefore, was to learn that the machine was no longer at the restaurant. It had long since been replaced with another model stored in a different area of the kitchen.

But we were told there was video. That ought to make up for a lack of in person investigation to at least some degree. Apparently, though, the video had been lost. Strike two.

So we asked to speak to whichever of the current staff members had seen the video so they could tell us what they saw and answer our questions. And of course we were informed that none of the current staff had actually seen it for themselves, either. Like us, they'd only been told of its prior existence. Strike three. This is starting to sound more like urban legend than an actual video which ever truly existed. But if any readers happen to actually know what happened to the real video—on the off chance there ever was one—we'd love to hear from you.

Through the rest of our investigations, we found several things to be of particular interest in between our long moments of watching and waiting while nothing happened (typical of any paranormal investigation).

We did detect very strong variable electromagnetic fields around the bar and the front desk. Strong enough and intermittent enough to catch our attention, at least. However, we were able to trace all of those readings to the presence and operation of computers, cash registers, freezers, refrigerators, and other electric equipment common in a commercial kitchen or business.

Loud noises are common in this building. We traced them to the large vents and fans on the building's roof. We're not saying these account for all of the claims of people hearing footsteps. There's simply no way to be sure. But under the right conditions, some of them did sound footstep-like, especially to a mind predisposed to thinking of ghosts stomping through the building.

There is also a low frequency hum of moderate volume which we determined to be caused by the building's HVAC system. On previous investigations and in our previous volumes of this series, we have noted that low frequency vibrations or noises can sometimes cause eerie and uneasy feelings or even hallucinations. Sounds of about 18.9 Hz are sometimes called the "ghost frequency" because it is near the resonant frequency of fluid in the human eye and also seems to (in some individuals) affect certain brain structures. It's below the range of human hearing (which usually has a lower limit of about 20 Hz) so it can't be consciously perceived, but it can play all kinds of tricks on one's experiences.[17] While this doesn't prove that any or all of the ghost experiences are mere hallucinations or the result of infrasound causing uneasy feelings, it does provide a potential explanation for at least some of the stories. Had the sound or vibration actually been below the level of human perception and detectable only on our gizmos, we'd be more convinced than we are. As it stands, the actual sound at the Melting Pot wasn't infrasound but only nearly so.

The best part of that? The low-frequency sound is most noticeable in the vent in the women's restroom. The very same women's restroom that's the subject of so many of the ghost stories.

Though we can't be sure whether the sound frequencies played a role, we did manage to solve at least some of the mysteries of that restroom. While we were measuring and recording the other sounds for further analysis, we heard the mysterious breathing sound that had been reported in the center stall. It gave us a momentary fright until we realized it was actually just an automatic air freshener which had been set to go off at a predetermined interval after no motion had been detected in the area by its sensor.

And then, right on cue, one of the water faucets turned on. We did note that they were operated by motion sensors—the kind that turn the water on when you wave your hand under them. But at the time this one went off, no one was waving a hand in front of a sensor. No one was even near the sinks. In the restroom,

17   In Volume 1, Chapter 10, we describe a case at the Denver County Jail in which one of our investigators experienced the frightful feelings induced by such frequencies firsthand. Not everyone is affected and of those who are, they're not all affected in the same way. But infrasound is no joke.

yes. We'd crammed a bunch of adult men into this women's room to track down the ghosts (yet another reason it's important we do our investigations after hours because we'd hate to try to explain that one). But none of us were anywhere near the faucets when that one turned on.

Figure 12.4. The haunted women's restroom, during our investigation (photo: Bryan Bonner).

The first thing we tried, just to see if it would work, was walking back and forth in front of the sinks at various distances. That didn't do it. Sensors on those sinks are designed to detect hands in the sink, not people just walking in the larger restroom. But we got lucky while we were marching about like fools trying to make a faucet activate. One of our cameras happened to have its infrared mode turned on and its field of view happened to pan across the sink. The water started again.

Mystery solved!

As it turns out, the sensor on the faucets is, itself, an infrared sensor and that's how it tracks motion to "know" when to turn the water on. When our own infrared light crossed its sensor, it "thought" it had detected a hand in the sink and dutifully ran the water. As usual, that doesn't mean a ghost can't turn the water on sometimes, but we have at least found a plausible explanation for the claim.

And so we conclude another investigation. This was a fun one because we actually were able to experience a good number of the claimed phenomena and were able to explain a sizable fraction of them. Of course, we can't say for certain the Melting Pot is not haunted and so as always, we keep this case open in our ledgers. But between the lack of documentation for some of the supposed historic events and our ability to find natural explanations for at least some of the ghostly manifestations, we can at least say that not everything you've read about the ghost stories here is true.

References & Further Reading

Brick, C. (2018). *Haunted America: Ghost & Legends of Colorado's Front Range*. Charleston, SC: The History Press

Denver Public Library Special Collections and Archives (n.d.). The History of the Denver Public Library. *The Denver Public Library*. <https://history.denverlibrary.org/exhibit/history-denver-public-library>

Harris, S. C. (Executive Producer). (2011). Lizzie Borden Took an Axe (Season 3, Episode 6) [TV series episode]. In Phillips, M. & Ayalon, H. (Executive Producer), *My Ghost Story*. Mark Phillips Philms & Telephision; Biography Channel

Hulse, D. F., Christensen, K., & Larison, P. (2021). Carnegie Library: Local Landmark—1973. *Littleton, Colorado*. <https://www.littletonco.gov/Building-Development/Historic-Preservation/Historic-Littleton-Buildings/Carnegie-Library>

Littleton Police Department (n.d.). Littleton Police History: A Brief History of the Littleton Police Department. *Littleton, Colorado*. <https://www.police.littletonco.gov/About-LPD/Littleton-Police-History>

Melting Pot (n.d.). History of Melting Pot. *The Melting Pot*. <https://www.meltingpot.com/history.aspx>

Reedy, A. (2019). Are these Colorado restaurants haunted? That depends on who you ask. *The Denver Post*. <https://www.denverpost.com/2019/08/23/haunted-denver-restaurants/>

Xu, H. (2024). Eat and Drink Your Way Around Historic Littleton. *Westword*. <https://www.westword.com/restaurants/eat-and-drink-in-historic-downtown-littleton-20846115>

Yoe, S. (2023). Haunted Dining in Denver. *Denver Center for the Performing Arts*. <https://www.denvercenter.org/news-center/haunted-dining-in-denver/>

13

The Gates of Hell and the Phantom Camaro: Riverdale Road

If you follow Colorado paranormal lore even a little bit, there's a strong chance you've heard of Riverdale Road. Famously one of the most haunted stretches of road in the world—if not *the* most haunted—it's collected tons of paranormal claims, ghost stories, urban legends, and other folklore.

Of all our cases, and of all the popular haunts people talk about, this is easily one of the most-requested. Because it's an eleven-mile stretch of road, it was also one of the most challenging for us to come up with a good means to investigate.

Figure 13.1. A sign marking Riverdale Road (photo: Bryan Bonner).

We're going to present this chapter a little bit differently than the others in this first part of the book. As with Third Bridge in Chapter 9, it doesn't make a whole lot of sense to present the history before the paranormal claims. Given that Riverdale Road is eleven miles long, we wouldn't even know which history to present and we certainly can't describe every building along the road. Unlike in Chapter 9, though, we're not even going to separate our investigation from the paranormal claims themselves because there are simply so many paranormal claims, we'd be sending readers flipping back and forth just to follow the thread of discussion. Instead, we'll present the entire story in one large block, offering our notes on each of the paranormal claims as we present them, and discussing whatever history seems relevant along the way.

As we've indicated, Riverdale Road is thought to be one of the most haunted roads—certainly in Colorado, maybe in the United States, and possibly even in the world.[1] When the rental car company SIXT conducted a (fairly extensive but admittedly unscientific) study of which roads in the United States people believe to be most haunted based on social media mentions, Riverdale came in second in the entire country.[2]

It's certainly attracted its share of paranormal enthusiasts over the years, ranging from those who read about it from their armchairs to those brave enough to drive its length late at night and see what they might encounter. One anonymous commenter on a haunted house forum even suggested that it's a bad idea to take wedding photos there because people will mistake the bride for one of the ghosts![3]

Riverdale Road finds its northmost (or "beginning") point at an intersection with East 160th Avenue in Brighton, Colorado and then runs for approximately eleven miles, following a somewhat winding pattern but in a generally southward direction until it ends where it hits Yosemite Street in Thornton. Because it's so long, it's difficult to describe the entire road as having any particular character. Parts look just like any other suburban street. Others seem dark, twisty, and foreboding. Lore suggests the entire road is haunted, but likely its reputation mostly stems from those creepier parts.

Long before we started embarking upon anything resembling an official investigation of the street, we started collecting all of the stories people have told

1   For example:
Sandmeier, A. (2023). One of the Most Haunted Streets in America Is Hiding Here In Colorado. *Only In Your State*. <https://www.onlyinyourstate.com/experiences/colorado/most-haunted-road-cd> (accessed August 19, 2025).
Bitler, D. (2022). Is this Colorado road haunted? *KDVR*. <https://kdvr.com/news/local/riverdale-road-is-this-11-mile-stretch-haunted/> (accessed August 19, 2025).
2   McKee, S. (2025). Colorado's 'most haunted road' known for 'Gates of Hell' and twisted past. *Denver Gazette*. <https://denvergazette.com/outtherecolorado/adventures/colorados-most-haunted-road-known-for-gates-of-hell-and-twisted-past/article_9dd294ce-9bfb-4085-8c63-1d23f71fc4f1.html> (accessed August 19, 2025).
3   Colorado Haunted Houses (n.d.). Riverdale Road – Thornton, CO Real Haunted Place. *Colorado Haunted Houses*. <https://www.cohauntedhouses.com/real-haunt/riverdale-road.html> (accessed August 19, 2025).

about it. They are legion, and we'll treat each one in turn shortly. Not to give away too much of our thinking on the subject before we even talk about the particular claims, but we started to think immediately in terms of how urban legend and folklore develop.

Sometimes—often—there's a hint of truth behind even a largely false urban legend, but one of the things that's always interested us and has been of great importance as paranormal investigators is the study of how these tales shift, morph, relocate, develop, and get passed on over the years. By the time an urban legend achieves the status *of* urban legend, as a matter of fact, often its origins have been completely lost to all but the luckiest and most dedicated of folkloric researchers. Sometimes an event which took place in one location and in one context gradually gets added to the lore of another location. Miniscule changes in each telling, whether intentional embellishments on the part of the teller or just accidental mischaracterizations, accumulate.[4]

In the case of something like Riverdale Road, one of the things we noticed is that some of the specific paranormal claims have also shown up in completely unrelated lore in other locations. Either these types of manifestations are more common than one might expect or perhaps one or both of the stories originally came from somewhere else and was just made to fit.

A useful concept here is that of the meme. Many of us know it now as a word to describe those humorous combinations of text and images passed around on social media. Indeed those are memes, but we're using the word in its original context to refer to a unit of cultural transmission analogous to how genes are units of biological transmission. The term was originally coined by the biologist and author Richard Dawkins but has since been further developed by numerous other scholars. In brief, a meme is a sort of idea. It can be passed on from mind to mind, or it can die out. It can also develop "mutations" that might allow it to pass on more effectively. A memeplex is a group of memes that somehow fit together. Individual memes are more likely to be passed on in the context of a suitable memeplex, whereas memes that find themselves lumped in with an ill-suited memeplex tend to be dropped by the wayside.[5]

Our hypothesis as we began looking into all the claims surrounding Riverdale Road was that at least some—perhaps not all, but at least some—of the paranormal claims didn't originate there, but got added to the Riverdale Road memeplex simply because they seemed to fit well with the overall mythology growing up around the location.

The earliest online reference to Riverdale Road being haunted we've been able to find is in a 2007 article which only describes one of the many ghostly manifestations commonly associated with the site (phantom dogs), though it seems to allude to pre-existing lore and mentions that the road used to be the Cherokee Trail before it was settled, developed, and paved by Europeans.[6] The

---

4   Brunvand, J. H. (2001, 2002). *Encyclopedia of Urban Legends.* New York: W. W. Norton & Company.
5   Dawkins, R. (1976, 2016). *The Selfish Gene.* Oxford, UK: Oxford University Press.
Blackmore, S. (2000). *The Meme Machine.* Oxford, UK: Oxford University Press.
6   Yellow Scene (2007). 13 Things That Go Bump in the North Metro Night.

latter part of the claim, at least, is true. The Riverdale Road corridor was indeed located along the so-called Cherokee Trail, which was used by the Ute and later by Americans seeking passage to the west, until it eventually developed (in roughly its current form) into Wolpert Road (named for the Wolpert Ranch, about which much more later) in the 1870s and 1880s and eventually into Riverdale Road when it was paved in the early 1950s.[7]

Most of the chatter we've heard about the road comes from more recent sightings and discussions. Our theory about its development, then, is that smaller-scale whispers and rumors about the road being haunted had been passed around for quite a while, including perhaps in the pre-internet days, and perhaps stemming at least in part from a mid-1970s fire about which we'll have a lot more shortly. It wasn't until the mid 2000s—and in particular the mid 2010s—that paranormal afficionados seemed to start taking Riverdale Road not just as *a* potentially haunted location but as one of *the* must-see spots. With that growing reputation, likely, new legends were added to the lore, either out of whole cloth or migrated from other stories set in other locations.

Obviously this is speculative, but it seems a plausible explanation for how Riverdale became as infamous as it did, given that some of the stories don't really otherwise have all that much to do with each other.

But on that note, and in no particular order, here are the most popular reports of paranormal activity on or near Riverdale Road, along with some of our thoughts on each.

Phantom Jogger of Jogger's Hill: One of the more prominent stories is that there's a phantom jogger who can be seen on Riverdale Road near its intersection with 120th Avenue. Supposedly the ghost is that of a jogger who was killed in a hit and run accident (or some variations of the story say a murder). His spirit still lingers. In addition to the occasional visual manifestation, witnesses report hearing running footsteps, hearing a heartbeat or taps on the car windows, and/or feeling an impact while driving even though they didn't actually hit anything. The location has become known as "Jogger's Hill."

Regarding the name of "Jogger's Hill," that's not an official moniker. It's a name applied to the scene only in the paranormal lore, though if the story turns out to be true, it seems like a fair enough way to honor the ghost. The trouble is, we haven't been able to find any records of the hit and run accident (or murder) at that location. We do know of several fatal crashes on the road, which we'll discuss later, but none of them match the specifics of the story as it's been reported.

That portion of Riverdale Road is not particularly hilly, so we're not entirely sure where the name comes from, though we do notice that 120th itself inclines up a small hill to pass over Riverdale, so perhaps that's the hill people are talking about.

If one were to go jogging down Riverdale Road in that location, though, one would indeed be taking one's life into one's own hands (or perhaps more

Yellow Scene Magazine. <https://yellowscene.com/2007/10/01/13-things-that-go-bump-in-the-north-metro-night/> (accessed August 19, 2025).

7   Pace, L. W., Fischer, W. R., & Nichol, A. J., et al. (2005). Riverdale Road Corridor Plan. *Adams County, Colorado.* <https://adcogov.org/sites/default/files/Riverdale%20Road%20Corridor%20Plan.pdf> (accessed August 19, 2025).

accurately, into the hands of all the passing drivers). While it looks like a wonderful place to go for a jog along a not-too-busy road, and it might be easy to find oneself there if one had been jogging in one of the nearby open spaces, there are no sidewalks along Riverdale for much of its length, including at the intersection with 120th Avenue. Therefore that would indeed be a dangerous activity.

So we'd consider the jogger being struck by an automobile entirely plausible even though we have no evidence it ever happened.

In terms of folklore, this can be seen either as an example of a ghost simply haunting the location of his own demise, which requires no explanation, or as a potential variant of the "vanishing hitchhiker" legend in which motorists (or, prior to automobiles, coach drivers as early as early as 1602) encounter someone who seems to be asking for help, only for that person to vanish without a trace.[8] Similar folkloric roots might be applied to some of the other stories we'll encounter shortly.

The Lady in White: Possibly a hitchhiker (see above) or possibly the wife of a man who burned down his mansion (see below), this ghostly figure is sometimes seen wandering the road, particularly on foggy autumn nights. In some variations, she looks like a normal person wearing a white dress (perhaps a wedding dress). In other versions, she's clearly ghostly from the start and appears luminous against the dark night. When seen, she appears to be searching for something. When approached, she disappears.

As with the jogger, we can find no records of any lady matching the description of this ghost being killed (intentionally or accidentally) on Riverdale Road. Some do connect her to the house fire we mentioned, but we'll get to that story in a moment; for now, you only need to know that the description of this Lady in White does not match the true history of that fire. But perhaps she could be an entirely different Lady in White. Anything is possible.

However, Ladies in White are a well-known and well-documented tradition in global folklore (and, indeed, horror literature and cinema). Cultural differences abound and so each region seems to have a different take on how the White Ladies came to be ghosts and what they might portend. Medieval legend suggested that the appearance of a White Lady could be an omen of a death in the family. Many cultures have variations in which these are the spirits of murdered wives. Others might be ghosts still pining for unrequited love.[9] The famous La Llorona of Hispanic American lore is also often depicted as a Lady in White. The prevalence of these stories across cultures suggests something about this type of figure may be archetypal, which could explain why it became part of the Riverdale mythology (obviously, assuming we're approaching this as folklorists instead of believers in each particular ghost—we'll gladly adjust our reasoning if we ever find direct evidence that she actually haunts the road).[10]

8   Brunvand, J. H. (1981, 2003). *The Vanishing Hitchhiker: American Urban Legends and Their Meanings.* New York: W. W. Norton & Company.

9   For another local variation on this theme, see Volume 1, Chapter 2 of this series, which connected a fabricated tale of a local ghost to folklore very much akin to the Ladies in White.

10   Brunvand, J. H. (2001, 2002). *Encyclopedia of Urban Legends.* New York: W. W. Norton & Company.

That said, knowing there's a White Lady legend on Riverdale, we can understand why the bride to be we mentioned earlier got some strange looks while she was taking photographs along the road!

The Gates of Hell: Easily the most frightening of all the claims of Riverdale Road is that it is home to the very gates of Hell itself. Religious beliefs about such things vary, and we'll tread carefully around those. In the lore connected to Riverdale Road specifically, a minority view seems to be that the road is home to the one and only physical gateway to Hell on Earth. Most instead believe that Riverdale Gates of Hell are just one among several locations boasting a gateway to the underworld. Either way, it's not a comforting thought to those driving the road or living in the area.

Amalgamating the variety of stories we've collected on the subject, the general view is that a pair of rusty iron gates on Riverdale Road do indeed lead to Hell. Supposedly these gates were part of a burned-down mansion where the owner conducted Satanic rituals before lighting his own house on fire, killing everyone inside (including his own family, which connects in some variations of the story to the aforementioned Lady in White). Two variants of the story emerge. Some think the Gates of Hell were already there and demonic influence caused the man to burn down the house; others think the evil act of burning the house is, itself, what caused the Gates of Hell to open at the location.

We'll get to Satanic rituals when we discuss another related claim in a moment, but for now, we do know a few things about the specific Gates of Hell story.

First of all, there are no rusty iron gates. If such gates were ever there, they've long since been removed. So if there is a gateway to Hell along the road, it's more of a metaphorical gateway than a literal pair of gates. That's largely unimportant, though. What's more important is that the fire really did take place.

Above, we made brief mention that Riverdale Road was once called Wolpert Road after the Wolpert Ranch. Well, the house that burned down was none other than the Wolpert House. And this story provides a perfect example of how real history can blend with folklore and become something quite terrifying.

The house in question was located at 9190 Riverdale Road, an address at which no structure currently exists. Though land plots and addresses have shifted slightly, its approximate location is still identifiable if you plug it into a GPS device or online map service. It wasn't just any house. It was one of the earliest in the area, built in or around 1864 by a settler named David Wolpert, who came to the region from Ohio in 1859 chasing the gold rush, but who ultimately chose to establish a ranch at the Riverdale location rather than taking on a career in mining. Wolpert, along with wife Catherine and children Lucille, Mary, and David, lived in the charming two-story brick house until David (Senior) and Catherine died in 1909 and 1915, respectively.[11]

Limos, M. A. (2019). The History of the White Lady: An International Ghost. *Esquire*. <https://www.esquiremag.ph/long-reads/features/white-lady-ghost-history-a00293-20191113-lfrm> (accessed August 19, 2025).

11   Rudolph, K. (2019). The David Wolpert House: Dissecting A Riverdale Road Ghost Story. *The Denver Library*. <https://history.denverlibrary.org/news/denver/david-wolpert-house-dissecting-riverdale-road-ghost-story> (accessed August 19, 2025).

By the mid-1970s, the Wolpert House was starting to fall into disrepair and a land developer set its sights on the property, prompting former University of Colorado Museum director Dr. Hugo Rodeck to prepare an application to list the building on the National Register of Historic Places and thus guarantee its ongoing maintenance and preservation as part of Colorado's heritage.[12]

In May of 1975, a chicken house on the property burned down. Reports don't indicate any significant damage to the main house or to any of the rest of the property except for the chicken house itself, but the local fire department determined the Wolpert House itself was a fire hazard and ordered the current tenants to evacuate, which they did. Leaving the house unoccupied invited vandalism. On November 28, 1975, the main house itself burned down, leaving only crumbling remnants of the brick walls and one smaller outbuilding in the rear. It's not clear what started the fire or why it so completely destroyed a brick house (which should be at least somewhat flame-resistant), and no one knows whether the fire was caused accidentally during more minor vandalism, whether it was deliberately set by arson, or whether it was just "one of those things."[13]

Figure 13.2. This dead tree marks the location of the Wolpert House (photo: Robert Lewis).

When we went on one of our investigative outings to Riverdale Road (one of many over the years, each with a different purpose), we wanted to find the location of the original house and see if any remnants might still exist. The property is now part of the Pelican Ponds Open Space, a publicly-accessible park, so we were able to wander around for a bit without fear of trespassing. One thing that immediately stood out to us is that there's a large dead tree lying on its side, with a fence running through it, at the approximate location where we knew the original house ought to have been. We wondered if that tree might have been a part of the original Wolpert Ranch.

12   *Ibid.*
13   *Ibid.*

Searches through archival literature, both paranormal and historic, didn't provide a direct answer. However, visual comparison with archival photos showing a tree standing in front of the burnt house's ruins suggest the tree we saw is indeed the same one that stood at the front of the Wolpert property. If that is so, it's an easy landmark to be able to identify exactly where the house stood, several yards into the Open Space from Riverdale Road itself.

We wandered a bit further from the road, toward a small lake or pond on the far side of the property. Among the grass, we found a few bricks just lying on the ground. We seriously doubted they were original bricks from the Wolpert House—the property has been cleaned and repurposed far too thoroughly for that. But we were just curious enough to document their location for future reference, just in case.

This led us down the rabbit hole of studying the history of brick manufacture. People often ask us what skills or knowledge are required to become a paranormal investigator. Well, as this demonstrates: anything and everything. Never in our wildest dreams when we started looking into ghost stories did we think we'd need expertise on the making of bricks, but here we are. As it turns out, there are important visual clues in one of these bricks that help solve the mystery. One of them has rectangular holes through its center. Though bricks themselves are an ancient technology, we learned that bricks with holes in them, designed to maximize strength and ventilation while minimizing weight and materials cost, though invented as early as the mid nineteenth century, only entered common use in the middle of the twentieth century. That would make it unlikely but not impossible the bricks we found on the site were from the original house.

Between that and the quality of the cleanup project to turn the land into an Open Space, we don't think there's anything left of the house.

Figure 13.3. Bricks at the Wolpert House site (photo: Robert Lewis).

All that is to say, there is indeed a kernel of truth behind the Riverdale Road legend. A house really did burn down. Contrary to the myth, though, it was unoccupied and so no one died in the fire, nor is there any reason to suspect any connection to Satanic or other forms of dark ritual. Nor are the gates still present on the property, if there ever were gates in front of the house (no photos we've seen show them).

Satanic Rituals and Evil Beings: A lot of lore suggests the property has been used for Satanic rituals over the years. Most of it is connected to the afore-mentioned (and debunked) story that the house fire was part of such a ritual, but some have dropped that part of the story and just insist Satanic rituals are done *somewhere* along the road. Cultists are said to even have performed human sacrific-es there. In the most extreme variations of the story, demonic beings themselves have manifested on Riverdale Road as a result of these rituals.

It should probably go without saying, but there are no reports of Satanic human sacrifices taking place anywhere in the area. Indeed, stories of such dark rituals, while common in lore, are almost always apocryphal.

Stories of Satanic rituals, sacrifices, cults, abuse, and more are not uncom-mon even today, but the lore surrounding those is more prominently associated with the so-called "Satanic Panic," beginning in the 1970s and reaching a fever pitch in the 1980s, in which countless communities became convinced Satanists were in their midst doing horrible things to their children. Entire volumes can be and have been written on the subject, giving it a much more detailed treatment than we can here, but it's worth mentioning that these panics even caused numer-ous individuals—all perfectly innocent—to be convicted of crimes. Crimes not only that they didn't commit themselves, but crimes which were never committed at all and occurred only in rumor, false memory, and delusion.

It would be a step too far to claim that Satanic rituals of these types have never occurred. There are plenty of crazy people in the world. But claims of Sa-tanic sacrifices or ritual abuse are, based on volumes of well-documented historic and psychological studies, generally considered to be the result of nothing more than baseless moral panic, groupthink, and mass social delusion, often worsened or amplified by supposed "professionals" who implant or reinforce the delusion instead of treating it.[14]

Our point is not only to suggest that it's unlikely any such sacrifices, rituals, or abuses took place on Riverdale Road, but also to suggest that the addition of these kinds of stories to the list of claims about the location is entirely consistent with how urban legends develop and how particularly salient ideas or memeplex-es can overtake stories they shouldn't "naturally" or truthfully have anything to do with.

Underground Chicken Coop with Strange Writings: Less commonly reported but in our opinion one of the most interesting stories (at least from a storytelling perspective) is the claim that somewhere on Riverdale Road, usually near the Wolpert House, is a chicken coop buried underground. Within the chick-en coop are collections of arcane, occult writings prepared by the house's owner

14   Kratetz, L. D. (2017, 2018). *Strange Contagion: Inside the Surprising Science of Infectious Behaviors and Viral Emotions and What they Tell UIs About Ourselves.* New York: Harper Wave.

Bartholomew, R. E. (2001). *Little Green Men, Meowing Nuns and Head-Hunting Panics: A Study of Mass Psychogenic Illness and Social Delusion.* Jefferson, NC: McFarland.

Carroll, R. T. (2003). *The Skeptic's Dictionary: A Collection of Strange Beliefs, Amusing Deceptions, & Dangerous Delusions.* Hoboken, NJ: Wiley.

Loftus, E. & Ketcham, K. (1996). *The Myth of Repressed Memory: False Memories and Allegations of Sexual Abuse.* New York: St. Martin's Griffin.

during his life.

Once again, we find an addition to the same story of a Satanic owner burning down the house, so we needn't give this one much additional attention. The one thing we'll point out is that, once again, there is a kernel of truth to the story but that it has been exaggerated beyond all recognition in the retelling. The true part is that there once was a chicken coop on the property. As we discussed above, it burned down a few months before the house did. Everything else is pure folklore. No credible reports describing any kind of arcane or occult writings on the property have ever come to light. We also don't know of anyone ever finding an underground structure along Riverdale Road, chicken coop or otherwise.

The Witch Tree or the Hanging Tree: At least two variants of this story exist. In the first, somewhere on Riverdale Road was once a dairy where accused witches were hanged from a tree on the property. In the other, it wasn't witches but slaves or other oppressed people who were lynched and hanged from the tree. Either case leads to roughly the same result in the paranormal lore: people have reported seeing ghostly visions of the bodies hanging from the tree.

There's not a whole lot to give the hanging tree story historic credibility. There are no records of witch trials or witch hangings in Thornton, or even in the entire State of Colorado. Witch trials in the United States (including the infamous Salem case) were predominantly a colonial-era feature of New England and the surrounding areas and had faded out prior to westward expansion and the founding of Colorado.

Lynchings are a tougher nut to crack when it comes to historic research, as not every such event likely resulted in surviving records. Because Colorado was never a Slave State (having been admitted to the Union after the abolition of slavery), the notion of slave hangings is unlikely. Other forms of lynching are possible but no records corroborate the story.

Our opinion is that this is likely lore that developed to explain the large dead tree lying along the road near the Wolpert property. There's absolutely no evidence at all to suggest it was ever used for hangings, but it is a creepy looking dead tree that seems a bit out of place to the casual observer who doesn't know the history. We envision a (completely made up and speculative but entirely plausible) story in which some people were visiting Riverdale Road to investigate the ghost stories. Maybe they came upon the old tree and one of them turned to the other, trying to produce a good scare, and invented the story that It was once an old hanging tree. Perhaps the other person, believing the story to have actually been part of the lore, then repeated it to someone else, and so on. Again we emphasize we're not saying that's what *did* happen, just that it's *one* plausible explanation for the story. This is how folklore often develops.

Ghostly Children: The heading pretty much says it all in this case. One of the claims is that people have seen ghostly children on the road or sometimes that they've heard children laughing late at night when no children are present. There's less lore surrounding this one, and it doesn't seem to connect in any way to the Satanic claims or to the Wolpert House in most versions, though a few people have suggested that maybe the ghosts are the spirits of children sacrificed in the Satanic rituals or who died in the fire. But that's a minority of the reports. One addition that sometimes gets added to this story is that if you drive along the road, the children might leave small handprints on your car windows.

Ghosts of children are ubiquitous in stories of many haunted locations, so it's unsurprising there should be such stories here. In terms of validating or debunking the lore, there's not a whole lot we can say. This is more of an empirical question and the only real way to investigate is to visit Riverdale Road and see whether you do or don't see anything. So far, we have not seen or heard these children. We would note that the sounds of children laughing, if the stories are true, at least hypothetically could have been the sounds of living children (or teenagers) giggling as they snuck around the haunted property.

With regard to the handprints on the car, that's actually a common piece of folklore that gets passed around from location to location. Often, including sometimes (but not always) in connection to Riverdale Road, the children are meant to be the lingering spirits of the victims of a school bus crash. We haven't found any records of school bus crashes on Riverdale Road, but there have been some fatal ones in Colorado over the years, and it's possible that ghost stories from those crashed might have just been "moved" to Riverdale Road due to its haunted reputation.

The most famous case we're aware of that matches that part of the story is the case of the San Antonio Ghost Tracks. According to that legend, there's a place in San Antonio, Texas where a school bus broke down on some railroad tracks in the 1930s or 1940s and was struck by a passing locomotive, killing all the students. Now, the legend goes, if you park your car on those railroad tracks,[15] the spirits of those children will push your car to safety. This can be verified, supposedly, both because the car rolls uphill and because if you sprinkle powder on the car, you can see their handprints.[16]

Though not directly part of our Riverdale Road investigation, we do have some thoughts about the San Antonio haunting, too. First, with regard to the car rolling uphill: it actually doesn't. It rolls downhill. The location just happens to be one of those spots where the topography of the road is such that it produces an optical illusion of rolling uphill. Second, with regard to the handprints showing up in powder on the car, we'd be surprised if handprints *didn't* show up. People touch their cars, and no one ever thought to wipe the cars down before conducting the experiment. People engaged in that exercise were just dusting for their own fingerprints. Finally, we have found no records of a train hitting a school bus anywhere in or around San Antonio during the period in question. There was a crash matching that description, however, in Utah. On December 1, 1938, thirty children were killed when a train hit their stalled school bus. Though it occurred in Utah, San Antonio press at the time covered the story, which makes it the most likely origin of the tale.[17] Fortunately, for safety's sake, the "ghost tracks" location

15   For the love of God and all that is holy, do NOT park your car on railroad tracks!

16   River City Ghosts (2020, 2025). San Antonio Ghost Tracks. *River City Ghosts.* <https://rivercityghosts.com/ghost-tracks/> (accessed August 20, 2025).

17   Pettaway, T. (2020). The Ghost Tracks: 82 years ago Tuesday, one of San Antonio's most famous ghost stories was born. *My San Antonio.* <https://www.mysanantonio.com/news/local/slideshow/The-Ghost-Tracks-82-years-ago-today-one-of-San-213626.php> (accessed August 20, 2025).

has been modified so the unusual "uphill" rolling will no longer occur.[18]

Closer to home, there was a crash in Colorado which also may have influenced some of the national urban legends. On December 14, 1961, a train crashed into a school bus in Weld County, killing twenty children. A 2007 series about this crash was the last "long" feature of *The Rocky Mountain News* before it ceased operations in 2009.[19]

Other "ghostly handprint" stories exist all around the country. Most either have similar origins or have been borrowed from one another. While we can't speak authoritatively on the presence of ghost children *per se*, we're pretty sure the handprint story came either from San Antonio or one of the other similar tales.

The Ghost Boy and the Bloody Handprints: Sometimes these are reported as a single story of a single ghost and sometimes the two elements are separated and disconnected. The first element is that Riverdale Road is haunted by the spirit of a young boy, perhaps murdered or perhaps accidentally killed (struck by a car, maybe) on his way to school one morning. The second element is that if you look at the street signs or stop signs, you can sometimes see bloody handprints on them.

Ghostly children and ghosts leaving bloody markings behind are common among many ghost stories, so we can't connect this one to any particular piece of folklore. If it did just develop as part of the urban legend, likely it was simply an addition of a classical kind of ghost story to the roster of spirits on this road.

Historically speaking, we can find no records of a young boy dying on Riverdale Road under any circumstances remotely similar to those described. Deaths occur just about everywhere in the world, and not all of them get written up in the newspapers, so it's possible a young boy could have died there, but had it been either a murder or a fatal car accident, we would have expected to find at least some report somewhere.

Separate from the supposed history, the presence of a ghost boy or the bloody handprints would be an empirical question. On that, all we can say is that we haven't seen them for ourselves yet. Given the number of reports, we wouldn't be surprised if some prankster went out and put some stage blood (or perhaps animal blood from a butcher shop) on the signs. Therefore if we (or anyone) ever were to actually see them, the next step in the investigative process would be to have the bloody handprints professionally analyzed to determine if it really is blood and, if so, if it really is human. From there, the next investigative steps, if any, ought to suggest themselves based on the results.

Phantom Camaro: Perhaps the most commonly reported story of Riverdale Road is that sometimes if you're driving late at night, you might encounter a ghostly black Camaro driving along the road, often with only one headlight working. Sometimes it is said to entice cars into a deadly race. Other times it follows the witness. Other times still it's just seen driving the road, sometimes vanishing rather than driving out of sight.

18   Moreno, J. (2018). San Antonio's 'Ghost Tracks' will be history after next week. *KSAT.* <https://www.ksat.com/news/2018/10/20/san-antonios-ghost-tracks-will-be-history-after-next-week/> (accessed August 20, 2025).

19   Vaughan, K. (2007) The Crossing. *The Rocky Mountain News.* Archived online at <https://thecrossingstory.com/> (accessed August 27, 2025).

Historically, there's not much we can say about it except that we have not found any evidence of a car crash along Riverdale Road involving a Camaro. Doesn't mean necessarily there never was such a crash, but we've looked pretty hard and we've never found it.

This is another one straight out of popular folklore. Ghost cars are fairly common. Many of them are black. Often, for whatever reason, they're Camaros. Rarer, but still not unheard of, they even have one headlight out. Similar stories have been reported along a lot of different roads. From folklore to horror literature, the phantom car is a perennial favorite, even going back to the pre-automobile days when the stories were of phantom stagecoaches.[20] Phantom hearses and taxicabs are also common in folklore.

That leaves us as usual with the empirical question of whether such a phantom Camaro exists. Also as usual, all we can say is that we haven't yet seen it. Important investigative notes of caution should one ever think one sees it: first, it's entirely possible a perfectly ordinary person might drive a black Camaro, and this must be ruled out first; and second, if one sees a vehicle with only one headlight, one must similarly rule out the possibility that it's just a damaged car or that it might turn out to be a motorcycle. Always make sure you get a good look before jumping to conclusions. We suspect at least some of the reported sightings may have such explanations.

Demon Dogs and Phantom Animals: This is a pretty broad category. The common factor here is that people have reported seeing supernatural animals on Riverdale Road. Some are thought to be demon dogs (or hellhounds), which may be an offshoot of the Gates of Hell story. Others are thought to just be ghosts of dogs, coyotes, or other animals. Occasionally, these are connected to Native American lore (see below).

There's simply no way to conduct a historical analysis of this one, except to the extent that we already ruled out the Gates of Hell story. All that's left, then, are the empirical and folkloric elements. On the former point, we again haven't seen them. A large stretch of Riverdale Road does run along the Pelican Ponds Open Space which is home to some wildlife, so we've seen plenty of animals, but none that ever seemed supernatural or beyond ordinary explanation. At least that's our personal experience.

On the folkloric side of things, once again we have a case in which phantom animals are a common occurrence. Even one of our private cases later in this volume (Chapter 18) involved a claim of a ghostly pet cat. So that's nothing to be surprised about. Hellhound or demon dog stories are less common than "ordinary" ghosts, but still a regular feature in the paranormal lore and in supernatural fiction.

We also think it may be possible that some of these stories could be connected to the "Black Dog" of American trucker lore which traces its roots to British tales of black dogs as death omens or evil spirits but has come to also be associated with a warning to exhausted truckers to get off the road before they fall asleep at the wheel.[21] A potential (skeptical) explanation for the black dog

20   Brunvand, J. H. (2001, 2002). *Encyclopedia of Urban Legends*. New York: W. W. Norton & Company.

21   Isler, D. (2024). The Haunting of the Highway: The Legend of the Black

phenomenon in trucker lore is that sleep deprivation and dark nights can cause certain types of hallucinations. We don't know if that necessarily explains the phantom animals of Riverdale Road or all of the Black Dog encounters on the American Interstate system, but we're convinced it accounts at least for some of them after experiencing similar phenomena for ourselves when exposed to long periods of dark in the Cave of the Winds (see Volume 1, Chapter 6).

Native American Burial Ground: Many supernatural stories claim that the source of the haunting is that a building was built on Native American burial ground. Likely we have the film *Poltergeist* (about which, see Volume 1, Chapter 5) to thank for that trope's popularity, but the notion certainly predates the horror movie. The idea is that, perhaps because of poor treatment of Natives by European (and later American) settlers, modern buildings constructed over Native Burial ground becomes haunted or cursed. In the case of Riverdale Road, of course, the claim is that at least portions of the road were in fact built over such burial ground and that this explains all the supernatural phenomena.

It is known that Native Americans made extensive use of the region that is now Riverdale Road. No archaeological evidence, though, suggests that the immediate vicinity of Riverdale Road was ever a burial ground. Go far enough back into pre-history, though, and you'll probably find that pretty much the entire country was once burial ground for one people or another, so there may be a hint of truth to the story, but we can't find anything more concrete than that.

General Paranormal Activity: Beyond the specific claims we've analyzed above, there are quite a slew of what we'd call "generic" paranormal claims about Riverdale Road. Multiple witnesses report unusual phenomena that don't necessarily follow any of the established stories. For instance, some people say they've heard unusual sounds. Car radios sometimes malfunction there. Photographers have captured images of "orbs." Headlights flicker. Fog suddenly covers the road. Drivers feel like they're being watched.

Some of these have straightforward explanations. For instance, orbs are most likely either insects, dust, or other particulate matter in the air reflecting a camera's flash back to the lens (see Volume 1, Chapter 24 for a much more detailed explanation of how orbs form). Flickering headlights or radio malfunctions could be supernatural in origin or could simply be machines displaying the malfunctions that machines sometimes display. Fog covering the road, while rare, is not an unheard of meteorological event. So this entire category of paranormal claims falls under the heading of ones we'd have to see (or hear or feel) for ourselves in order to make anything resembling an informed pronouncement.

Car Crashes: One thing we have confirmed is that there have been quite a few car crashes, some of them unfortunately fatal, along Riverdale Road. People inclined to believe in the paranormal might attribute these to malevolent entities on the road. We can't deny that, but we would offer as an alternative hypothesis the idea that, much as we found with Third Bridge in Chapter 9, the road might simply be treacherous if a bunch of people are out there having a good time and looking for ghosts. Doubly so if they're drinking when they do it.

In just the last several years, we know of at least several traffic fatalities on

Dog. *Forward Thinking.* <https://www.ftsgps.com/resource-center/the-haunting-of-the-highway-the-legend-of-the-black-dog/> (accessed August 20, 2025).

the road:

*October 22, 2017: A fourteen-year-old girl was driving on Riverdale Road when she lost control of the car and crashed. She died later in the hospital. Three other passengers were injured but not killed. Police said speed and alcohol were factors in the crash, and that none of the passengers wore seatbelts.[22]

*January 31, 2021: One person killed.[23]

*August 23, 2022: One person killed and another seriously injured.[24]

*September 3, 2022: Three people killed (a twenty-eight year old man, a twenty-three year old woman, and an eighteen year old woman) and one injured (a twenty-five year old man) after their car went airborne and landed in a ditch.[25]

On-site investigations: Given what we've just described, you can understand why we wanted to be cautious about our on-site investigations. Nevertheless, we wanted to do everything we could to look into the claims. Most of our findings, we've already described. Which is to say, we've never seen any of the paranormal phenomena other people have reported. It's not for want of trying. In addition to days we've visited with a specific purpose in mind (for instance, trying to find the location of the Wolpert House as described above), we've spent a lot of time driving the road, both during the day and at night. Individually and as a group, all of our investigators have driven the road more times than we can count, often with dashcams running, and once with a spherical camera mounted to the hood of the car. Never did we see any of the phantoms or other paranormal happenings.

Resources preclude us from being able to set up 24/7 monitoring to maximize our chances of finding anything, so that's all we've been able to manage so far. Perhaps one day in the future we'll be able to put the entire road under constant surveillance for a period of time and determine more conclusively whether there really are such frightening things happening over there. Until then, this remains yet another open investigation in our books, albeit one we think we have a pretty good skeptical handle on by this point.

---

22   Jensen, K. (2017). 14-year-old drives car off Riverdale Road in fatal crash, CSP says alcohol involved. *9News*. <https://www.9news.com/article/news/local/14-year-old-drives-car-off-riverdale-road-in-fatal-crash-csp-says-alcohol-involved/73-485374243> (accessed August 20, 2025).

23   Bitler, D. (2022). Deadly crashes on Riverdale Road over last 5 years. *KDVR*. <https://kdvr.com/news/local/deadly-crashes-on-riverdale-road-over-last-5-years/> (accessed August 20, 2025).

24   CBS Colorado (2022). One killed, one seriously injured in overnight single-car crash in Thornton. *CBS News*. <https://www.cbsnews.com/colorado/news/one-killed-one-seriously-injured-in-overnight-single-car-crash-in-thornton/> (accessed August 20, 2025).

25   CBS Colorado (2022). 3 dead, 1 hurt after car goes airborne, lands in ditch in Adams County. *CBS News*. <https://www.cbsnews.com/colorado/news/adams-county-riverdale-road-rollover-deadly-crash/> (accessed August 20, 2025).

References & Further Reading

Bartholomew, R. E. (2001). *Little Green Men, Meowing Nuns and Head-Hunting Panics: A Study of Mass Psychogenic Illness and Social Delusion.* Jefferson, NC: McFarland

Bitler, D. (2022). Is this Colorado road haunted? *KDVR.* <https://kdvr.com/news/local/riverdale-road-is-this-11-mile-stretch-haunted/>

Bitler, D. (2022). Deadly crashes on Riverdale Road over last 5 years. *KDVR.* <https://kdvr.com/news/local/deadly-crashes-on-riverdale-road-over-last-5-years/>

Blackmore, S. (2000). *The Meme Machine.* Oxford, UK: Oxford University Press

Brick, C. (2018). *Haunted America: Ghost & Legends of Colorado's Front Range.* Charleston, SC: The History Press

Brunvand, J. H. (1981, 2003). *The Vanishing Hitchhiker: American Urban Legends and Their Meanings.* New York: W. W. Norton & Company

Brunvand, J. H. (2001, 2002). *Encyclopedia of Urban Legends.* New York: W. W. Norton & Company

Carroll, R. T. (2003). *The Skeptic's Dictionary: A Collection of Strange Beliefs, Amusing Deceptions, & Dangerous Delusions.* Hoboken, NJ: Wiley

CBS Colorado (2022). One killed, one seriously injured in overnight single-car crash in Thornton. *CBS News.* <https://www.cbsnews.com/colorado/news/one-killed-one-seriously-injured-in-overnight-single-car-crash-in-thornton/>

CBS Colorado (2022). 3 dead, 1 hurt after car goes airborne, lands in ditch in Adams County. *CBS News.* <https://www.cbsnews.com/colorado/news/adams-county-riverdale-road-rollover-deadly-crash/>

Colorado Haunted Houses (n.d.). Riverdale Road – Thornton, CO Real Haunted Place. *Colorado Haunted Houses.* <https://www.cohauntedhouses.com/real-haunt/riverdale-road.html>

Dawkins, R. (1976, 2016). *The Selfish Gene.* Oxford, UK: Oxford University Press

Isler, D. (2024). The Haunting of the Highway: The Legend of the Black Dog. *Forward Thinking.* <https://www.ftsgps.com/resource-center/the-haunting-of-the-highway-the-legend-of-the-black-dog/>

Jensen, K. (2017). 14-year-old drives car off Riverdale Road in fatal crash, CSP says alcohol involved. *9News.* <https://www.9news.com/article/news/local/14-year-old-drives-car-off-riverdale-road-in-fatal-crash-csp-says-alcohol-involved/73-485374243>

Kratetz, L. D. (2017, 2018). *Strange Contagion: Inside the Surprising Science of Infectious Behaviors and Viral Emotions and What they Tell UIs About Ourselves.* New York: Harper Wave

Limos, M. A. (2019). The History of the White Lady: An International Ghost. *Esquire.* <https://www.esquiremag.ph/long-reads/features/white-lady-ghost-history-a00293-20191113-lfrm>

Loftus, E. & Ketcham, K. (1996). *The Myth of Repressed Memory: False Memories and Allegations of Sexual Abuse.* New York: St. Martin's Griffin

McKee, S. (2025). Colorado's 'most haunted road' known for 'Gates of Hell' and twisted past. *Denver Gazette.* <https://denvergazette.com/outtherecolorado/adventures/colorados-most-haunted-road-known-for-gates-of-hell-and-twisted-past/article_9dd294ce-9bfb-4085-8c63-1d23f71fc4f1.html>

Moreno, J. (2018). San Antonio's 'Ghost Tracks' will be history after next week. *KSAT.* <https://www.ksat.com/news/2018/10/20/san-antonios-ghost-tracks-will-be-history-after-next-week/>

Pace, L. W., Fischer, W. R., & Nichol, A. J., et al. (2005). Riverdale Road Corridor Plan. *Adams County, Colorado.* <https://adcogov.org/sites/default/files/Riverdale%20Road%20Corridor%20Plan.pdf>

Pettaway, T. (2020). The Ghost Tracks: 82 years ago Tuesday, one of San Antonio's most famous ghost stories was born. *My San Antonio.* <https://www.mysanantonio.com/news/local/slideshow/The-Ghost-Tracks-82-years-ago-today-one-of-San-213626.php>

River City Ghosts (2020, 2025). San Antonio Ghost Tracks. *River City Ghosts.* <https://rivercityghosts.com/ghost-tracks/>

Rudolph, K. (2019). The David Wolpert House: Dissecting A Riverdale Road Ghost Story. *The Denver Library.* <https://history.denverlibrary.org/news/denver/david-wolpert-house-dissecting-riverdale-road-ghost-story>

Sandmeier, A. (2023). One of the Most Haunted Streets in America Is Hiding Here In Colorado. *Only In Your State.* <https://www.onlyinyourstate.com/experiences/colorado/most-haunted-road-cd>

Vaughan, K. (2007) The Crossing. *The Rocky Mountain News.* Archived online at <https://thecrossingstory.com/>

Yellow Scene (2007). 13 Things That Go Bump in the North Metro Night. *Yellow Scene Magazine.*   <https://yellowscene.com/2007/10/01/13-things-that-go-bump-in-the-north-metro-night/>

14
The Unknown Sound: Mattie's House of Mirrors

We close this first part of the book with one of our favorite investigations to tell people about: Mattie's House of Mirrors. Though it no longer exists as a stand-alone venue, it was, at the time of our investigation, a bar and restaurant located at 1946 Market Street in Denver. Before that, it was one of Denver's most popular brothels.

This story's got everything: brothels, duels, murders, ghosts, and even some rather unusual findings during our own investigation. Oh, and rumors of a half-naked sword fight, just to top it all off.

Figure 14.1. Mattie's House of Mirrors (photo: Bryan Bonner).

The History

The history of Mattie's House of Mirrors, at least in its earliest incarnations, is largely the history of Market Street. And the history of Market Street is the history of prostitution in Denver. It's no secret that Denver's history is full of the kinds of behaviors seldom spoken of in mixed company. Just cracking a book on the city's past would be enough to make any self-respecting Puritan blush. Greed, gambling, violence, and prostitution were far from uncommon occurrences. Don't get us wrong; there was a lot of more wholesome history, too, but we can't erase the more deviant aspects of our past, which were prominent enough for prospector William Hedges to suggest there was not "a place on this continent where a greater amount of evil to the square acre was so spontaneously and openly developed."[1]

At least when it came to prostitution, most of the activity was centered on a specific stretch of road. It was once called Holladay Street but was renamed Market Street, and we all know just what kind of "market" we're talking about. The name change came about following a request from the Holladay family to not have their name associated with a street on which such scandalous practices took place. Market Street was chosen officially because the stretch was home to a variety of shops and grocers in addition to the brothels, but most people—ourselves included—can't help but wonder if the city's planners had their tongue firmly planted in cheek when they chose the new name.[2]

Obviously there were plenty of brothels in the neighborhood, but arguably the most prominent (and certainly the focus of our attention here) was the House of Mirrors, named for the floor-to-ceiling mirrors (framed with carved nudes) once prominently featured in the building's ballroom.

Our story begins with Jennie Rogers (1843 – 1909), the self-proclaimed Queen of the Colorado Underground. She was born Leeah Weaver[3] in Allegheny, Pennsylvania to poor farmers and spent her early years helping sell her family's produce and warding off suitors until she finally gave in and married a physician named G. Freiss. Unfortunately, Dr. Freiss was often away on business for long periods of time and his wife, quite a social creature, was left alone. Not the best circumstances for a happy and lasting marriage. Thus, Jennie left her husband and took off with the captain of a steamship, on which she lived for a time before taking a service role for the Mayor of Pittsburgh. But the mayor quickly tired of her willingness to openly discuss her numerous affairs and dismissed her with enough cash to start her own business in St. Louis, Missouri, where she started her career as a madam.[4]

1   Quoted in: Secrest, C. (2002). *Hell's Belles.* Boulder, CO: University of Colorado Press.

2   Trembath, B. K. (2019). Market Street's troubles are nothing new. *The Denver Library.* <https://history.denverlibrary.org/news/market-streets-troubles-are-nothing-new> (accessed August 21, 2025).

3   For simplicity and clarity, we will use her chosen name of Jennie Rogers throughout.

4   Enss, C. (2018). Wild Women of the West: Jennie Rogers. *Cowgirl Magazine.* <https://www.cowgirlmagazine.com/wild-women-jennie-rogers/> (accessed August

From St. Louis, Jennie was lured to Denver by the mining rushes of the late 1800s and arrived in either 1879 or 1880 (we think the former), leaving behind her latest affair with the police chief of St. Louis to open a new brothel on Market Street (then still called Holladay, though we'll consistently refer to it as Market Street regardless of year to avoid confusion). She took up residence in several different buildings on "The Row" for a few years before she rented a building at the location in question from another madam named Minnie Clifford in 1888.[5]

Brothels on Market street at the time were something of an open secret. Everybody knew they were there, but the general sense seemed to have been that as long as they kept to their own district, it was best to leave them alone. Enough prominent politicians and business leaders frequented such houses of ill repute that any legal action against them could have resulted in mutually assured destruction. However, while they were barely tolerated, they were not quite legitimate businesses. Even real estate transactions in that industry were often characterized by sly manipulations and under the table transactions.

The same was true of the business arrangement between Minnie Clifford and Jennie Rogers. Minnie sold the building out from under Jennie to another woman named Mary Leary for $10,000 (equivalent to about $350,000 in 2025—not an exorbitant price for such prime real estate, but also ample evidence that business must have been good if the madams had that kind of money to throw around given all their transactions needed to be cash as most banking institutions would want to avoid entanglements with their line of work). Not to be outmaneuvered, Jennie bought the property back from Mary for $12,000, demolished it, and built her own brothel—the current House of Mirrors building—in the same location.[6] According to unconfirmed but widely-believed rumor, funding for the new construction came courtesy of the chief of police himself, who helped cover up a potential homicide for a local businessman and would-be politician and then blackmailed the same in order to fund Jennie's brothel.[7]

The building itself is something to behold. It came from a time when architecture, even for a brothel, was meant to be beautiful and elegant. This one certainly is. It was designed by architect William Quayle (1835 – 1906), who ironically also had a knack for building churches, including the Trinity Methodist Church and the First Congregational Church.[8] Saints and sinners alike deserve quality architecture, we suppose. The brothel in its new building resumed operations in 1889.

21, 2025).

5   Sneesby-Koch, A. (2018). A Brief Walk Along Denver's Notorious Market Street. *History Colorado*. <https://www.historycolorado.org/story/going-places/2018/11/26/brief-walk-along-denvers-notorious-market-street> (accessed August 21, 2025).

6   *Ibid.*

7   Collins, J. M. (2007). *Brothels, Bordellos & Bad Girls: Prostitution in Denver's Red-Light District.* Albuquerque, NM: University of New Mexico Press.

8   Sneesby-Koch, A. (2018). A Brief Walk Along Denver's Notorious Market Street. *History Colorado*. <https://www.historycolorado.org/story/going-places/2018/11/26/brief-walk-along-denvers-notorious-market-street> (accessed August 21, 2025).

The very same year, Jennie learned of the whereabouts of a long-lost love named John A. "Jack" Wood and began a correspondence which led by the end of the year to her second marriage. Jack died of unknown causes eight years later and is buried at Fairmount Cemetery. Throughout the following years, Jennie would travel extensively while still maintaining the House of Mirrors brothel. During her travels, she met and fell in love with Archibald T. Fitzgerald and they were married in 1904 but Jennie quickly realized he was a bigamist and separated. Somehow, he always managed to talk her out of divorce. Historians think his motivations were mostly financial, and his spending brought Jennie near bankruptcy. Stress from financial troubles, legal issues, and marital strife finally weakened her health to the extent that she died in 1909. The local madams and "working girls" attended her funeral. Archibald Fitzgerald did not. She's buried in Fairmount Cemetery next to her second husband, Jack, under a headstone reading "Leah J. Wood."[9]

Following Jennie's death, the brothel came into the hands of Mattie Silks (1845/6 – 1929). She was born Martha A. Nimon in Fayette County, Pennsylvania, also to poor farmers, and spent at least some of her early years working on wagon trains as a freighter, during which time—though it's not clear exactly when—she also began working as a prostitute.[10] What is known is that she already owned her own brothel in Springfield, Illinois by the time she was nineteen years old.[11]

It's not clear how she acquired the name "Mattie Silks." Some historians think she loved silk material and took the name "Madam Silks" for that reason, with "Mattie" as a diminutive form of Martha. Others think the Silks name came from a man she knew in Kansas.[12] Others still say she took the name from a man named either George or Casey Silks, who she met when she came to Georgetown, Colorado in 1875 and either married or maintained a long relationship with.[13] Either way, it stuck and that's the name she's known by far more than her real one.

As you might guess from the above, there's a great deal of confusion about Mattie's life, particularly her relationships. Another important relationship she had was with Cortez "Cort" Thompson, with whom she'd had a long relationship prior to their marriage. But Cortez was already married, so it wasn't until his wife passed in 1884 that he was free to marry Mattie. According to at least some of the stories, they were married within four days of his first wife's passing, but it's certain at least that it was within a year of his receiving the news.[14]

9   Enss, C. (2018). Wild Women of the West: Jennie Rogers. *Cowgirl Magazine.* <https://www.cowgirlmagazine.com/wild-women-jennie-rogers/> (accessed August 21, 2025).
10   Collins, J. M. (2019). *Good Time Girls of Colorado: A Red-Light History of the Centennial State.* Guilford, CT: Twodot.
11   Enss, C. (2015). Wild Women of the West: Mattie Silks. *Cowgirl Magazine.* <https://www.cowgirlmagazine.com/wild-women-wednesday-mattie-silks/> (accessed August 21, 2025).
12   *Ibid.*
13   Collins, J. M. (2019). *Good Time Girls of Colorado: A Red-Light History of the Centennial State.* Guilford, CT: Twodot.
14   Enss, C. (2015). Wild Women of the West: Mattie Silks. *Cowgirl Magazine.*

One of the most colorful chapters in Mattie's life involved a romantic rivalry over Cort and an infamous duel. Almost everyone tells the story differently, but the one thing they all agree on is that there was a duel of some sort, that it involved Mattie and a woman named Kate Fulton,[15] and that it was over Cort in some way. Some think Kate and Cort had a prior relationship before Mattie entered the picture. Others think Cort and Kate maintained a long-running affair that Mattie tolerated until she simply couldn't take it anymore.

One version of the story is that when Mattie and Cort announced their engagement, Kate burst in and began an argument which began to turn violent. Eventually the two women agreed to settle their differences with a duel. They stood back to back in the middle of the street as crowds gathered to watch, marched the requisite number of paces, turned, and fired. But the only person shot was Cort himself, who'd been non-fatally struck by one of their bullets. Kate was arrested and Mattie and Cort were married after the latter recovered from his injury.[16]

In another version of the story, it wasn't a duel with pistols at all. Instead, the two women engaged in a topless sword fight. This is by far our favorite version of the tale and we genuinely wish it were the case, but it's almost certainly untrue.

The more likely and best-documented version (though none of these claims are conclusive) says there wasn't a formal duel at all. Rather, the two women simply had a fistfight after an argument, with Kate apparently taking the worst of the beating. Later, in a separate incident, a carriage pulled up alongside Cort's buggy and someone shot him—again, non-fatally—in the neck. Sometime later, Kate left town, returned again, and the two women had another fistfight. It's not clear whether the shooting was in any way connected to the women's fight at all.[17]

After Jennie Rogers' death in 1909, Mattie purchased the House of Mirrors brothel in either 1909 or (more likely) 1910 and continued to operate it as the premier brothel on Market Street. Both under Jennie Rogers' and Mattie Silks' ownership, it developed quite the reputation. The first floor was a restaurant known for its quality food as well as a "viewing area" where gentlemen could select which of the working girls they might want to spend the night with.

Upstairs were the bedrooms, and they had quite an interesting design. Gentlemen had two kinds of rooms they could choose. Spending a night in one of the smaller rooms, measuring about nine by seven feet, cost less than the larger rooms which had a wood stove to supply heat. Instead of normal locks, the rooms had a unique design. When the girl and her gentleman for the evening entered the room, a bed would drop down against the door, thus "locking" it

---

<https://www.cowgirlmagazine.com/wild-women-wednesday-mattie-silks/> (accessed August 21, 2025).

15   Well, almost everyone agrees on that. We've read a few reports that indicate the duel was actually with none other than Jennie Rogers, but this claim is easily debunked if you look at the timelines.

16   Enss, C. (2015). Wild Women of the West: Mattie Silks. *Cowgirl Magazine*. <https://www.cowgirlmagazine.com/wild-women-wednesday-mattie-silks/> (accessed August 21, 2025).

17   Collins, J. M. (2019). *Good Time Girls of Colorado: A Red-Light History of the Centennial State*. Guilford, CT: Twodot.

in place. This also allowed the building to fit more rooms into a smaller space. Quite ingenious, if you ask us. Employees and customers alike seemed to enjoy how everything worked. Mattie even offered her working girls generous benefits including higher than average pay and medical care.

Like Jennie Rogers before her, Mattie Silks kept things running smoothly most of the time while somehow managing to simultaneously navigate a complicated personal life. Cort had a child from his previous marriage. When he received word that his daughter had died and left behind a young daughter of her own named Rita, Cort wanted nothing to do with the child and refused to take her in. Mattie wouldn't let that stand. Though she couldn't force Cort to house the child in their own home, she took it upon herself to adopt Rita and have her placed in a respected (and expensive) boarding house. When Cort died four years later in 1900, she arranged for the most extravagant funeral she could afford—and she could afford a lot.[18]

Even at the age of seventy-seven years, she was still the "Queen" of the district, but age and exhaustion meant she needed a bit of extra help. She hired "Handsome" Jack Ready as her brothel's bouncer and financial advisor shortly after Cort's death. With his help, she was able to keep Mattie's House of Mirrors running until shifting moral sentiments in Denver forced the closure of most or all of the brothels—including Mattie's—in 1915.[19]

Figure 14.2. Mattie Silks' grave (photo: Bryan Bonner).

Though Jack Ready was initially just an employee, the two quickly fell in love and were married in 1923. They lived a quiet retirement together until Mattie died from complications following a fall in 1929. Once boasting a fortune of over half a million dollars (and that's not inflation-adjusted), she left an estate of $4,000 in real estate and $2,500 in jewels, divided equally between Jack Ready and Rita. She was buried next to Cort Thomson's unmarked grave in Fairmount Cemetery under a headstone reading "Martha A. Ready." Her funeral was poorly attended

18   Enss, C. (2015). Wild Women of the West: Mattie Silks. *Cowgirl Magazine.* <https://www.cowgirlmagazine.com/wild-women-wednesday-mattie-silks/> (accessed August 21, 2025).

19   Parkhill, F. (1951). The Scarlet Lady: Mattie Sils made over half a million dollars from the world's oldest profession, but a good-for-nothing broke her heart. *The Denver Post.* <http://blogs.denverpost.com/library/files/2012/11/MattieSilks.pdf> (accessed August 21, 2025).

and without benefit of clergy. Ready died penniless about two years later. His funeral expenses were paid through collections in local bars, and he's also buried in Fairmount Cemetery, but not near Mattie's own grave.[20]

After Mattie's House of Mirrors ceased to be a brothel, it went through several other incarnations we'll talk about in a moment. Before we get there, we do have one unfortunate incident to report: the only confirmed death in the building, which occurred in 1894.

Ella Wellington was either owner (leasing from Jennie Rogers), co-owner (with Jennie Rogers), or manager (under Jennie Rogers) of the House of Mirrors at the time. Prior to becoming a madam, she'd been married to Fred Brouse, but took up her life in the world's oldest profession after growing bored of married life and running off with another man. Her new life seemed to treat her well, at least financially. On the fateful day of July 27, 1894 she was at the House of Mirrors wearing an elegant dress and a necklace valued at $2,000 (equivalent to about $75,000 in 2025 money) among other expensive pieces of jewelry. But when she ran into some visitors who brought news that her former husband had remarried and was doing well, she's reported to have stormed upstairs shouting "Oh, I'm so happy! So happy that I'll just blow my God damned brains out!" When she reached her bedroom where Arapahoe County Clerk William R. Prinn was lying (some reports say sleeping and others say waiting for his "date" for the evening), she's reported to have drawn her firearm (some say a .32 revolver) and shot herself in the head.[21]

Figure 14.3. Interior of Mattie's House of Mirrors; the suicide room (photo: Bryan Bonner).

20   *Ibid.*
21   Ordway, K. (2021). Shady Dames of Denver. *The Denver Westerners Roundup.* <https://denverposse.org/wp-content/uploads/2021/05/Roundup4-2021.pdf> (accessed August 21, 2025).
Collins, J. M. (2007). *Brothels, Bordellos & Bad Girls: Prostitution in Denver's Red-Light District.* Albuquerque, NM: University of New Mexico Press

Some rumors have persisted that her death may not have been a suicide. We've heard people claim that she was found not by Prinn but by another of the working girls, lying dead in her bed of a gunshot wound to the head. This cannot be the case, because if she was lying alone in her bed with no one else in the room, it would have been impossible for the other girl to have opened the door thanks to the unique locking mechanism. Others have speculated that perhaps her death was a murder and that it was covered up because the gentleman she was with was a local leader. This is sometimes told in relation to the ghost stories we'll come to in a moment and is admittedly more plausible but also inconsistent with available documentary evidence. The real story, sad as it is, seems to be indeed just what it looks like on the surface: a troubled woman took her own life in a fit of depression.

Her story ends on a further sad note. Apparently her funeral was quite well-attended. One of the attendees was a man named Fredrick Sturges, one of her regular customers or admirers who was so devastated by her death that he spent several nights clinging to her headstone, where he eventually committed suicide by overdose. In his pocket was a photograph of Ella bearing the inscription, "Bury this picture of my own dear Ella beside me."[22]

Though not strictly related to Mattie's House of Mirrors, as none of the murders took place at the property, there was also an active serial killer on Market Street at about this time. The so-called "Denver Strangler," who was never identified or apprehended, murdered three prostitutes (Lena Tapper, Marie Contassolt, and Kiku Oyama) in 1894, and is rumored (but not confirmed) to have also murdered professional psychic medium Jula Voght in 1898 and prostitute Mabel Brown in 1903.

One other suicide, near the restrooms on the second floor, is the subject of rumor but has never been confirmed. According to the story, the deceased had been a boyfriend of one of the working girls and hanged himself, perhaps distraught that his lover was engaged in sex with other men. It's probably apocryphal. Ella Wellington's is the only confirmed death in the building we're aware of.

Returning to Mattie's House of Mirrors, the building went through several owners after the brothel closed. After just a couple years without an occupant, it was purchased by recent Japanese immigrants and turned into a Buddhist temple in 1919. The new owners didn't know about the building's history as a brothel and weren't pleased when they found out, but thanks to the Great Depression and then Japanese internment during World War II, it wasn't until 1947 that they finally managed to move their temple to its current location on Lawrence Street.[23] After that, it was used as a warehouse. There are some conflicting reports from that point. We've been told that it has also been used for a variety of purposes including a barber shop and a bicycle repair shop after it was a warehouse. However, the only formal reports we can find indicate it was a warehouse from 1947 until it became a restaurant in 1998. Perhaps the other businesses leased part or all of the space *from* the warehouse. We can't be sure.

Regardless, we do know that it was purchased in 1998 by restauranteurs

---

22   *Ibid.*
23   Tri-State/Denver Buddhist Temples (n.d.) Our History. *Tri-State/Denver Buddhist Temples.* <https://tsdbt.org/our-history/> (accessed August 21, 2025).

Javier Juarez, George Mannion, and Chris Myers, who turned it into a restaurant called Mattie's House of Mirrors and maintained its interior décor in such a way as to honor its history.[24] It was during this period that our investigation took place.

Since then, much of the interior has been unfortunately lost and all that's really left of the original building is the façade. Between 1998 and 2018, it was merged with the adjacent LoDo's Bar and Grill. Since 2021, the entire property (LoDo's and Mattie's combined) has been Dierks Bentley's Whiskey Row, another bar and restaurant.

Paranormal Claims

Denver is home to quite a few historic brothels. Many are still standing and used for different purposes. Of those, quite a few have a haunted reputation, so it really ought to be no surprise that Mattie's House of Mirrors is also reputed to be home to at least a few distinct ghosts.

Sometimes the tales connect to the death of Ella Wellington in one of the upstairs rooms. Occasionally people talk about the ghost of the boyfriend who may or may not have committed suicide in the building. Most of the stories, though, are of ghostly experiences that can't be pinned down to any particular person.

Several staff members working there at the time of our investigation reported hearing people talking and walking about on the top floor of the restaurant after closing time. Of course, whenever they went up to check, the found the place empty.

Long gone now but still present during our investigation was a piano, also on the second floor (where the bulk of the paranormal claims seem to have taken place), which multiple people said they've heard playing itself when no one was near it.

Guests have also had experiences. One of our favorites occurred during a dinner service. A young girl excused herself from her family's table to use the washroom. She returned a little sooner than she should have and told her parents she couldn't go to the bathroom because of all the smoke. Waitstaff who overheard the conversation got understandably concerned that there might be a fire in the restaurant and went to investigate. There was no fire. Indeed, there wasn't any smoke. But there was an antique matchbook resting on the lid of one of the toilets.

Several guests and staff alike have reported seeing unusual faces reflected in the building's numerous mirrors.

Footsteps have been reported throughout the building, coming even from locations known to be empty of any people.

Sometimes the elevator would go from floor to floor without being called. Staff told us this was a regular occurrence and that repairmen had been unable to find any cause.

A former manager told us about one evening when he was alone in the

24   Wagner, K. (1998). Call Me Madam. *Westword*. <https://www.westword.com/restaurants/call-me-madam-5059394> (accessed August 21, 2025).

building taking care of the day's last paperwork late at night. He noticed the lights were dimming and brightening as he worked. Once he started paying attention to them, he further discovered not only were they changing brightness, but they seemed to be doing so in time with his own breathing. To experiment, he held his breath for several seconds. The lights maintained their brightness. But as soon as he started breathing again, the lights resumed their strange behavior. We've heard a lot of ghost stories in our day, but that one was new even to us!

Our Investigation

We were fortunate enough to visit Mattie's several times for investigative purposes, generally focusing on the second floor because that's where the only known death in the building occurred (we were told which room it was supposed to be and we have every reason to believe that information was correct—the room's location had been converted into the bar on the second floor) and because that's where the lion's share of the paranormal experiences have taken place.

After initial introductions, tours, and baseline readings were done on the first investigation, we set up monitoring and surveillance equipment according to our usual methodology in the following locations.

Cameras (capable of recording in color when there is sufficient light but which switch to infrared mode automatically in low-light conditions) were placed in the first floor dining room, on the main staircase, in the main second floor hall-way facing the restrooms, in the men's room on the second floor, and throughout the upper-floor dining area, affording a view of the entire room from multiple angles.

Microphones were similarly placed throughout the entire building: at the back of the first floor dining room near the kitchen, at the second floor bar, at the top of the back staircase, in the middle of the second floor dining room, and at the front of the same dining room.

Remote thermometers were placed near our base of operations at the bottom of the main staircase, at the top of the stairs, in the southwest corner of the second-floor dining room.

Perhaps we sound like a broken record for saying this in just about every chapter, but in between our periods of silent monitoring, we also took EMF readings at intervals ranging from forty-five minutes to one hour at three points in the kitchen, in the hall by the first floor restroom, in front of the elevator on the first floor, at the base of the rear stairs, in the first floor dining room, at two points in the second floor dining room, at the bar on the second floor, at the top of the rear stairs, and in the men's room on the second floor.

To avoid contaminating our data, we asked our hosts to disable the heater in the building. This was another one of those cold and snowy nights typical of our investigations, so it was a decision we didn't take lightly, but the thing was just too noisy. So we bundled up and proceeded without it. Temperatures therefore did drop considerably throughout the evening, but we didn't measure any thermal anomalies inconsistent with the environmental conditions.

Early in the evening, it looked like we were in for another long and boring night of watching and listening to nothing. Electromagnetic fields were steady and consistent with the kinds of appliances and utilities in the building. Nothing was happening on the cameras. It was just another one of those days.

Then, at about 2:00 in the morning, the service elevator started to move to a different floor. No one was on it. No one was even near it. For a while, we thought maybe one of our team members was playing a joke on the others. It's the sort of thing we might do to each other when bored, but we also take data integrity seriously, so we'd always fess up after a few minutes. Everyone denied any knowledge of it. Then, at 2:14, it moved again, to a different floor. And again at 2:25. By this time, we had gathered everyone into the same place, so we knew with absolute certainty none of our people nor our staff hosts were pressing buttons. It went to all three floors in a random order without any rider or operator.

Staff confirmed for us that repairmen had previously found no fault with the elevator machinery, but pointed out that sometimes it does that. It's been attributed to ghostly activity in some of the stories, so we were excited to have seen one of the claimed manifestations. Though not elevator mechanics, we did our best to check it out for ourselves and similarly could find no fault. We didn't detect any EMF anomalies around it, and though we couldn't disassemble it to look at all the parts, we didn't see any evident cause.

Can we rule out a mechanical cause? No. Even a machine that's functionally sound can sometimes develop certain "quirks," and that's a plausible explanation for this one. Neither, though, can we rule out the supernatural explanation. It's one of those mysteries we're just glad we had a chance to see.

Mattie's wasn't done with us yet. About half an hour later, at roughly 3:00 in the morning, one of our investigators claimed to have heard footsteps from the bar area on the second floor. We knew that to be the location of Ella Wellington's death. It didn't look like a bedroom anymore, but the site was marked by a framed newspaper article on the wall describing her suicide. Things were starting to get more interesting.

Before anyone could go upstairs to check it out, we heard one of the most unusual sounds we've ever recorded. And we did, in fact, record it. We wish print could convey audio, and we'd play it for you now; in lieu of that, we'll do our best to describe it and invite you to check out our website or ask us about it at one of our public lectures so you can hear it for yourself. It sounds like two adult male voices carrying on a conversation. None of the words are intelligible (one of us insists one of the words sounds like "spoon," but wouldn't swear to it, especially since it makes no sense without the rest of the context). At one point there's even the sound of one of the speakers sniffling as if fighting off a cold or allergies.

Here's the thing. There was nobody in that room. We know that because the restaurant was closed and our entire team and our staff hosts were downstairs. We also know there were no radios or televisions on. Nor could it have been our own conversation carrying up through the floor and just being picked up by the second-floor microphone because we were all in "silent monitoring" mode at the time. Even if we had been speaking, we only picked up the sound on our upstairs microphone, not the one near our base of operations.

After a few minutes of deliberation, one of us thought maybe we'd picked up a conversation from outside on the street below. There are windows overlooking an alleyway in that room. Perhaps that's where the sound came from. This was one of those occasions on which we ended up being thrilled it was a snowy night. Because there was snow on the ground, we could run outside and check for tracks. Nothing. Not a single footprint or bicycle track or anything of the sort. No one had been below that window for at least a few hours before the recording.

So we honestly have no idea who or what we recorded. In the years since, we've tinkered with that audio using professional audio processing software on our own and we've also sent it to just about every linguist and audio engineer we can think of. Between the lot of us working on it, that audio's been sped up, slowed down, played backwards and forwards at all different speeds, and we can't make any sense of it. All we or any of the experts can come up with is that it sounds like two distinct male voices speaking an unknown or indecipherable language. One of the linguists we asked said he thought it sounded like Cherokee language, but we personally disagree. The cadence, to us, sounds more like English even though we can't understand what's being said.[25]

That audio track is one of the reasons this is one of our favorite stories to tell at our public lectures. We always like to play it two or three times for our audience and watch as everyone (ourselves included) strains to try to make out any of the words.

Throughout the same evening, we had some uncharacteristic equipment failures. We've mentioned before that we often suspect equipment failures reported by ghost hunters as a supernatural manifestation of really being nothing more than someone forgetting to change the batteries or something, but when it happens to ourselves we do pay a bit more attention because it's so rare and we're so careful with our methodology. In this case, several different pieces of equipment had problems for no apparent reason. The audio pre-amplifier burned out. Then a video camera placed in the corner of the "suicide room" repeatedly turned itself off (and this was by means of the physical on/off switch, not just a battery dying or an electronic failure). We don't know the cause of either of these failures.

That was our first investigation, and it was intriguing enough we went back again. Our second investigation followed largely the same protocol so we won't repeat ourselves. But we did have one more unusual experience this time.

The manager at the time who was hosting us asked if she could bring a Ouija board to use during the investigation. Normally, this is not part of our procedure for reasons we've explained elsewhere, but we didn't have any objection. We just asked her to be cognizant of what we were doing and to use it only during the times when people were moving around or taking a break rather than during silent time. Maybe she'd get a result we could look into (though all our caveats about cognitive biases that occur when using Ouija boards would still apply). If nothing else, it would give her something interesting to do while we were doing fundamentally boring work.

25   If you're interested in what it means for something to "sound like" a particular language without actually carrying any linguistic meaning, look up the 1972 song "Prisencolinensinainciusol" by Italian singer and entertainer Adriano Celentano. The entire song (including the title) is complete gibberish, but it was written so that it would sound like English—and it does, even to native English speakers. Not only that, it sounds like English spoken with a specifically American accent.

At one point, she was using the Ouija board on the first floor with one other person. We had them on camera just in case anything interesting happened, and it's a good thing we did. Just as she started asking about the rumored suicide—that is, the unconfirmed suicide of the boyfriend, not Ella's—the lights began to flicker.

Figure 14.4. The mysterious flashing light, top (photo: Bryan Bonner).

Weirder still, at the same moment the lights were flickering above the Ouija board downstairs, a different light upstairs, near the alleged suicide location, also started to flicker. Now that's an unusual result!

Never ones to leave well enough alone, we set to work trying to figure out what might have caused the lights to flicker. Assuming the timing (coincident with the Ouija board) could have just been a coincidence, we went looking for a loose bulb or switch or some bad wiring. Though we can't rule that out with 100% certainty, we didn't find anything that would suffice to explain the lights' behavior. Instead, we found quite the opposite. The two lights that were flickering were on completely different circuits. There wasn't a switch in the entire place that could operate both of them, nor a wire directly connecting them. That is, two completely isolated lights both started flickering at the same time, independently of one another, at the very moment someone started using the Ouija board. You know us pretty well by now if you've read this far, so you'll be unsurprised that we're still not prepared to call that sufficient evidence to confirm a paranormal claim. But it's one of the stranger things we've seen. Coincidences do happen, but we don't like relying on them as an explanation.

Recall also that flickering lights was one of the pre-existing ghost stories. In our case, they didn't flicker in time with anyone's breathing like they (allegedly) did for the former manager. Still, whether the cause turns out to be natural or supernatural, we do have to wonder if our experience and the former manager's might not have a common cause.

We did return for a third investigation sometime later, but by this time the building had been completely remodeled as part of the process of merging with the adjacent LoDo's bar. At this point, the entire interior had been changed from a restaurant to a rental banquet hall. The flooring had been replaced, ceiling fixtures had been rearranged, the piano had been removed, the second floor lounge had become an empty space for lecturers to give presentations, the upstairs kitchen was converted into storage, and the tables, booths, and chairs had all been removed from the first floor. In other words, pretty much nothing except the building's interior remained.

We did stick around for a while and try to conduct an investigation. Though conditions had changed, that didn't mean we shouldn't be able to continue our work. However, being so connected to the neighboring bar made it noisy enough that our work was extremely limited (audio recording became essentially useless). Nothing else unusual happened while we were there.

And so there we have it. Yet another great case full of unsolved mysteries that we get to keep in our open cases file. Likely this one will be a permanent feature of the open cases file because subsequent renovations have made it all but impossible to recreate the circumstances in which our anomalous experiences of the past occurred. That doesn't mean we wouldn't be happy to go back for further investigation, but we have relatively little hope we'd be able to conclusively address the weird experiences we had in its prior incarnation.

References & Further Reading

Collins, J. M. (2007). *Brothels, Bordellos & Bad Girls: Prostitution in Denver's Red-Light District.* Albuquerque, NM: University of New Mexico Press

Collins, J. M. (2019). *Good Time Girls of Colorado: A Red-Light History of the Centennial State.* Guilford, CT: Twodot

Enss, C. (2015). Wild Women of the West: Mattie Silks. *Cowgirl Magazine.* <https://www.cowgirlmagazine.com/wild-women-wednesday-mattie-silks/>

Enss, C. (2018). Wild Women of the West: Jennie Rogers. *Cowgirl Magazine.* <https://www.cowgirlmagazine.com/wild-women-jennie-rogers/>

Lamb, K. (2016). *Ghosthunting Colorado: America's Haunted Road Trip.* Covington, KY: Clerisy Press

Ordway, K. (2021). Shady Dames of Denver. *The Denver Westerners Roundup.* <https://denverposse.org/wp-content/uploads/2021/05/Roundup4-2021.pdf>

Parkhill, F. (1951). The Scarlet Lady: Mattie Sils made over half a million dollars from the world's oldest profession, but a good-for-nothing broke her heart. *The Denver Post.* <http://blogs.denverpost.com/library/files/2012/11/MattieSilks.pdf>

Rutter, M. (2005). *Upstairs Girls: Prostitution in the American West.* Helena, MT: Farcountry Press

Secrest, C. (2002). *Hell's Belles.* Boulder, CO: University of Colorado Press

Sneesby-Koch, A. (2018). A Brief Walk Along Denver's Notorious Market Street. *History Colorado.* <https://www.historycolorado.org/story/going-places/2018/11/26/brief-walk-along-denvers-notorious-market-street>

Trembath, B. K. (2019). Market Street's troubles are nothing new. *The Denver Library.* <https://history.denverlibrary.org/news/market-streets-troubles-are-nothing-new>

Tri-State/Denver Buddhist Temples (n.d.) Our History. *Tri-State/Denver Buddhist Temples.* <https://tsdbt.org/our-history/>

Wagner, K. (1998). Call Me Madam. *Westword.* <https://www.westword.com/restaurants/call-me-madam-5059394>

PART TWO:
Private Residences

Though the public venues we've discussed so far (and will continue to discuss in future volumes of this series) are arguably of the greatest general interest, they don't represent the entirety (perhaps not even the bulk, though it varies from year to year) of our work. On a regular basis, people write to us or approach us at one of our public lectures and ask for our assistance with their personal paranormal experiences.

Private cases such as these come in all shapes and sizes. Some of them never proceed past brief initial email discussion or perhaps a quick forensic analysis of a single photograph, while others may involve substantial on-site research on par with the kind of work we've done in some of the public cases.

For this second part of the book, we'd like to take you through a tour of some of these private cases, but there are a couple of caveats to get out of the way before we do so.

First, while we're attempting to provide a complete record of our case files in these books (though it will take several volumes to do so, especially since we're still actively conducting research on new cases), we're not going to document every communication we ever receive. Many cases simply don't go anywhere, and while we maintain records of these cases in our private files, we wouldn't presume to bore the reader with repetitive details of cases we were never able to fully investigate. On the other hand, neither are we presenting only the most outstanding and extraordinary cases. We do want to provide a reasonable overview of our activities.

Second and most importantly, because these are not public venues or public figures, we will not identify the individuals into whose homes and lives we've been invited. Most of the time, we can accomplish this simply by omitting their names or assigning pseudonyms. Every once in a while, some feature of the case narrative itself might be potentially identifying. When that happens, we will disguise the individuals' identities in a variety of ways, but never in such a way as to alter the relevant data we're trying to present.

With regard to photographic evidence, we have similar concerns. Occasionally, we may have a photograph that's restricted enough in its contents (or can be cropped to be so restricted) that it can illustrate the relevant scene without risk of identifying or embarrassing the client(s). When that's the case (both in this volume and in the series at large), we'll use genuine photographs. The rest of the time, we'll use artists' renditions and written descriptions to illustrate the cases.

In the pages to follow, you'll meet a diverse cast of characters, and we hope you enjoy their paranormal stories and our research into the same.

Note: though the chapters in Part One were (for the most part) divided into sections for history, paranormal claims, our investigation, and further reading, the private cases will not be. Most of these residences don't have much history to speak of, and further reading on these particular cases is typically nonexistent. Instead, we'll present the entire story in narrative form.

Essay
The Demons Go Away: Alternative Explanations for Paranormal Experience

On occasion, we've mentioned our work to people of our acquaintance who aren't involved in the same kind of research. Reactions to the claims made by our clients are not always positive. It's easy to dismiss some of these claims as being those of crazy people. But that has not been our experience. While there certainly are people in this world who suffer from a variety of mental health conditions ("crazy" is not a technical term) and occasionally they cross our path, most of the people who reach out to us are fundamentally sane and normal. Paranormal lore is a part of our culture and plenty of reasonable people believe in it, just as plenty of reasonable people disbelieve in it.

Often, perfectly sane and normal people who simply happen to have a belief or interest in paranormal phenomena have some unexplained event in their lives, and we consider it our duty to help them understand what they're dealing with. If we can find a natural explanation so they don't have to be afraid of their own homes anymore, we consider it a good day's work. On the other hand, if we could document a legitimate paranormal event to validate what they thought they were experiencing, we'd also consider that a good day's work.

When it becomes difficult is when we find ourselves working on cases in which the individuals involved are seriously struggling. We've seen marriages hanging by a thread, people about to walk out from their house to their financial ruin, families terrified to even sleep in their own homes. We've seen all kinds of psychological distress, either directly related to the paranormal claims or coincident with them. Any paranormal investigator needs to be careful when negotiating these kinds of territories. Some of our past experiences have involved people who were severely traumatized by prior investigation teams not taking due care in their practices or their reporting. Some of those cases have already been described in prior volumes and one of the worst cases of family trauma we've ever seen is described later in this one.

Even putting aside the extremes, there are several common things we've noticed among paranormal claimants at private residences. There's no person to whom all of these apply and there are some to whom none of them apply, but they are fairly common.

Marital distress is one of the most common issues, and it comes in multiple "flavors." A common one in the case of paranormal claimants is that one spouse (most often but not always the wife) believes there is something paranormal in the house, while the other spouse is skeptical. Marriages can easily survive some disagreements, but it becomes a serious problem when the skeptical spouse starts to question the other's sanity, or when the believing spouse starts to *think* the other questions their sanity even if that's not really the case. Other times, the believer thinks the skeptic is just being blind or stubborn.

How should a paranormal investigator approach such a situation? The first thing is to make it clear that beliefs can differ without either party questioning the sanity or motives of the other. Should a serious mental health problem present itself, obviously it needs to be dealt with and paranormal belief can sometimes be a symptom of those illnesses, but such cases are rare. Most people just happen to have different opinions. A good paranormal investigator (the kind like Rocky Mountain Paranormal which insists upon philosophical neutrality in every case) can be a disinterested third party to help mediate the dispute.

When it comes time to report findings, though, this can be fraught. If the investigator found nothing at all, the couple can be left in the same place they started. On the other hand, if the investigator finds proof (or even suggestive evidence) to support one side or the other, it needs to be made abundantly clear that just because one party was wrong (or seems to have been wrong) doesn't mean that party is foolish.

Professional psychologists, psychiatrists, marriage counselors, or members of the clients' own clergy might be called in to help with this process. That's important because paranormal investigators, though we may play a role in determining what's really happening in the house, are not qualified to play the role of doctor, counselor, or therapist.

Some cases involve spousal abuse which gets blamed on demons or other paranormal entities. Professional care is doubly important in those cases.

Substance abuse is another thing we see relatively often. A reader of one of our earlier books even commented to us that a lot of the people who see such unusual things must have been on drugs. And indeed, we've written about such cases. But that's certainly not all of them, and one needn't be high as a kite to experience something unusual. We're as sober as they come during our investigations and have still seen some weird things ourselves. When drugs (whether prescription or illicit) or alcohol are a factor, sometimes they precede the paranormal experience and might be the cause of hallucinations; in other cases, they follow the paranormal experience and were the person's coping mechanism. Our job is to determine which of those is the case, offer whatever guidance we can, and again to refer the people to professional help as needed.

Cleanliness can sometimes be an issue. Many of the residences we've visited have been messy. Sometimes just with clutter lying around and sometimes with actual dirt and filth everyone. On at least one occasion, unfortunately, we found black mold was the culprit and the house had to be condemned. But when it doesn't get to that extreme, often our advice involves cleaning up a bit both for its own sake and for reasons we'll explain in a moment.

One of the things we need to be careful with is how we describe our findings to the individuals involved, especially if they are in one of those situations where they're experiencing psychological distress. We don't want to push people into believing their house is inhabited by something evil just because we couldn't solve all of their mysteries anymore than we want to make someone feel foolish because we could solve them. Careful phrasing is paramount.

Worse are the situations when we really don't know. Either we didn't see the weird thing they've been experiencing or we couldn't solve it. Honesty is important and we have to admit when we're uncertain, but it's also important to help the clients cope with the unusual circumstances in which they find themselves.

Alternative explanations with common solutions are often a good way to go.

Imagine a hypothetical client who has many of these issues we've been describing. Maybe the wife believes it's a demon causing their problems and the husband is a skeptic. They've been at each other's throats over the whole thing. Maybe they're on the verge of divorce. Perhaps they're about to move, at great expense, because they can't stand being afraid of their own house. Maybe they're hoarders or close to it and the whole place is messy. And to top it all off, one (or both) of them is self-medicating with drugs or alcohol. Then we come in and maybe we solve some of the mysteries but not all of them so the whole case is still one big question mark in our ledgers. What in the world are we supposed to tell the clients and what are they supposed to do so they aren't spending the rest of their lives in a sort of Hell on Earth?

Our first advice, as always when there's that much psychologically going on, is to consult professionals. We'll help them find a good doctor, therapist, or clergyman to help along the way. People with that many problems aren't going to fix them overnight or just on our say-so.

But the way we report our findings can help the process along. Problems like these, nested one inside the other, don't all spring up at once. They feed into each other. And the way we would describe it to these clients is that it almost doesn't matter what the answer is to the paranormal question.

In the psychological literature and the more skeptical paranormal literature, it's well-established that paranormal experiences can stem from psychological or social problems. People with tons of stress and substance abuse can easily convince themselves there's a demon at the heart of their problem even if all their troubles are truly self-inflicted. Plus, people who believe they have demons might start to adopt a ton of unhealthy coping mechanisms. Or the stress just of dealing with that can cause psychological issues. This can result in a self-perpetuating feedback loop.

On the other hand, in the paranormal lore on the believers' side, it's equally well-established that living in conditions of negativity can attract negative beings. Those who believe in demonic possessions often suggest the demons have to be invited either by some psychological weakness or some great sin. People who believe in hauntings often think negative events can create negative "energies" which attract malevolent spirits or demons.

The solution, as it turns out, is the same whichever you believe. Put your life right, clean up your house, get your marriage in order, add some meaning back into your daily routine, and the demons will go away. If you believe in the demons literally, they go away simply because you have made your life and home inhospitable to them. On the other hand, if you believe the demon was only psychological in nature, getting your life in order still makes the demon go away because you've made your *mind* inhospitable for it.

That's the real reason we suggest people seek professional help when they think they're experiencing the paranormal. It's not because we think they're crazy, as we've mentioned. It's because whether the paranormal turns out to be true or false, getting that kind of professional help is usually exactly what people need to start putting things right and to save themselves from a downward spiral.

So, yeah: the title of this essay pretty much says it all. Clean up your house, put your life in order, start thinking clearly and soberly, and the demons really will go away. Regardless of their nature or origin.

15
The Friendly Farmer

We're going all the way back to the beginning for this one. This was one of our earliest experiences with private paranormal investigations just when we were first starting out and had only begun to develop enough of a reputation that people reached out to us with their paranormal experiences and questions. It involves a woman who lived alone and started receiving nightly visits from a ghostly figure while she was trying to sleep.

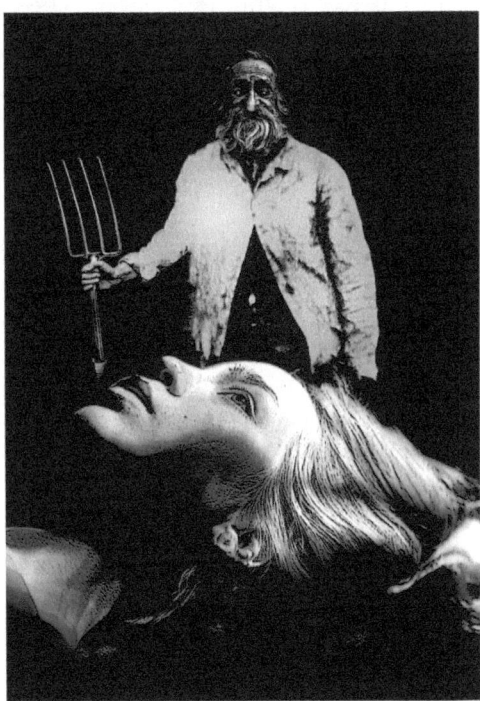

Figure 15.1. Artist's rendition of the Friendly Farmer ghost (image: Elderlemon Design).

As we said in the introduction to this chapter, this is one of our earliest investigations, and it really gave us an opportunity to cut our teeth on this sort of private investigation. Fortunately, this wasn't one of those really extreme cases like we alluded to in the essay at the beginning of this part. Our client here wasn't so much terrified as just confused and in search of answers.

This story involves a woman we'll call Alice (as we said earlier, all the names of the private clients have been changed to protect their identities). She lived alone in a single-family house she'd purchased in the 1990s. Prior to moving into her new house, she'd never experienced anything paranormal before.

Just after she moved in and finished unpacking, though, she had her first experience that very night. Exhausted from the work of moving, she decided to go to bed. After crawling under the covers, she rolled onto her side to click off the lamp on her nightstand. But when she rolled back over, she saw a disembodied face hovering just inches above her own.

Most people would probably have panicked and either turned on the light to investigate or run screaming from the room. Some might have moved right back out of the house as soon as they were able. Not Alice. She had one of the best responses we've ever heard to a paranormal encounter. Though she'd never had any significant experiences of her own before, she was a lifelong believer in the paranormal and so she wasn't afraid of it.

"I'm just too tired," she said to the ghost. "Go away."

Away the ghost went! The face disappeared, never to be seen again.

The house remained quiet for about the next year, then the paranormal happenings started up again.

Once again, Alice was in bed, about to go to sleep, when a figure, she says, materialized in the room, walked across her bedroom floor, and stepped right through the wall. She described the figure as an older man dressed in what she thought looked like farmers' attire. Nothing else happened for the rest of the night.

The same experience, though, started to repeat itself. Just before Alice went to sleep—not every night, but often enough—the ghostly farmer would materialize, walk across the room, and disappear through the wall. Alice said he always took exactly the same path across the room, no matter how she rearranged the furniture. The only deviation from his routine was that he would occasionally stop at the end of her bed and gently squeeze her foot before continuing on his usual path. She said this was done in a friendly manner rather than a threatening one, which is why we've taken to calling this ghost the Friendly Farmer.

According to Alice, the ghost could and sometimes did interact with her, so she felt like he was sentient and aware of her presence. But she also said most of his behaviors were so repetitive that he felt more like a recording.

In the paranormal lore, people draw distinctions between active, intelligent, or sentient hauntings, which are conscious spirits aware of their surroundings and able to interact with the material world in some way, and residual hauntings, which really are thought to be more like supernatural tape recordings. In the latter case, the ghost isn't aware or sentient but is just a supernatural reenactment of something that happened (usually at the location where the haunting occurs) in the past.

For our purposes, though we're aware of the distinction, it doesn't make much difference to us most of the time. Once we could prove a haunting is taking place, we'd start to care a lot about determining its exact nature. But until that time, we're just looking an answer to whether or not the haunting is happening at all, and only once that question is answered are we going to start worrying about whether it's conscious or residual.

Still, we do note that Alice's story is a bit unusual in that it seems to have some elements drawn from both varieties of paranormal lore.

Two additional incidents, both occurring shortly before Alice contacted us, prompted her to reach out. In the first, she saw a male figure who appeared to be standing by the foot of her bed. Except "standing" might not be quite the right word because he was only visible from the waist up. His legs, if he had any, were invisible. Alice said even though she had a clear view of his head, she couldn't see a face. It did not seem to be the same Friendly Farmer, though. This was someone or something different.

Then, on another night not long after, she saw yet a third spirit. This one was a woman wearing what Alice thought looked like attire appropriate to the 1930s or 1940s. It consisted of a white blouse and a dark skirt. This ghost seemed to have a sadness about her, Alice said. It felt like she was trying to communicate but couldn't find the means to effectively get her message across.

Once we got involved, we set a date for an investigation, scheduled to begin in the late evening and go late into the night, covering the entire timespan during which Alice experienced her phenomena, with plenty of margin on either side. Since this one wasn't an emergency case, we scheduled it a short time in the future and used the interim to figure out anything we could about the property.

Alice was not the first owner of the house, but it didn't go too far into the past, either. Just two decades or so. She moved in during the 1990s and it had been built in the 1970s (we do know the exact years but we approximate the dates here as one more way to help preserve the client's anonymity). Before the current house was built, the property was a vacant lot in an open field. Not a whole lot of history to go on, and we couldn't find any useful information about any of the former residents that might tell us anything important. As far as we could learn, though, the house had never been the site of any major newsworthy tragedies or anything like that.

When we arrived for the on-site investigation, we toured the property. This was a nicely but simply decorated home that didn't have any signs of trouble. Neither, for that matter, did Alice herself. She seemed sane and levelheaded, credible in her storytelling, and didn't display any signs of mental illness or delusion. So far, so good.

We checked around the house for any environmental concerns that might be relevant—low frequency vibrations, strong EMF fields, gas leaks, mold, et cetera—and didn't find anything that would either be cause for concern or a potential explanation for any paranormal experiences.

The one thing we did notice is that a bird cage hung just off to the side of her bed. We don't think that could explain any of the paranormal experiences Alice described, but we noted it just in case perhaps a weird combination of low-light and exhaustion might have caused her to perceive a reflection off the bottom of the cage as the floating disembodied head in the first experience. That's a real stretch and we don't think it's right, but we always try to account for every possibility.

Paranormal investigation can put one in some awkward situations. When that happens, it's important to just be completely frank with people and to not only *seem* trustworthy but to actually *be* trustworthy. The next step in our investigation was just such a situation.

Investigation under conditions as close to those in which the paranormal phenomena were experienced is paramount. In this case, those conditions would be Alice going to bed. We explained our thinking and told her that because that's when the ghosts seem to show up, she should probably go to bed at the usual time and we would watch and see if anything unusual happened. Furthermore, because she was always alone in the room at the time, we didn't want to be in there to potentially confound the experiment. Instead, we wanted to sit in the other room and watch remotely on cameras we'd place around her bed. Admittedly, filming someone while she sleeps is a horribly creepy thing to do under normal circumstances, but as we said, sometimes we just have to deal with awkward situations. Fortunately, Alice understood and didn't mind participating in the experiment.

She went to bed and we watched and we waited.

Unfortunately, this was one of those cases in which not a thing happened while we were there.

So what should we make of a case like this? What did we tell Alice? Well, because the phenomenon didn't manifest while we were there, everything we can say is purely speculative, but we gave Alice a couple of ideas.

First, if we assume the ghosts really were visiting her on some evenings, we noticed that none of them ever seemed threatening. With that in mind, we said not to be afraid of it and just to be glad she got to experience something a lot of us spend our entire lives chasing.

Some skeptical explanations were also in order, though. We did note that Alice was a lifelong believer in the paranormal. For such people, even minor anomalies that just happen to all of us from time to time can be mistaken for something paranormal. To minimize the risk of making those mistakes, we just cautioned Alice (as we caution everyone) that it's okay to believe but it's also wise to question. We also noted that exhaustion can sometimes fuzz people's thinking.

There's also a well-known phenomenon called sleep paralysis. Generally, during sleep, the brain paralyzes the body so you don't accidentally act our your dreams and hurt yourself. But this mechanism can malfunction. Some people experience sleep walking or somnambulism in which the body fails to properly paralyze. Others experience sleep paralysis, which is the exact opposite. When this happens, the mind starts to wake up while the body is still paralyzed for sleep. In this half-awake state, people often hallucinate culturally relevant monsters. By that we mean medieval Christians tended to see demons, while science fiction nerds tend to see extraterrestrial aliens, and people who believe in ghosts might see ghosts. These are often terrifying experiences both because the visions themselves tend to be frightening and because the body is paralyzed so the victim feels helpless.

Could that be what was happening to Alice? We don't think so. We did mention it as a possibility she should be aware of, but she didn't seem to have the feelings of terror or paralysis common to those experiences. That said, could her experiences have been ordinarily hallucinations brought on by exhaustion or a half-awake/half-asleep state of mind? It's entirely possible, but we just don't know.

So at the end of the day, we parted ways and told Alice to contact us if things worsened or if she ever wanted us to investigate any future phenomena. She never called back.

We're grateful this was one of our earliest private investigations, though, because it gave us a lot of practice in reporting our findings while carefully navigating the client's beliefs. That would become quite relevant in some more serious cases years later.

16
The Cricket Demon

This is another favorite of ours. Sometimes people ask us about the most unusual claims we've ever investigated. There are a few contenders for the title, but "giant cricket demons" has to be at least in the top five. It's also a perfect example of why it's important to take every claim seriously and investigate it properly rather than just assuming a crazy-sounding claim is actually crazy, because this also ended up being one of those cases where our methodology really paid off.

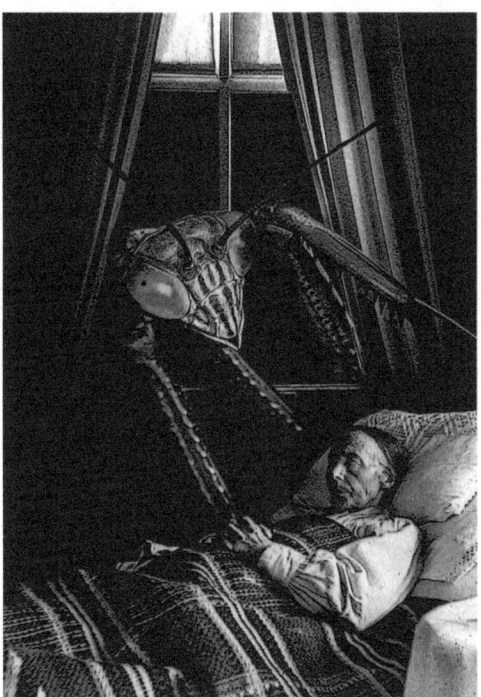

Figure 16.1. Artist's rendition of the cricket demon (image: Elderlemon Design).

Most of our cases end up being ghost stories. You've been reading this book (and presumably at least a good number of you have read the first two volumes), so you know that's true. The vast majority of claims we get called to investigate are ghost stories. Occasionally someone thinks it's a demon instead of a ghost. And from time to time we get to look into a UFO or cryptid sighting. But rarely do we get a call for something truly out of ordinary (at least if we define "ordinary" by paranormal standards. This is just such a case, though.

One of Rocky Mountain Paranormal's claims to fame is that we're one of relatively few groups that will investigate all manner of paranormal phenomena. A lot of people focus specifically on either ghosts (or maybe ghosts and demons combined), *or* aliens, *or* Sasquatch, and they don't show much professional interest in the others. Not us. We always say "both and" instead of "either or." We love it all.

We'll call our clients in this case Betty and Clarice. They were two elderly sisters aged ninety and ninety-two years, respectively, living together in a single-family home. Betty, the younger, was the one who reached out to us and described several things she'd been experiencing while Clarice slept. Sometimes she'd hear voices coming from the kitchen even though her sister was asleep and no one else lived in the house. Odd lights would occasionally come on in the basement. But most frightening of all, she said she saw giant monstrous crickets wandering through the house.

We asked what she meant by giant crickets, thinking maybe these were just large insects and she ought to be calling pest control instead of paranormal investigators. No, she said, these were crickets the size of large dogs or small people. Demonic, monstrous crickets. Now we were intrigued.

Every paranormal investigator has to choose his or her cases. We do get more letters and inquiries than we could ever possibly investigate in person. So if someone reaches out to us with a story of a ghost they saw one time twenty years ago, we'll file the story in our archives, but there's really no point in us showing up for something that hasn't recurred in two decades. We thank those people for the story and tell them to contact us again if anything new happens. It's just the reality of this kind of work. Similarly, there are some cases we can crack just on the basis of photographic evidence without ever needing to show up. So we do have to be selective. A lot of people, probably even most paranormal investigators, upon hearing "giant cricket demons," would have passed on the case, dismissed Betty as just a crazy person, and moved on.

This is not the right approach.

For us, the question isn't how unusual the claim sounds. Honestly, the weirder the better sometimes, if for no other reason than some day we might get to write a chapter entitled "The Cricket Demon." Our only real questions when deciding whether to take a case are: can it be done safely? Can we afford to get to the location (because we don't charge for our services)? And is there any chance, even a remote one, we might be able to witness the phenomenon and/or solve the case?

Here, all three answers were yes. Safety was of no concern in a single-family house occupied by two elderly women. Cost was no concern because they were local. And Betty said these things happened with such regularity we were almost guaranteed to be present when one of them took place. We agreed to the investigation.

When we arrived, we started off with a quick interview of both sisters followed by a tour of the house. During this time, we learned a bit more about their claims and also had a chance to start forming some preliminary opinions about the case.

The way their story went, only Betty was experiencing any of these bizarre happenings. Mostly they took place at night, when Clarice was sleeping.

Both Betty and Clarice seemed quite sharp and credible. Age may have slowed them physically, but neither showed any signs of mental illness or dementia. Both were also quite friendly and accommodating. We asked if either had any medical conditions and they said the only things were that Clarice was extremely hard of hearing and partially blind. They said this was probably one reason only Betty had seen the paranormal things.

Another thing we noticed was that the house was well-kept. These weren't the kinds of people whose lives seemed to be falling apart. In other words, nothing about the sisters or their house screamed "crazy" to us. We shudder to think what other might have said when they heard "cricket demons," but to us, this was just another case.

After concluding a perfunctory tour and our brief interview, we decided to start the investigation, before even setting up any equipment, by just sitting silently for a few minutes and acclimatizing ourselves to the sounds of their house. This is a useful technique on any paranormal investigation because it gives us a chance to get to know the building's perfectly natural quirks so we can more easily distinguish them from any anomalous sounds later on.

This time, it paid off even more directly.

While we were listening, we all heard—well, all except Clarice, whose hearing was too poor—what sounded like a faint, muffled conversation in the kitchen. Exactly like Betty had described! We managed to experience one of the claims before even starting the investigation properly.

We asked if they had any radios or televisions. Yes they did, but they were all turned off and none were in the kitchen. And this was a detached single-family house, so we didn't think it should be coming from neighbors (though we resolved to take a sound pressure meter around the neighborhood and check if it should become necessary). Before doing anything else, we went to the kitchen and found the sounds seemed like they were coming from the refrigerator. Upon closer inspection, we found the culprit. The fridge's freon compressor had a bad bearing. When it turned on, it made a noise which really did sound eerily like muffled voices.

One mystery solved already. We explained what was causing the problem. Both—Betty in particular—seemed relieved to have solved at least one of the problems, though obviously we hadn't yet gotten to the elephant in the room (or in this case, to the cricket demon in the bedroom). We also explained how it would be simple for a skilled repairman to fix the problem. Fortunately, they said the refrigerator was still under warranty and that they'd call for repair the next day.[1]

After that bit of luck, we went for a more detailed tour of the house while we started to figure out how we'd like to set up our overnight monitoring system. One of our investigators took an EMF meter around to start collecting baseline readings. This also turned out to be quite a fortuitous decision because we immediately found an unusual electromagnetic anomaly. While most of the house had EMF fields within expected ranges, Clarice's bedroom had remarkably high levels of EMF radiation. Almost dangerously high.

We pause for a moment to explain that, contrary to some popular myths, electromagnetic radiation (at least of the non-ionizing variety we're talking about) is not dangerous in and of itself under most circumstances. The point at which EMF can become medically dangerous is a complicated question which depends both on the frequency and strength of the field. For our purposes here, all you really need to know is that the electromagnetic fields commonly encountered in a home or around town are well below the limit of being dangerous at all.

However, even a field weak enough not to be dangerous can still be relevant to a paranormal investigator. Moderate to strong electromagnetic fields can, in some individuals who have a certain sensitivity to them, cause uneasy feelings or even hallucinations. Temporal lobe epileptics commonly have this sort of sensitivity, though it's not exclusive to those people.

That there was a strong electromagnetic field in this house was an important piece of the puzzle and we wanted to investigate that further. We exaggerate only slightly when we say the EMF levels were almost dangerous (they were still below public health exposure limits), but they were honestly the highest we've seen on any investigation, and certainly strong enough to produce hallucinations if Betty might turn out to be one of those people with a sensitivity.

After a bit of exploring with our meters, we traced the EMF source to a CPAP (Continuous Positive Airway Pressure) machine, used to prevent upper airway collapse in sleep apnea patients or some heart failure patients. These are the devices consisting of a machine connected to a mask the patient wears at night to ensure proper breathing. And this one was putting out way more electromagnetic radiation than any medical device ever should.

We asked about the device and they told us Clarice used it when she was sleeping and that they'd gotten it about three months prior. We then asked when the paranormal phenomena started happening and they said also about three months prior. Curiouser and curiouser.

As it turned out, Betty would often sleep in Clarice's room because she stayed up later and didn't like to be alone at night. Particularly when the paranormal things started happening, she'd get frightened and go to her sister's room just to be around someone. Completely understandable.

But now we were starting to develop a hypothesis that the CPAP machine might be playing at least some role in her experiences. If that were the case, she might have inadvertently been making things worse and frightening herself all the

---

1   Some of us have a bit of the handyman gene and like to tinker. Had the fridge not been under warranty, there's a good chance one of us would have volunteered to try the repair ourselves. It's probably a good thing this didn't happen, because that easily could have led to a whole different comedy of errors.

more by moving herself closer to the electromagnetic field. Such fields follow an inverse square law, such that the strength of the field is inversely proportional to the square of the distance from its source. In other words, the closer you get, the strength of the field increases not linearly but by the square of how much closer you've gotten.

We followed up on this a bit more, and Betty confirmed basically what we were thinking. She'd get a little frightened or lonely and go to Clarice's room, where she'd sit in the recliner near the bed. While there, she'd start to see the giant crickets and even a sort of cricket-human hybrid walking through the house, though fortunately that monster never came into the bedroom. For her part, Clarice never saw any of it. She was sleeping through most of it and even when awakened she didn't see these things, but just figured it was because she was almost blind.

At this point, we were pretty sure the CPAP machine and its EMF radiation were to blame. Either due to genetic disposition or age, it seemed like Betty was sensitive to the electromagnetic radiation and that it was causing her to hallucinate. The precise nature of the hallucinations may have been influenced by cultural beliefs, past experiences in her life, and other experiences (like the voices from the refrigerator's compressor) she'd been having around the house.

It's a plausible story. But so far, it was just that: a story. How could we know for certain? We designed a simple experiment the sisters could do over the next two nights. On the third night, we'd return to get their results and tell them about what we thought was going on. We didn't specify all the details at this time—only enough to convey our instructions—but said we'd explain everything when we came back. That way, we hoped to be able to do the experiment without unduly altering the results due to expectancy effects.

For both nights of the experiment, both sisters should stay in Clarice's room, and each should have either a notebook or a tape recorder so they could note anything and everything they saw, heard, or experienced during the night. Regardless of what they saw, they should stay put and just write it down. On the first night, we said they should keep the CPAP machine running all night. But on the second night, they should try to turn the CPAP machine off all night (and Betty would need to stay awake and monitor Clarice's breathing and only turn it on if it were medically necessary).

In truth, we had to go even a step further. We discovered the CPAP machine *always* emitted electromagnetic radiation, even when turned off. The only part the switch turned off was the low-voltage controller, not the high-voltage functional part of the device. Therefore, we advised Betty and Clarice to completely disconnect the machine from the power source during the second night. So neither of them would have to bend to the floor to reach the outlet, we connected it to a power strip before we left.

They did exactly what we suggested. When we came back, they shared their results. Betty said that first night was the worst night yet. Those cricket demons were all over the place. Worse, even some human figures joined them this time. All the monsters just wandered the house all night and neither of the sisters got a wink of sleep. But on the second night, they said, it was exactly the opposite. They didn't see a single monster or cricket, didn't hear anything unusual, and it was the most peaceful night they'd had in months.

Such evidence may not be 100% proof beyond all doubt, but that was more than good enough for us. We explained about electromagnetic radiation and said they needed to get a new CPAP machine.

Experience has taught us, though, that you need to be careful when you call up your doctor or insurance company about anything paranormal in nature. Though we knew what was happening and that there was nothing psychologically wrong with either of the sisters, we also feared that if they called up for a replacement and told their insurance company they needed a different model because Clarice's current one was causing them to hallucinate cricket demons, they might find themselves being fitted for a straitjacket. And if they mentioned our names, we might, too!

So, we said, they should call up the insurance company and just tell them they needed a new CPAP machine because this one seemed to be malfunctioning and that it made too much noise. They understood our reasoning and said they'd do exactly that.

We told them to contact us a week or two after they got the new machine and let us know if that solved the problem or if they were still having any trouble. If the latter, we'd come back and keep investigating. They got the new machine and said everything went perfectly back to normal and they never had any more issues or paranormal phenomena.

The only thing we didn't specifically solve in their case was the odd lights from the basement, but we assume that was either just a bad lightbulb or another hallucination. We didn't see it while we were there, and the problem went away after we helped them solve the bigger problem.

We love this story for a lot of different reasons. First, any story that's a bit unusual catches our attention, and this was the only cricket demon case we've ever heard of. Second, we always love when we can solve a mystery and help people get their lives back. As much as we're often hoping to find evidence of a genuine paranormal phenomenon, cases like this show that even negative results can be a good thing. These two sisters were terrified of their own house, and we consider it nothing but an honor that we had the opportunity to solve that problem for them. Third, this was about as good a demonstration as we've ever seen that sometimes EMF can cause hallucinations. We'd read about such things in the scientific literature before, but here we were able to see it firsthand, and that's a really useful piece of information to keep in our toolbelts for future investigations.

Most importantly of all, though, this case shows how important it is to investigate every claim with an open mind, no matter how outlandish it might sound at first. Most skeptics would have hung up or not written back as soon as they heard "giant cricket demons." Even most believers probably would have dismissed this as a crazy person or a prank instead of investigating. But had we done that, Betty and Clarice might never have found the solution they so desperately needed to what turned out to be a simple, if highly unusual, problem in their lives.

There are crazy people out there. In the paranormal world, we meet plenty of them. But not everyone with a crazy-sounding claim really is crazy. The only way to tell is to roll up your sleeves and get to work investigating each and every case you can.

17

The Possessed Man and the Stinky House

While ghosts are by far the most common paranormal claims we get the opportunity to investigate, we've also heard our share of demon stories—especially in private cases in which clients have become afraid of something in their own homes. This case, however, was a rarity for us. Not only did the client believe his home was being invaded by a demonic force, but he claimed to actually be possessed and under demonic control.

Figure 17.1. Artist's rendition of the client's demon (image: Elderlemon Design).

Thankfully, we were the first investigators called in on this case. Other times, as we've described in prior volumes, less-diligent paranormal teams have preceded us to a client's home and done more harm than good by leaving the individuals convinced they were in grave danger from demonic forces, even in cases wherein we've later been able to find natural explanations for all of the phenomena and then had the difficult task of trying to un-ring the bells rung by our predecessors.

This case didn't take us very long on site and won't take us long to describe here, but it does contain some important lessons.

We were contacted by a man who believed he was under demonic control. He didn't specifically use the word "possessed," but his description is consistent with what people consider to be claims of demonic possession. We'll explain that in more detail in just a moment. The client was a single man in his thirties who we'll call Damien (because we can't help ourselves but to make a horror movie reference as we attempt to preserve client anonymity) who lived alone in an average-sized two-story single family home.[1]

In addition to the client's belief that demons were attempting to control his thoughts and behaviors, he said he often saw things moving around his home, heard voices, and seemed to be growing increasingly paranoid about life in general as time went on. We knew this was going to be an important case for us to investigate because Damien was clearly struggling. As we alluded to in the essay at the beginning of this part, whether we wanted to jump to a paranormal explanation or to assume he was instead experiencing some kind of mental health episode, we knew it was going to be important for someone to intervene in the case.

Plus, we have to admit, we were intrigued to have our first case of alleged demonic possession and were grateful for the opportunity to see what might really be going on.

Taking a step back for a moment, we should talk a little bit about how demonic possession is viewed in general terms. As is often the case, our discussion will be brief and somewhat superficial as the subject merits entire volumes on its own, but we do think it's important to at least have some grounding in terms of what people actually believe about demonic possession versus what is merely the stuff of Hollywood and horror fiction.

Almost all cultures, peoples, and religious orientations have a concept of demonic or spirit possession, though different terminology might be applied in different cases. Especially in pre-scientific cultures, mental illnesses (and even certain physical ailments) were often thought to be the work of malevolent entities and a variety of ritualistic practices and forms of exorcism were developed to combat these forces. You might be inclined to think such practices vanished entirely (or were at least relegated to fringe religious groups) as scientific understanding of mental illness started to develop. Rather than vanishing, though,

---

1   If you're curious as to why we also go to the trouble of specifying whether the residence is a single-family detached home, a condominium, or an apartment, the reason is simple. Success in paranormal investigations often hinges on the researchers' ability to notice potential environmental causes of apparently anomalous phenomena. Neighbors with loud televisions, for instance, could explain unusual sounds if the client lives in an apartment or condo, but this is less likely in a detached home.

many of the early beliefs and rituals were demystified and incorporated into the developing field of psychiatry.[2]

Demons that once possessed people, in this new view, were thought not to be supernatural entities but the "ghosts" or "demons" that metaphorically haunted individuals' unconscious or subconscious minds. Sometimes doctors with interpretations from dynamic psychiatry rather than from religion would nevertheless request exorcism rituals for their patients because they thought the patients' own belief in the efficacy of the ritual would make it an effective treatment. Often—but not always—it worked.

It was in this spirit that we approached our study of the case. We insist on philosophical neutrality so we'd never dare say the client were mentally ill rather than possessed, but we do admit we had our suspicions along those lines right from the start. Still, getting a client appropriate treatment—whatever that treatment might end up being—often requires starting off by working within the client's own belief system. As we said in our earlier essay, when one puts one's life in order, the demons often go away regardless of whether one chooses to believe in them literally or only metaphorically.

However, because different religious faiths have different concepts of demons, diabolic influence, possession, and exorcism, it's important to have some idea where the client is coming from before we start. In this case, we knew the client to be a Christian, though we never learned exactly which denomination (if any denomination in particular) he belonged to.

Protestants and Catholics don't always agree on everything, but their views of demonic possession are similar, with the main differences being that many Protestant faiths base their spiritual warfare more on the deliverance prayers of individuals while Catholics use those types of prayers in a more subsidiary role to the formal Rite of Exorcism which is reserved only to the priesthood (and then, only with the express consent of the local ordinary or bishop). Because the Catholics have a well-established literature of demonology and possession (and because often even Protestant clergy will sometimes refer suspected possessed people to the Catholics for this exact reason), we'll briefly explain their view of the matter, albeit with all the caveats that we're only scratching the surface, that these ideas are highly nuanced (and not all theologians even within a particular faith agree), and that other faiths might differ in some or all of the details.

Demonic influence over human lives is generally divided into the ordinary and the extraordinary. Ordinary influence consists of temptation into sinful activity and of snares (external influences meant to drive individuals to the same end) and are experienced by all persons just as a matter of course. Extraordinary influences are divided into the categories of oppression (which are direct personal demonic attacks external to the individual and may manifest in the form of noises, objects moving, infliction of pain, etc.), obsession (which are internal psychological attacks wherein one is intellectually and emotionally tortured by the demons but is not under their explicit control), and possession (in which the demons assert actual ownership over the individual's body, which may be partial

---

2   Ellenberger, H. F. (1970). *The Discovery of the Unconscious: The History and Evolution of Dynamic Psychiatry.* New York: Basic Books.

or full and temporary or chronic).[3]

Each stage of the diabolic attack is characterized by certain signs clergy can use in confirming the case or determining whether it would be better referred instead to mental health professionals (who are, these days, supposed to be consulted before proceeding with any ritual exorcism anyway).

In the case at hand, the symptoms described to us were more in line with what one would call a demonic obsession. As far as we could tell, Damien seemed to still believe he was under conscious control of his own faculties, but that the various manifestations he was hearing and seeing were *attempting* to gain control of his faculties. He was looking for whatever help he could get. We explained quite clearly that we're not clergy or exorcists but that we'd gladly take a look at his case and then help him find the professionals who'd be able to help.

Because of the nature of the case, we didn't wait long before visiting Damien's house for our initial interview and investigation. This case seemed like something of an emergency.

During our interview, we learned that Damien had been on psychiatric medication but that he'd stopped taking it because he believed it was, in his words, "letting the demons in." The house also had a strong and almost unbearable smell of chemical cleansers or disinfectant.

After chatting for a while, we conducted an initial tour of the house and made a couple additional discoveries.

First, the upstairs bathtub was filled to the brim with hundreds upon hundreds of tree-shaped car air fresheners—the kind people hang from their rearview mirrors. The smell of these (combined with some other cleansers we couldn't identify) made it impossible to stay in the house for any prolonged period of time. When asked, Damien explained he was trying to purify the air in his home to keep the demons from gaining any stronger hold.

Second, while we swept the house for electromagnetic radiation, low frequency vibrations, and other environmental contaminants, we discovered there was indeed a very strong EMF emanating from just behind the headboard of Damien's bed. It wasn't from any supernatural source. The main power bus for the home ran right through that wall.

At this point, we had three plausible hypotheses for what might really be going on. Perhaps Damien, by stopping his medication, was suffering a mental health episode.[4]

Or perhaps prolonged exposure to fumes from all the air fresheners might have had some kind of effect on his mental health. At the time, we didn't know of any specific information linking air fresheners to hallucinations, but just our common sense suggested that exposure to *that much* of any kind of fume could pose a health risk. Subsequent research solidifies this as a plausible explanation. Air fresheners like the ones people hang from their car mirrors contain numerous volatile organic compounds including formaldehyde, benzene, toluene,

---

3   Ripperger, C. (2022). *Dominion: The Nature of Diabolic Warfare*. Keenesburg, CO: Sensus Traditionis Press.

4   He did tell us what medications he was taking, but we'll withhold that information except to say they were psychiatric drugs.

and xylene, which have been associated with a variety of health risks.[5] Worse, chronic exposure to toluene, which Damien was certainly experiencing given the condition of his home, has been linked to both hallucinations and psychosis.[6] Fortunately, certain classes of antipsychotic drugs have been found effective in treatment of solvent-induced psychosis.[7] So though these kinds of questions are above our paygrade, we figured we at least had some useful information to pass along to the relevant professionals.

Finally, though prescription noncompliance and solvent exposure seemed like more important issues, we also couldn't rule out potential hallucinatory effects from chronic EMF exposure due to the power main behind the bed (see Chapter 16 for a more detailed description of that phenomenon). Sensitivity to electromagnetic radiation is common in patients with certain types of brain injuries (or pre-existing conditions like temporal lobe epilepsy). Once again, these questions were a bit beyond our paygrade, but we thought it might be plausible that Damien's other issues might have created or exacerbated such a sensitivity.

The big question at that point was: how do we proceed? Telling people who believe they're experiencing something paranormal that we think it might have natural causes can cause some people to think we're calling them crazy. That's certainly not what we intend, but we didn't want to push a client away or cause him to retreat into potentially dangerous behaviors. So we consulted with him and—without using any words like "psychosis" or "hallucination"—suggested we thought the conditions under which he was living might be making his situation worse.

Following our own advice from the essay at the beginning of this part, we suggested that whatever the initial cause of his condition, we thought three changes would likely help matters. First, he should see his doctor or therapist and work on resuming his medication (and we helped him to schedule an appointment to do exactly that). Second, he should try living without the air fresheners for a while. And finally, he should consider moving his bedroom downstairs to get away from the power bus behind his bedroom wall.

He agreed to try those things and we said we'd follow up after a few weeks to check in and see how things were progressing and if any further assistance was needed. About three weeks later, we spoke to Damien again and were gratified to hear he was doing much better.

Does this all mean Damien was just hallucinating and that there were no demons? We can't say that with absolute certainty, though it does seem like the best explanation given the evidence. But it almost doesn't matter. Getting his life back in order solved the problem whether it was natural (as we think) or supernatural in nature.

---

5   Steinemann, A. (2017). Ten questions concerning air fresheners and indoor built environments. *Building and Environment, 111:* 279 – 284.

6   Lee, M., Lin, B., Chan, M., & Chen, H. (2020). Increased behavioral and neuronal responses to a hallucinogenic drug after adolescent toluene exposure in mice: Effects of antipsychotic treatment. *Toxicology, 445:* 152602.

7   *Ibid.*

18
Ghostly Pet Cat

Depending on who you ask, humans aren't the only beings who experience an afterlife. In this case, we were contacted by a woman with a whole slew of claims, but the one she most wanted us to look into was a ghost cat which she said leapt into her bed almost every night and wanted to play with her when she was trying to sleep. She wasn't so much afraid of the cat; she just wanted us to document it. But the other things happening in and around her apartment…those, she found a lot more frightening.

Figure 18.1. Artist's rendition of the ghostly pet cat (image: Elderlemon Design).

Philosophers, scientists, and theologians will probably never stop debating about the nature of life and death, whether or not there's an afterlife, and—if there is—what it might be like and who or what might get to go there. We won't presume to settle those debates here, but we will point out that human ghosts aren't the only ones we get called out to investigate. Animal spirits are a minority of claims we've encountered, but they're common enough in both our cases and in the paranormal lore that they shouldn't be ignored or dismissed entirely.

The case at hand involved a middle-aged woman we'll call Elizabeth who lived alone in a suburban apartment. She reported several paranormal phenomena which didn't seem to necessarily be connected with each other, but she wanted us to investigate the lot of it and see if we could document what had been happening.

It all started with the ghost cat. According to Elizabeth, she'd been dealing with the ghost of a cat for several months. Every night when she went to bed, the cat would jump on her bed and annoy her all night. It wasn't that she was afraid of it. This wasn't an evil cat, or at least it didn't seem so. It seemed just like a friendly pet which happened to be a ghost.[1] Our first order of business as far as Elizabeth was concerned was simply to document her experiences with that cat.

But she also had several other claims she wanted us to look into while we were there, to see if we could figure out what was going on. At sundown, she said, she would see strange lights in her home, like little rainbows that appeared all around the room. More troubling, she said "something nefarious" was happening in the unit next to hers. She didn't know what they were up to but sometimes she'd hear screams and chanting at all hours of the night. To us, that sounded like something right out of Ira Levin's *Rosemary's Baby* (or Roman Polanski's film adaptation of the same title).

Because she said the strange lights usually showed up around sundown, we arrived early in the evening to begin our investigation. After an initial interview during which we didn't gain much new information except just a reiteration of what we've already described, we toured the apartment and figured out our plan of attack. Mostly, this case was going to just involve waiting and watching to see what might happen. Most of the reported phenomena occurred in the bedroom, so we put most of our monitoring equipment there, with just a couple cameras covering the other rooms. We did put a microphone right against the wall to the adjoining unit so we might be able to record any of the mysterious screaming or chanting Elizabeth had described.

Once everything was set up, we retired to the room we'd chosen as our base of operations and continued chatting with Elizabeth for a while until it got late enough for the spooky stuff to start happening.

Right around dusk, sure enough, we also saw the mysterious lights. Just as described, little rainbows started appearing all over the bedroom. It didn't take us very long to figure out the cause, though. A crystal chandelier hung in the

---

1   Those of us who are animal lovers but suffer from certain pet allergies are quite intrigued to learn of the possibility of ghostly pets. We can't be sure but we certainly hope allergens don't experience an afterlife. If that's the case, ghost pets might just be the next big thing!

room. At dusk, the angle of the setting sun was such that direct sunlight entered an open window and struck the chandelier's crystals, splitting the white light into rainbows. Because the chandelier was fairly elaborate, each of the crystals had an opportunity to cast a separate rainbow, essentially filling the room.

Many people are tempted to say they'd never be fooled into supplying a paranormal explanation to something as simple as a chandelier acting as a prism. But it's not that easy. Often, people who can rationally evaluate claims made by other people struggle to evaluate their own. They're too close to the phenomenon in question, and especially if it coincides with another pre-existing belief system (in this case, related to the other paranormal claims), it's difficult to remain levelheaded and search for all the potential explanations. Even with all of our own experience—and we do think we're pretty good at these things by now—if one of our own people started to experience a paranormal phenomenon in his or her own home, we'd insist on a disinterested third party to investigate for exactly this reason.

At private investigations, we often have to remind people to maintain their usual routine to the greatest extent possible. We understand it can be difficult with all of us hanging around, but it's important to keep conditions as close as possible to the circumstances under which the initial experiences happened. We'd told Elizabeth the same thing and after our interviews were all done, she went about her usual business while we just sat and watched.

Around 10:00 in the evening, she rejoined us for a few minutes and said this was about the time she usually went to bed. She hadn't taken her medication yet but she always did so right before bed. She asked if we needed anything before she turned in—we didn't—so she bid us good night and went off to take her pills and retire.

Here, we need to make a confession. We snooped. Yes, yes, we know it's inappropriate, but we considered it highly relevant to this paranormal investigation, so one of our members took a glance at the medicine cabinet and found her medication included both potent sleeping pills and antipsychotics. No judgment, and we don't like poking around in other people's business, but that is the kind of information we need to be able to solve paranormal mysteries. We noted the discovery and went back to silently monitoring as Elizabeth slept.

The night passed uneventfully. In fact, the only thing we noticed out of the ordinary was just *how* uneventful the night was. Elizabeth didn't toss or turn or roll over even once. Most people will at least move around a little bit at night, so this was a bit strange, but we figured it was just the effect of the sleeping medication. Over the years in this unusual field of work, we've spent more than our fair share of time watching people sleep, so we're perhaps the most uniquely qualified people in the world other than actual sleep scientists to note her sleeping pattern seemed strange.

We didn't hear anything from the neighbors. No screams. No chanting. Nothing. The only thing that we did discover was that on the opposite side of the wall where she'd heard all the unusual sounds was the neighbor's television. This was the most likely source we could think of for any strange sounds from the adjoining unit. Obviously that doesn't mean they don't scream and chant on other evenings, but we can only report what we saw and heard, and all in all, it was a pretty boring night.

When Elizabeth awoke in the morning, she rushed back to our little monitoring station and said she wanted to review the video we'd taken during the evening. The cat, she said, had been even more active than usual. It was by her pillow all night, wanting to play. She said she had to keep pushing it away to get any rest at all.

This caught us a bit off guard. We'd been up all night watching the monitors and hadn't seen anything. Not only had we not seen a ghostly cat, but we hadn't seen Elizabeth moving at all, and here she was claiming to have been shoving the ghost cat away all night. We said so, and she just couldn't believe it.

Both for our sake (just in case we'd missed something during the night) and for hers, we replayed the video. It was incontrovertible. There was no cat on the video. Nor had Elizabeth stirred all night. She acknowledged what she saw on the tape but still had trouble accepting it. She'd experienced something firsthand and the tape showed something different.

There's a phenomenon called cognitive dissonance. It's the term psychologists use to describe the mental discomfort (in some cases extreme) when one discovers one has held two or more beliefs which turn out to be contradictory. In this case, Elizabeth's personal experience conflicted with what she saw on videotape. This is an unpleasant experience. There are different ways people can try to deal with cognitive dissonance. They can abandon one or the other of their beliefs, they can modify one or both of them, or they can try to develop a mental framework to accommodate the beliefs and explain away the contradiction.

Our most plausible explanation for these contradictory experiences, which we explained in some detail, was that likely the medication Elizabeth had been taking was keeping her in a very deep sleep and that she was just dreaming the cat, but that for whatever reason the dreams were particularly realistic and believable and so she couldn't easily differentiate dreaming from reality without external confirmation.

Alas, there wasn't much more we could do in this case. We suggested that if these phenomena kept bothering Elizabeth, she might consider talking to her doctor about adjusting her medication if necessary or checking in with a sleep clinic for more expert overnight monitoring than we'd be able to provide. She said she'd consider those options. We also said she should feel free to reach back out to us if her issues persisted or worsened or if she needed anything else. She didn't contact us again, so we can only hope things finally settled down.

19

The Girls and the Demon in the Closet

Many children believe they have monsters in their closet. It's a common childhood fear. But when an adult claims there's something monstrous—perhaps even demonic—living in her closet, then it's time to call in the paranormal investigators. Though this was only a short investigation, it covers quite a bit of important paranormal lore as well as human psychology.

Figure 19.1. Artist's rendition of the demon in the closet (image: Elderlemon Design).

It's a terrible thing to be afraid of one's own home. "Safe as houses" is a known phrase for a reason. Home is supposed to be our place of refuge against the troubles of the world. Whether due to family conflict or something paranormal, breaches of the sanctity of the home cut to the heart of our human experience. That's a big part of the reason why we're so eager to take on these private paranormal investigations. In some cases, we've encountered people so afraid of their own houses they were about to walk out from under their mortgage, to their potential financial ruin. This case, fortunately, hadn't gotten to that extreme yet but it did seem to tend slightly in that direction, so we responded as quickly as we were able.

This case takes place at a two-bedroom apartment permanently occupied by two female roommates in their twenties who we'll call Frannie and Giselle. Additionally, the apartment was often occupied part time by their boyfriends who we'll call Henry and Isaac.

Both girls were starting to get terrified of the apartment, but only one of them had directly experienced anything. Frannie said she'd seen and heard something in her bedroom closet. It was always dark and shadowy enough during those experiences that she could never really get a close look at it to be able to describe it in any detail, but she said she'd started to suspect it was something malevolent or dangerous, perhaps even demonic. For her part, Giselle was also getting frightened of it just based on what Frannie had told her, even though she'd never seen anything unusual.

By the time they reached out to us, things had gotten so bad that Frannie refused to stay in her bedroom with the door closed (which we imagine may have put some strain on the relationship when boyfriend Henry was staying overnight, though we didn't press that subject). Nor would she sleep with the lights off.

Our on-site investigation began as they all do, with an interview and tour of the property. At the time, only the two girls were present. At one point, a rather confused Isaac showed up and asked why a bunch of strange people were moving what looked to all the world like an unholy hybrid between a television studio and a mad scientist's laboratory into the apartment. Appropriate explanations were offered and he left without another word. We're not sure if he was spooked by the paranormal or if he just didn't want anything to do with something so strange. Either way, that worked out to our advantage because the fewer people present the better, especially when we were going to be trying to collect data in an already cramped apartment.

We set up our video and audio monitoring throughout the apartment but with a particular emphasis on Frannie's bedroom. We arranged the cameras so we could see the entire bedroom and the hallway leading to it, but also had multiple camera angles on the supposedly haunted closet so we knew we wouldn't miss anything that occurred in the most likely location.

During our initial walkthrough of the apartment, we scanned for EMF, as we pretty much always do. We did notice an unusually strong field near the head of Frannie's bed. There weren't any electrical outlets or appliances nearby so we weren't immediately sure what could possibly be causing it. As we watched our EMF meters, though, we noticed that it didn't seem to be a static field. It was pulsing or oscillating periodically. One of our team members suggested that pattern was reminiscent of the pattern produced by the transmission of wireless data.

We pause our story for just a moment to explain something important. Electromagnetic field meters you pick up for just a few dollars from an online shop or your local general store are unlikely to be sensitive enough to resolve the oscillating patterns of data transmission. But, while we do use those meters too, we also use much more expensive meters designed for industrial uses which are sensitive enough to pick up such signals. Determining which is better for paranormal investigation is a tradeoff. The more sensitive the device, the better your odds of noticing something like—as we did here—data-like signals. But with greater sensitivity comes a greater probability of a false positive. We've seen one of our meters detect the changing electromagnetic field produced by a person casually strolling down a hallway at the opposite end of a large building, as well as (on another occasion) an incoming thunderstorm several miles away. Less sensitive models are better for handheld use or general monitoring, but not sensitive enough to provide the full picture of the situation. This is why we use both—and importantly, why we know exactly what the limits are of each of our pieces of equipment.

Getting back to our story, we wanted to know if there could be a source of data transmission in the area and asked if there might be a cellular tower or something along those lines near the apartment building. Either Frannie or Giselle (alas, we can't remember which) opened the window, and sure enough, there was a cellular tower just outside.

We noted it in our records and explained some of our past experiences with electromagnetic radiation while we waited for nightfall and the remainder of the investigation. Importantly, we were not saying this cellular tower was the cause. It was a piece of the puzzle which might go some way toward potentially explaining some of what might have been happening, but in cases like this, one piece of data does not a mystery solve.

Later, as evening wore on but still well before bedtime (which we expected to be the most important time on this investigation), Giselle came running into the room. She'd finally had an experience of her own, and was now even more terrified because it wasn't just that she'd seen or heard something unusual.

No, she said something had just scratched her on the back.

Physical harm from paranormal experiences is exceedingly rare. Horror fans might be misled into thinking people are constantly getting hurt—or even killed—by demons or evil spirits. Lore does suggest these kinds of things happen from time to time, but our experience has been that they're almost unheard of in legitimate cases. In all our years, in all our cases, only a couple times have we found someone claim to have been physically harmed by something paranormal. More common (but still relatively rare) are cases in which less violent physical contact is made—a spirit grabs an ankle or touches a shoulder or something like that. Almost always, though, the experiences are visual or auditory rather than tactile. So this was quite an unusual claim for us.

With that in mind, and since Giselle hadn't been personally experiencing anything for herself until that point, we wondered if perhaps she'd just imagined it. Personal experience has shown that when you sit in a supposedly haunted location while everyone's telling the ghost stories, your mind can start to play tricks on you. Even with all our experience and our skeptical mindset, it still happens to us from time to time, so we always have to be cautious about these things.

Here, though, Giselle insisted this wasn't the case. Something had most definitely scratched her shoulder and upper back. She pulled down the back of her shirt to demonstrate the scratch marks. Sure enough, she'd been scratched. These weren't deep scratches, but we could clearly see the irritated skin. But we quickly spotted the culprit—there was a stiff tag sewn into the shirt's collar which looked quite scratchy. We asked if this was a new shirt. She said it was. We then asked her to rub the collar against her upper back and see if that felt the same as what she'd experienced. It did. So the whole affair turned out to just be a false alarm. Perfectly understandable under the circumstances, but this is why it's so important to always follow up every experience with further inquiry.

When bedtime approached, Frannie was reluctant to go to sleep. She was still terrified of what she might find lurking within—or worse, emerging from— her closet. We suggested an experiment in the interim. Instead of going to sleep, she could sit silently in front of the frightening closet so we could see if anything happened even while she was still awake. She was reluctant but she agreed to do so, and with the lights off and the door closed, as these were the circumstances in which the initial experiences occurred. Never to fear, though. We would be watching the whole time on our video monitors and if even the smallest thing out of the ordinary happened, we'd all come a-running.

A note here on methodology. Most ghost hunters (as distinct from paranormal investigators—a distinction we've explained in more detail earlier) tend to investigate in the pitch dark. It's creepier, for sure, but we think it's usually a bad idea. People rarely report having experienced anything in pitch dark. They certainly can't have actually *seen* something with no light. So we like to leave the lights on. This case was different. Though it wasn't *pitch* dark, the initial claim occurred during the night while people were sleeping, so the lights were off. As such, we wanted the lights off for our investigation, too. Therefore, we used infrared cameras so we could still see clearly despite the lack of light. Some lore suggests that infrared cameras should be used on paranormal investigations because ghosts are more likely to show up on that kind of video. We've never seen any evidence—on our own investigations or anywhere in the literature—to suggest this is the case. When we use IR cameras, it's purely and simply because they're able to see in conditions of low light or no light.

Frannie sat in the darkened room on the floor by her closet while we all huddled around the video monitors in the other room. As she sat down, she looked toward the closet door and it gently swung closed, just a little bit.

One of our investigators got up and ran into the room to see what had happened. Perhaps he was a little over-excited by the prospect of having witnessed something unusual because he sort of burst into the room.

"Did you just move the door?" he asked.

"No," she said. All she'd done was sit there. But we'd certainly seen the door move. So we went back, now with Frannie joining us, to review the tapes.

Initially, we'd noticed the door close on the main camera (whose view is shown in Figure 19.2), but fortunately we also had other angles on the scene. In one of them, we clearly saw Frannie sit down, glance toward the open closet door, and gently push it closed. For a minute, she couldn't believe her eyes!

Figure 19.2. This door closed during our investigation (photo: RMP archives).

You see, she was being completely honest—or so she thought—when she denied closing the door. When our investigator came into the room, he interrupted her train of thought so suddenly she legitimately had no memory of what she'd been doing.

This is a well-known phenomenon in psychology. Imagine you're in school, listening attentively to your professor's lecture, when someone suddenly drops a large book on the desk next to you. The interruption is so sudden and so disconnected from what you'd been consciously attending to that it completely breaks your train of thought. Even just getting called on in class—as many students will attest—can have a similar result. You're sitting there listening attentively and the professor unexpectedly calls on you to answer a question. Maybe you're the best student in the world, but in that very moment, you'd be lucky to remember your own name, much less the solution to a differential equation or the ultimate and proximate causes of the Second Messenian War or whatever the professor asked.

Memory is not like a tape recorder that linearly records everything that happens to us. It's episodic and is processed and stored according to relevant topical groupings. When something unexpectedly shifts us from one topic to another, we often forget what we were doing in the former context. There's even a related phenomenon called (funnily enough, given our subject here), the "doorway effect," in which people forget what they were doing when they cross a boundary like a doorway. Shifting from one location to another is sufficient in some cases to make people forget what they were up to in the prior location, even if that memory is directly related to why they went to another room in the first place.[1] We

1   Radvansky, G. A., Tamplin, A. K., & Krawietz, S. A. (2010). Walking through doorways causes forgetting: Environmental integration. *Psychonomic Bulletin & Review, 17*(6): 900-904.

all know the feeling of stepping into a room and immediately asking, "what did I come in here after?" Frannie's experience next to the closet is a closely related phenomenon. That she was also afraid of the closet at the time probably made forgetting her action all the easier.

No harm done, though. We were able to figure out what happened and moved on with our investigation. Unfortunately, nothing else of interest happened.

At the end of the day, though we managed to debunk a couple tangentially-related claims like the scratches on Giselle's upper back, we didn't fully prove or disprove the underlying paranormal claim. The best we could offer were some hypotheses we thought might be worth testing. One such idea was that EMF radiation or even just simple exhaustion could be causing hallucinations. Once a belief in a paranormal entity is established, even minor perfectly normal sights and sounds can get blown out of proportion. On the other hand, though, just because we didn't witness a paranormal event doesn't conclusively mean there never was one.

To test these conflicting hypotheses, we suggested keeping a journal of any unusual experiences so Frannie (and Giselle, if she were so inclined) could examine each phenomenon with fresh eyes. Sometimes a pattern emerges. Other times, simply looking at it again when one isn't sleepy is enough for one to realize what the real (natural) cause might have been. Further, we suggested Frannie might consider sleeping in a different room for a while to see if putting more distance between herself and the electromagnetic radiation might have an effect.

She agreed to try these for a few weeks, after which we'd follow up and see if things had improved or if we needed to come back out for further investigation. When we followed up, she said she'd followed our advice and the paranormal issues had ceased.

Another case solved. But even though we immediately knew the moving closet door was not a paranormal phenomenon, this does teach an important lesson about the reliability of eyewitness testimony. Often, people will say that if a paranormal claim isn't really true, then the witness must either be crazy or lying. Frannie's experience at the closet door proves this wrong. She certainly wasn't crazy, nor was she lying. She just genuinely didn't remember moving the door. Other people with more outlandish claims may similarly be perfectly sane, credible, and completely honest but simply victims of faulty human memory or mistaken interpretations of events. Skeptics in particular would do well to remember that just because they don't believe in the ghost or the demon or the alien, that doesn't mean the person who does is crazy or lying.

20

The Noisy Demons

A constant source of fascination for us is the way people of different religious or cultural backgrounds come to such different interpretations of events. Show the same fact pattern to a scientific skeptic, a Christian, a Pagan, and a Buddhist, and you're likely to get *at least* four wildly different stories, all of which could be seen to explain the facts from within those individuals' own philosophies. In this case, we were called by a Christian household who believed their home was being overtaken by demons.

Figure 20.1. Artist's rendition of the noisy demon house (image: Elderlemon Design).

The clients who called us on this case were a devoutly Christian middle-aged couple we'll call Joseph and Katherine. They'd just purchased a new single-family home in a suburban Colorado city and almost immediately started hearing strange sounds throughout the house. They told us based on their religious background that they thought their house was under some kind of demonic attack and they were terrified of the whole thing and wanted whatever help they could get.

The specific sounds they reported including knocking on the doors and walls, even though no one was there to knock, the sounds of footsteps throughout the house and the attic, and just generally odd sounds all through the house almost all the time.

Though we personally don't like leaping to the conclusion of "demon" just because someone has been experiencing something frightening, we did take note of the kinds of things they were claiming. If indeed a demon were to blame for the noises they were hearing, it would be consistent with what's called demonic "oppression," wherein the demonic forces attack people through external phenomena (in this case, such as odd noises throughout the house).[1]

Whenever someone suggests there are demonic attacks in their home, we have to ask about the condition of their home life. It's not that we're being nosy. It's just that experience has taught us sometimes these can be fraught relationships. If one believes in and wants to accept the case as one of demonic possession, it would be important to make sure spousal abuse wasn't a resulting behavior. Similarly, from a skeptical perspective, we want to make sure this isn't a case in which some violent form of mental illness is just being blamed on a demonic influence (e.g., the "Demon Made Me Beat Her Up" case in Volume 1, Chapter 21 of this series; see also Chapter 22 of the present volume for an even darker case).

Fortunately, nothing of the sort was the case here. By their own description, Joseph and Katherine were a happy and loving couple and their problem was purely and simply with the house itself.

By the time they contacted us, things had gotten so bad that Katherine said she would no longer even set foot on the property. She'd moved in with her parents to get away from the house. The day after they initiated contact with us, Joseph called to schedule a date for the investigation. He said it needed to be soon because he, too, wanted to flee the house and that he planned to leave the morning after our investigation never to return. We tried to caution him to let us check it out before jumping to any conclusions, but his mind seemed made up. We hoped we might be able to find natural explanations for the phenomena sufficient to convince the couple to at least wait until they could sell the house at a profit rather than fleeing to their financial detriment.

When we arrived, though, we knew they were already on their way out. Upon parking out front, we noticed all the windows and doors were open and they were blasting loud music that could be heard from the street. As we got closer, we found Joseph had tuned all the radios to a Christian broadcast and seemed to be trying to fight the demons or to drive them out with the religious messaging

1   Ripperger, C. (2022). *Dominion: The Nature of Diabolic Warfare*. Keenesburg, CO: Sensus Traditionis Press.

and music. But they were clearly not planning on staying regardless of the out-come. Other than those radios and televisions and a bare mattress on the floor of the master bedroom, the home had been emptied of the couple's possessions.

Joseph gave us a quick tour of the house. We asked which locations seemed to be the worst in terms of paranormal phenomena so we'd know where to focus our attention, but he said they'd been hearing things throughout the entire house. It didn't matter which room they were in; they'd hear something evil.

Each of the doors we came across on our tour seemed to have some kind of a substance on it. Joseph explained he and his wife had previously had clergy to the home to apply a blessing. While there, the clergyman had also anointed each of the doors with oil. Apparently, it had not had any effect on the demonic noises.

When night fell, Jospeh said he was going to try to get some sleep and that we could have free run of the house to investigate however we pleased. He retired to the bare mattress on the master bedroom floor and fell asleep almost immediately, and we set about on our usual practice of alternating between silent monitoring and more active measuring and exploration.

We heard the noises, too. They started almost immediately. Indeed, there were sounds coming from all over the home.

The first sounds we noticed were the noises of the heater and air condition-ing ducts throughout the house expanding and contracting as the temperature changed. We immediately knew what those sounds were, but we can also under-stand how they can trouble someone in a new house. Every building sounds a little bit different and if you're predisposed to think in terms of demons, you're likely to jump to conclusions. That's doubly true in this house because these duct noises were *loud*. We might have recognized them for what they were because we've been exploring so many buildings for so long that we're used to just about every noise a house can make, but even we were surprised that this house just liked to be loud for some reason.

Then we heard the sounds from the attic. Every horror fan knows when you hear sounds in the attic of a supposedly haunted (or in this case, demon-infested) house, you probably shouldn't open the hatch and check the attic because you might find yourself face to face with something big, green, and evil. Well, despite being lifelong horror fans, we never got that memo. Once the noises started, we immediately went up the ladder to check them out.

One thing actually did give us a momentary fright. It was obvious that at some point someone had been living in the attic. Fortunately, whoever it was seemed to be long gone and was not the source of the noises. No, the current noises came from a family of birds who'd decided to roost up there. What we'd heard were the sounds of these avians walking and fluttering about in their make-shift home.

Noises didn't just come from inside the house. Exterior sounds, also quite loud, were common. Per Jospeh's request, the windows were still open. Plus, the insulation in the home seemed poor. Between those two factors, we got quite the earful of everything that was happening nearby in the neighborhood. Among those outside noisemakers were an incredibly busy intersection, nearby railroad tracks, a Denver International Airport flight path, and a nearby bar whose patrons loudly poured onto the streets and started revving their car engines when the bar closed.

We're not exaggerating when we say this was—bar none—the single noisiest house we've ever encountered. But none of the noises we heard seemed to have any paranormal source. It was just noisy.

Further, we whipped out our measuring equipment and determined that both the railroad tracks and the airplanes flying low overhead produced low-frequency sounds and vibrations which, as we've explained in several chapters, can sometimes cause odd feelings or hallucinations. In this case, because all the claims were related to sound, we didn't think hallucinations were to blame, but we did wonder if perhaps the odd sounds combined with odd feelings resulting from low-frequency vibrations might be partly to blame for Joseph and Katherine thinking the house was so evil.

By morning, we figured we had a pretty good handle on the situation. We didn't think anything paranormal was happening. Instead, it was a combination of several natural but annoying factors. When Joseph awoke, we explained our findings, hoping we might convince the couple to just invest in some better insulation instead of fleeing their own home. Joseph didn't reject any of our findings out of hand, but also wouldn't let go of his belief that something demonic was afoot.

They left the house that morning and never returned.

21
Demonic Attack

It's one thing to think your house is haunted or inhabited by a demon. It's quite another thing to believe you've been victim of a sexual assault by the demon. Such was the case here. In many ways, both because of the severity of the claims and because of our investigative findings, this is one of the most serious cases we've ever worked. Ironically, its early stages were also among the funniest, and parts of this story would not feel out of place in a theatrical farce or a comedy of errors.

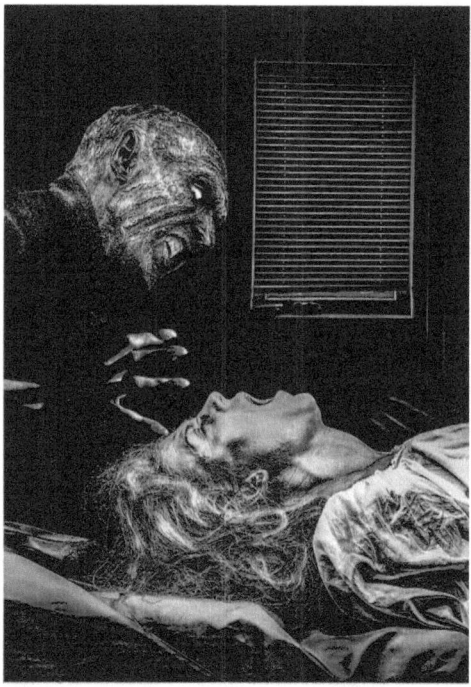

Figure 21.1. Artist's rendition of the demonic rape (image: Elderlemon Design).

Throughout these books, we've mentioned how much we prefer to be the first investigative team on scene. It's not that we're saying every other team is incompetent. Far from it. There are teams out there we would readily invite to join us on any investigation and to whom we would gladly refer clients if we weren't in a position to take the case on ourselves. However, there are also a lot of shady fly-by-night operations in the ghost hunting world, and choosing the wrong group can end up doing a lot more harm than good.

This is one of the cases in which a different group showed up first, and our description of what happened with them will be the funny part of the story. Unfortunately, things quickly took a darker turn and this ended up being one of the more tragic cases in our files (though the singular most tragic case will be the subject of Chapter 22). We don't think the other team really caused too much harm here, but the confusion of having different teams involved (some more competent than others) probably didn't help matters, either.

Our involvement in this case began later than usual in the process because we were referred to the clients by another team who thought they'd gotten in over their heads. The prior ghost hunters had conducted an "EVP session"[1] and determined the case to be "worse than Amityville," so they just couldn't deal with it themselves and thought we might be of more help to the clients.

With regard to Amityville, we're generally of the opinion that the case is overblown. The murders were real, and the books and movies resulting from the tale are wonderful horror stories, but the story of the paranormal phenomena there has largely been debunked, both by skeptical researchers[2] and by later confessions by some (not all) of the (alleged) hoaxers themselves.[3] That said, it has such a reputation that another ghost hunting group telling us a case was even worse than that one certainly caught our attention and we immediately started reaching out to the relevant parties to see what we could do.

This case involved a single-family home occupied by a couple in early middle age who we'll call Lawrence and Mary. They also had two children who we'll call Nancy and Oliver. Both children were teenagers and the daughter was slightly older than the son.

When it came time to write this chapter, we struggled quite a bit concerning how to present the narrative. Usually, we like to lay out all of the paranormal claims and then describe our investigations. Here, though, the claims didn't come to us all at once, so we've decided to present the case in a linear narrative, describing events in the order they came to us regardless of the order they happened in reality.

When the former team of ghost hunters called us, we happened to be out of town on another investigation, so we agreed to take the case but said the best

---

1   Recall, this is the practice of listening for ghostly voices on audio recordings.

2   For instance: Kaplan, S. & Kaplan, R. S. (1995). *The Amityville Horror Conspiracy: The dramatic true story of an incredible twenty-year investigation.* United States: Belfry Books.

3   For instance: Snopes (2005). Was 'Amityville Horror' Based on a True Story? *Snopes.* <https://www.snopes.com/fact-check/the-amityville-horror/> (accessed August 24, 2025).

thing to do would be to start by reviewing the evidence they collected remotely. They said they'd conducted quite an eye-opening and terrifying EVP session and also collected numerous photographs, all of which we wanted to examine. We gave them a good email address and asked them to pass everything along.

Here begins the comedy of errors.

Every time they tried to send us an email, it bounced back because the files they were attaching were too large. In return, they received the standard bounced email error message from "Mailer Daemon." Computer-savvy people know exactly what that is and what to do with it. But these were earlier days on the Internet, and the other team were emphatically *not* tech-savvy people. They took these error messages to mean that a literal demon—presumably the one torturing the clients—was stopping the emails from getting through to them.

Occasionally, we have a bit of a prankster streak when it comes to dealing with certain other ghost hunters. We would never mislead a client, but sometimes we will stoop as low as to have some fun with the ghost hunters. So we told them that to confuse the demon, they should break the emails into smaller messages and rename their files as "cute puppy." The demons wouldn't take notice of such files and so their messages would go straight through. Maybe it was a mean-spirited joke, but it did manage to solve the communication problem and we started receiving their files.

To say we were unimpressed by their evidence would be an understatement. All we could hear in their EVP recording were their own voices and their own footsteps as they shuffled around the house. All we could see in their photographs were images of the interior of the house. Nothing even hinted at anything paranormal. Worse than Amityville indeed! We hadn't yet gotten to the point at which things actually did take some turns for the worse.

The first hint of that came in a follow-up communication from the other ghost hunters. They started describing some of the family's behaviors and it sounded to us a bit too much like the "Demon Made Me Beat Her Up" case (described in Volume 1, Chapter 21) for our comfort. In that case, a husband, believing himself to be influenced by a demon, assaulted his wife on multiple occasions. We didn't have any direct evidence of that here, but our antennae were starting to perk up and we knew we needed to check up on this family just in case.

Fortunately, we'd already scheduled an on-site meeting and preliminary investigation with the family to be held not too long in the future. The evening before we were to head over, we called ahead to confirm the appointment. It was going to be an extremely long drive for us, so we just wanted to make sure the family remembered the appointment and would be there to greet us. They didn't answer the phone. Our antennae started to twitch even more. We tried to reach them several times by phone and they never answered, so we started to become concerned something might have happened to the family.

Because it was such a long distance away and we couldn't seem to reach anyone, we contacted their local police department to see if we could request a simple welfare check just to make sure they were okay. We explained to the officer or dispatcher who answered that we were with a local paranormal investigation group and that our clients weren't answering the phone so we wanted to see if they could—and we cannot emphasize this part enough—just swing by to perform a simple welfare check. While on the call, we heard the person on the other end mute the line just a little bit too late to mask the laughter. Apparently they couldn't quite process the unusual phone call. Nevertheless, they agreed to send an officer over to make sure everything was okay and we said we'd call back the next morning to check in.

Morning arrived and we called the police to follow up. No one had idea what we were talking about. No one there remembered our call and they had no record of either our call or of a welfare check. Maybe they thought it was a prank call and just ignored it, but we are extremely unimpressed with that kind of result. They apologized and said they'd send someone over (again).

But it turned out we didn't need to wait for the police to report back. While all of this was happening, we checked in on the social media pages of the other ghost hunting group. They posted that a family they were working with was upset because they'd missed a meeting and had called the police instead of showing up. According to their post, when the police arrived, officers told the family that a ghost hunting group had called them because they thought the family was "crazy." The gist of their post was that they'd never do such a thing and that they were only there to help with the demons. We're still not sure why no one in the family answered our call the night before, but the whole situation started to make at least some degree of sense.

We called the family again—successfully this time—and explained the whole situation. It was us rather than the other group and we certainly never called anyone crazy but just asked for a welfare check because they weren't answering their phone. By this time, it had gotten too late in the day to carry out our planned investigation, so we re-scheduled for the following weekend, and everything seemed to have been smoothed over. Now, we'd be able to just conduct our investigation without any more crossed communications or misunderstandings. We figured we might have a few choice words with the local police department later but ended up getting too distracted by the events that followed.

When we arrived for the investigation, we began the same way as we always do, with introductions and a tour of the house. The entire family was present at the time, so we could collect any stories they all had. For reasons which will become obvious in a moment, though, we chose to conduct our more in-depth interviews with the clients separately so we could see what each might say without the others in earshot.

Of course we asked where the main points of interest were and they said they had unusual experiences everywhere, but there were several spots they thought we should focus on primarily: the master bedroom, the daughter's bedroom, the son's bedroom, the back yard, and an unfinished closet. We'll take each location in turn.

Teenage son Oliver reported that he'd seen mysterious shadows in his room. We marked that room for monitoring during investigation, but we also pointed out to Lawrence and Mary that it appeared at just a glance through his room that their son had a drug problem which might be affecting his experiences. All throughout his room were makeshift pipes and empty plastic baggies. But they refused to believe their son could be partaking of anything illicit and wouldn't hear any more about it.

As for Nancy's bedroom, it contained a set of bunk beds even though she was the only one using the bedroom at the time. She told us that one day while she was in bed, her boyfriend came into the room and the blanket tightened around her and pulled her off the bed. She was adamant her boyfriend hadn't done this. It was more like some spirit or demon did it because it got upset that he'd entered the room.

The back yard was also a point of interest. Everyone in the family reported having seen shapes or figures moving in the yard at all hours of the night. Lawrence and Mary further said they'd seen someone looking into their bedroom from a door (near the foot of their bed) which went directly to the back yard.

The unfinished closet only had one story attached to it, as far as we could learn. While sorting through some things in the closet, Lawrence said he found an old book he didn't recognize. When he opened it, he was immediately knocked out and didn't wake up until several hours later, somehow now on the living room floor. We never learned what became of the book.

The parents' master bedroom, though, was the location of the most troubling stories, and there are several of them.

According to Mary, Lawrence would often sit up in the middle of the night and start babbling in a language neither of them spoke. Speaking in tongues is an established belief in a number of religious faiths and denominations, and it comes in different flavors. Sometimes speaking in tongues is a spiritual experience in which one babbles in a language which doesn't really exist as a spoken language on Earth. Other times, individuals are claimed to actually speak in a specific foreign language—one they don't know how to speak. Reports vary, but often the former is considered a positive spiritual experience or even a gift of the Holy Spirit, while the latter is often taken as a potential sign of demonic possession. When asked which was the case with Lawrence, Mary wasn't certain.

Lawrence also sometimes walked in his sleep. Regularly, Mary said, he would walk into the back yard and just wander around all night. In the morning, he never remembered any of it. Mary said she'd seen him leave and come back, but wanted to know what he'd been up to and where he walked all night so she went through his phone's GPS data. Specifically, she was looking for evidence of an extra-marital affair, but all she found was that it showed the path he took up and down and around the back yard on his walks.

We examined the phone for ourselves and we don't believe this story. For one thing, it sounded a bit unusual to us that a sleepwalker would bring his phone with him into the back yard in the first place. Weird but not impossible. However, the phone wasn't like modern smartphones with highly accurate GPS navigation functions. It would have been incapable of producing the kind of report Mary described. The accuracy of consumer grade GPS in phones of that era was around 20-300 feet depending on the equipment. The yard was only around thirty feet at its widest area. Whether she was lying to us or just misunderstood something she saw is not entirely clear, though we suspect the latter.

By far, the most troubling claim about the master bedroom was the one they saved for last. They said Mary was regularly raped by a demon, and they had the video to prove it. Obviously we need to take these kinds of claims seriously. Whatever the cause, these are traumatic claims and people need to be referred to whatever kind of professional help seems the most important. But determining what kind of professional help is required depends on the true nature of the phenomenon, so we continued our investigation as gently and politely as possible.

They showed us the video proof of these assaults. It had been taken with the camera on a flip phone, so it was of low quality. Even still, we could clearly see the entire scene. All we could see in the entire video was Mary lying in bed asleep, occasionally rolling over. No demon, no flailing about, nothing that seemed to corroborate the story we'd been told. But that both Mary and Lawrence believed the story and thought the video proved it suggested that whatever was really going on, this was all too real for both of them.

At this point, we'd seen enough that we knew which areas we wanted to monitor, but we also thought before we even started with any of that, we needed to sit down with Lawrence and Mary to discuss some potentially sensitive topics. We all gathered in the living room for a polite but frank and open conversation.

We asked, other than all the paranormal stuff, what life was like in the household. What were the family dynamics like? They confessed there was a lot of strife and major fights were a regular occurrence. Divorce had come up on several occasions, but they'd agreed to wait until their youngest child turned eighteen. Furthermore, Lawrence admitted they'd recently had to kick out a friend of his who'd been living in their basement and caused a lot of trouble for them. Said "friend" was apparently manufacturing and selling methamphetamine out of their home!

We explained they should have their house checked for contamination from the drugs because those could be causing any number of health or mental health issues. We refrained from adding that they should also warn paranormal investigators before inviting us in to such a house, but the thought crossed our minds. We also asked if the figures they'd been seeing in the back yard might be the "friend" they'd kicked out, either spying on them or looking for a way back in. Or perhaps it was some of his customers who didn't realize he wasn't there anymore. They admitted it could be. So for that, we recommended setting up some kind of a security system with lights and cameras.

While on the topic of drugs, we again mentioned all the drug paraphernalia we found in Oliver's room, but they still wouldn't hear any of it. They did admit that they were both addicted to some illegal substances and that Mary took a variety of prescribed medications for different conditions. She'd recently stopped taking the legal drugs, though, because she thought the illegal ones were helping more.

At this point, we said we thought the drugs, stress, potential lurking drug dealers, and other family dynamics were likely the root of the paranormal experiences. Even if something paranormal were afoot, we said, it was probably brought on by these issues.

We asked if they were attending any sort of counseling or therapy. They weren't, but they said they did sometimes attend a local church. We asked if they'd spoken to their clergy about any of these issues. Lawrence said a few weeks prior he'd driven to the church to do exactly that. While in the parking lot, though, he found more drugs in his car and took them instead.

No one wants to hear a lecture on morality from a paranormal investigator.

Still, we gave them the same argument we made in the essay which began this part that whatever the cause of their unusual experiences, the most important thing they needed to do in order to solve the problem was to clean up their own lives. We said we could refer them to any number of qualified professionals—doctors, therapists, clergy, whatever seemed most appropriate to them—who specialized in cases just like theirs. Until they did that, there probably wasn't going to be very much we could do.

Instead of listening, they got angry and threw us out of the house.

We dutifully left but told them to feel free to reach out whenever we might be of assistance.

A few weeks later, we received a follow-up letter from Lawrence. He said Mary had to be admitted to a hospital because she'd severely hurt herself while trying to remove the "worms" she saw moving under the skin of her arms. Eventually the children came of age and moved out of the house. Mary and Lawrence divorced. We've had no further contact from any of them.

Sometimes when we share our stories with skeptical people, they say the claimants must all either be crazy or on drugs. We always correct them. Many of the cases in this volume show that perfectly sane and sober people do have unusual experiences. Sometimes, though, the drugs really are the problem. We hope they all got the help they needed.

22

Prison Time for the Clients

We'll warn you right from the start. This is the worst case we've ever seen. Indeed, in all the paranormal literature through all of history, there are only a handful that have bothered us more than this one. It started off much like any case involving a family who believed their home was under demonic attack, but it ended in tragedy for all involved.

Figure 22.1. Artist's rendition of the levitating child (image: Elderlemon Design).

We gave serious thought to omitting this case from our books. When we say it's the worst case we were ever involved in, we're not kidding. It's really that bad. Still, we decided it was important to present the dark side of the paranormal world in all its ugliness as a sort of warning to readers. But just be forewarned, this is going to be a tough one to read. It's certainly a difficult one for us to write.

We should also note that because certain elements of this case are the subject of public records and of news reporting, we also went back and forth in a debate regarding whether to present the entire case—media reporting included—or to anonymize everything to the extent readers would never be able to figure out which case we were talking about. In the end, we decided the best course of action would be to present the case fully and truthfully, but to change all of the names and omit dates, locations, and certain unimportant details out of respect for the innocent individuals involved.

It all started much like any other case. We got a call from a family who believed their family home was haunted and perhaps even infested by demons. There were four people living in the house at the time—the two parents, an older son (roughly in his preteen or early teen years), and an infant daughter. We'll call the parents Paula and Quentin, the son Roger, and the daughter Sarah for the purposes of this discussion.

Per the clients' request, we conducted our initial consultation and interview at their home. Normally, we prefer to exchange emails for a time, then perhaps meet at a neutral location such as a library or coffee house, but this was what the clients preferred so this is what we did. When we arrived, all four members of the household were present. Our initial impression was that the home was extremely dirty. Boxes of God-knows-what had been stacked up in just about every corner of the house. But Quentin and Paula seemed reasonably friendly, if a bit "out there" in their beliefs and demeanor. And while the house seemed unpleasantly dirty, it didn't seem bio-hazardously so.

We asked what they'd been experiencing that led them to call a paranormal investigator to their home. They had a laundry list of claims.

They had seen shadows walking around the door leading from the kitchen to the garage. When asked for a bit more clarification, they described what might be called "shadow figures" in the paranormal lore, sometimes thought to be mere spirits but often considered to be malevolent entities.

Sometimes they'd see "demonic eyes" peering in at them in the bathroom window.

Recently, they'd installed a security system and said they saw "orbs" all throughout the house. The security system also recorded a shadow going down the hallway from the kitchen toward the bathroom.

A friend of theirs was a self-professed psychic who came to the home on a regular basis to speak to the ghosts and demons alike and to give the family warnings about things she thought would happen in the future. When you get to the end of this chapter, you'll realize she apparently forgot to mention the big one.

Quentin told us he'd done some research about the home's background and history and discovered it was on ancient "killing ground" where there had been some kind of a Native American massacre. He thought perhaps that might explain why the home was haunted, cursed, or attractive to demons.

Several times, infant daughter Sarah woke her parents up in the middle of the night. When they went to check on her, they found she'd somehow gotten out of her crib and was now sitting in the middle of the room. They could never figure out how she managed to escape.

On one occasion, Paula even walked in to see Sarah literally levitating above her crib!

Finally, they said they regularly had trouble with their electrical equipment. Lights would turn on in rooms where they'd already been turned off. Sometimes the microwave would run itself without any food in it.

All in all, they were convinced something bad was going on in their house and they were desperate for someone to shed some light on everything.

Though they didn't volunteer this information immediately, we later learned through various correspondences with the family themselves and with other ghost hunting groups they'd brought in that they often conducted "investigations" of their own. Sometimes Quentin would use his son Roger as bait, sending the lad into the haunted room to see if the ghost or demon was there. While we do use ourselves as "bait" regularly on our investigations, we're adults who are experienced in such matters and are disinterested third parties rather than members of the household in question. Suffice it to say, we don't recommend using your own young children as "demon bait."

At another point during their saga, Paula apparently contacted a psychic from out of state who, for a not-so-small fee, claimed to have been able to remotely cleanse the house of demons and evil spirits. It didn't work.

We arranged for an on-site investigation to take place a couple weeks later. When we arrived, we were in for one of the most humorously shocking sights we've ever seen. Not only was the entire family in attendance to greet us at the door, but so were most of their neighbors and some of their friends. They'd even put out deli trays to keep everyone fed during the event. We had to very politely explain that there was no way we could collect quality data with what looked like half the town joining us in the house. Therefore, the deli trays could stay—with our thanks—but everyone else except us and the family had to go. We needed to be able to monitor the house under conditions as close to normal as possible, and the paranormal phenomena didn't occur when so many people were around.

We assumed we'd been the first paranormal investigation team to look into the house, but at this initial investigation we were told we were actually the fifth! One other group had been working on the property for at least two years already, but the clients weren't happy with their results. They told us this because they also gave us several discs containing the notes, data, and reports from one of the other teams, which they wanted us to examine. We'll get to that when we discuss the rest of our findings in a moment.

Later on, well after our own investigation, we came across some correspondence between Paula and a representative from one of the other paranormal teams. According to the investigator, they were backing out of the investigation because they didn't appreciate the family bringing in other teams while their work was ongoing, the family dismissed their concerns about elevated electromagnetic radiation, uncleanliness, and cigarette smoke, they weren't finding anything

terribly interesting, the clients were only sporadically responsive in their correspondence, and the family (Paula in particular) simply wouldn't accept that there
weren't ghosts and demons lurking around every corner.

Most troubling of all from this report was a claim that Paula had said Quentin had served time for beating her but that the ghost was making him do it.
Nothing of the sort was said or done during our own investigation, but that the
other investigators took it seriously is somewhat concerning to us. Had we been
given this information before our investigation instead of much later, we might
have proceeded differently.

We also happened across Paula's response. It's a multi-page rant full of mis-
spellings, non-sequiturs, and idiosyncratic capitalization and punctuation in which
she accuses just about every other group of not caring, but praises Rocky Mountain Paranormal for being caring, thorough, and logical. As much as we'd like to
say this is because we genuinely are those things (and we certainly try to be), it
seems more like we were just the paranormal group that happened to be in her
good graces that week and hadn't yet given her answers she didn't want to hear.

During our one night at their home, we monitored all the relevant locations
as usual following our standard protocols and also spent some of our down time
examining the data from the prior team's disks. One thing we didn't take any
notice of at the time but which seems odd to us in hindsight is that infant Sarah
didn't wake up or make a sound all night, despite being an infant with a bunch of
strangers traipsing through her house (including her nursery or bedroom).

By the end of it all, we thought we had a pretty good explanation for at least
most of the phenomena they'd been reporting. We'll start with the videos we
examined from the other team and then talk about our own findings.

One video was supposed to show a mysterious figure in the kitchen. We
determined it was actually a reflection of the camera's infrared light bouncing
off a wall next to the refrigerator. Its apparent motion was caused by the actual
motion of the camera.

Another video purported to show a bolt of light shooting from young Roger's hand. At the time, he was sitting next to Sarah's crib (again, she didn't move).
At one point, he moved his hand, and once again, the infrared beam from the
camera bounced off his hand and back to the lens, causing a momentary bright
spot.

A third video was taken after the prior team had performed a "smudging"
or "cleansing" ritual in the house (see Volume 1, Chapter 16 for more on smudging). Supposedly some kind of energy flow was captured in the video. In reality, it
was just a bit of smoke from their own ritual wafting back into the camera's view
(smudging generally involves the burning of sage, as it did in this case).

Yet a fourth video claimed to show "weird stuff" and a "bright light." It
was actually just a hair stuck to the bottom of the camera's lens, appearing as an
ill-defined bright spot. The video also contains a lot of orbs, which were just dust
particles, as to be expected particularly in a dirty house (see Volume 1, Chapter
24). A fifth and final video purported to show "lots of activity" consisted of
nothing but large numbers of these supposed orbs, all of which share the same
explanation.

We shared these analyses with the clients and they seemed to at least hear us out, but they also said they weren't sure we were correct. They weren't yet pre-pared to give up their paranormal explanations so easily. Honestly, that's pretty common. Sometimes if you've believed something for a long enough time, it takes longer than just one conversation to change your mind.

But we had plenty more. By the end of our overnight investigation, we found that we hadn't experienced anything we couldn't explain, and we had come up with a lot of either definitive or plausible explanations for some of the specific things the family had reported.

We actually saw the demonic eyes in the bathroom window. It caught one of our team a bit off guard while he was using the facility, but he quickly realized what it was—taillights from a car on nearby roadway. Close enough to still see the lights but far enough to be unable to see the rest of the car in the dark.

Similarly, the shadow figures moving down the hallway turned out to be shadows cast by car headlights passing in front of the house. The home was lo-cated at a busy intersection, so these were quite common occurrences.

Prior to our on-site investigation, we'd gone into the archives to study Quen-tin's claim that the land had been the site of a Native American massacre. We found absolutely nothing. All we found was the home's real estate record and that it had been an empty field before the home was built. Admittedly, there have been a lot of violent incidents in the past, so we asked where he'd gotten his informa-tion so we could do some more digging on follow-up. He was unable to answer.

The incidents involving young Sarah were more speculative, but we think we figured it out even though we didn't want to wake her up or cause a disturbance attempting any experiments. The first story was that she'd managed to somehow get out of her crib and onto the middle of the floor. We didn't see it happen (again, she didn't stir all night), but we did explain that often babies learn to climb a lot earlier than parents would expect. Maybe she'd just found her way out by purely natural means.

That she'd been seen levitating was tougher, but we thought we had an ex-planation for that, too. Over the years, we'd heard stories of people having epi-leptic seizures being reported as having levitated. Obviously they were really just flopping about, but a major seizure can be surprising in how violent of move-ments it can produce. Not taking the time to measure the actual distance between body and bed under such circumstances, a panicked parent could easily believe a child was "levitating" a lot higher than in reality. So we suspected young Sarah might have epilepsy and told her parents they needed to get her examined by a doctor as soon as possible. If it turned out to be something else, we'd rethink our conclusions and try further investigation, but if we turned out to be right, she would require medical treatment.

They didn't particularly like our answers, though they didn't react negatively either. They just continued trying to argue that there must be something super-natural going on. Every time we knocked one idea down or suggested a natural explanation, they came back with another one. Something we've observed over the years is that if people genuinely believe they have a paranormal occurrence, or sometimes if they just need someone to talk to, they will (consciously or un-consciously) keep building the paranormal mythology and increasing its severity until they get an investigator's attention. After receiving our report, they told us another group had brought a device in which could talk to the dead. There are several devices on the market claimed to be able to do that, and we happened to be familiar with the one in question. It was just a broken radio.

As we packed up, we reminded them to seek medical attention for Sarah's potential epilepsy but also suggested that given the amount of stress they were under, they should probably seek some kind of professional counseling. Whether or not their house was haunted, we thought they could probably benefit from that kind of help. They said they'd consider it, but we now realize they probably never did. We said we'd review all of our own recordings as well as what we'd been given by other teams and get back to them with any additional findings.

After the investigation, Paula started calling us several times per day to fill us in on all the new phenomena and evidence she'd noticed. Sometimes she'd even call just to discuss the news of the day. Other times, she wanted to complain about how "mean" Quentin was. Eventually we had to ask her to stop, to only contact us if she had real business for us, and that if she just needed to talk to someone, she should contact the professional help we'd recommended. The next phone call came from Quentin, who said Paula told him we'd advised her to leave him (we hadn't) and he wanted to challenge us. We told him the same thing—we never told her to leave him, but we recommended for both of them to seek professional counseling. The phone calls continued and escalated to the point of harassment (at one point we considered petitioning for a restraining order) but after about two months the calls finally stopped.

Fast forward several years. One of our team members was watching television when a news story came on telling viewers that Paula and Quentin had been arrested and charged with felony child abuse. Child welfare agents had asked the police to perform a welfare check. When police saw Sarah (no longer an infant) she appeared malnourished and with her head flopping forward. She had a seizure in their presence.

In the investigation that followed, police learned she had indeed been diagnosed as epileptic but was never given adequate treatment. Instead, her parents poured salt on the floor around her to keep the demons away. According to a copy of the arrest warrant we obtained, the parents stated that they did not believe that the issue was medical but spiritual in nature. When she was finally stripped from their care (and charges against them filed), she was seven years old, weighed only thirty-seven pounds, had no communications skills other than the ability to say yes and no, was no longer toilet trained, and was estimated to have the cognitive abilities of a two-year-old.

When we saw the report, we reached out to the detective in charge of the case and provided our own case notes. Our testimony was not required at trial, though, as the case really spoke for itself. These were parents who were so invested in the paranormal explanation for their problems that they wouldn't listen to the doctors, they wouldn't listen to us, and they wouldn't even listen to the warnings of other groups who are often known for being more willing than Rocky Mountain Paranormal to assume the supernatural explanation.

Both Quentin and Paula were convicted of felony child abuse and, as of this writing, remain in prison. By the time of their arrest, Roger was an adult living his own life, albeit likely traumatized by the whole affair. Sarah was treated in a hospital and eventually released into a foster home. Because her records are sealed, we've lost track of what happened to her after the fact, but the last report we heard claimed that several months later she had gained a more healthy weight and grown several inches.

Everyone reads cases of child abuse and neglect in the newspapers. Never did we want to be involved in one. We're not sure what else we could have done better at the time, but we wish we could have found some way to convince these people that their problems were medical and psychological rather than supernatural. Or even if they refused to give up the supernatural beliefs, we wish at least we, the other paranormal groups, the doctors, or anyone else had been able to convince them to at least pursue medical treatment alongside their quest for a paranormal explanation.

PART THREE:
Media Analyses and Other Activities

Activities of the Rocky Mountain Paranormal Research Society are not and have never been limited to investigations alone. While those do take the bulk of our time and represent the flagship of our work, we have our hands in a lot of other things as well. We now turn our attention to those cases that are somehow different from the standard investigations (whether public or private) we've been discussing so far.

In many ways, this third part of the book (and the corresponding parts in each volume of the series) represents something of a grab bag of different and often unrelated activities, but most of them fall into a couple of categories.

First are the media analyses, and those can be subdivided into our analyses of media that has been published and received enough attention to merit our commentary and our analyses of photographs or videos sent to us along with an invitation to perform an investigation but which never develop into a full-fledged investigative activity (typically because our analysis of the photo or video is sufficient to explain the phenomenon in question).

Then there are the experiments or other forms of academic research. From time to time, independent of any specific investigation, we perform a variety of experiments whose results may illuminate some aspect of paranormal lore or phenomena. We've done several such experiments in the past, one of which was included in Volume 1, though our experimental "department," so to speak, is beginning to pick up more steam in recent years, so later volumes will likely contain more of those cases.

Sometimes we also get involved in cases that aren't so much paranormal investigations as they are public services or civil or political actions. Clarity is in order here. Politically, Rocky Mountain Paranormal is committed to neutrality. We're an investigative and educational organization. Members through the years have been Democrats, Republicans, and other. We don't espouse any particular political or economic philosophy, we require no political test for membership, and we're committed to remaining non-partisan in our work. However, every once in a while, a political, social, or legal issue emerges that is specifically related to the paranormal. When that happens, we feel compelled to take action to ensure that everyone involved has the highest quality information available and behaves ethically. Such will be the topic of our essay at the beginning of this part, and one such action was described in Volume 2.

Finally, sometimes there's a case that just doesn't seem to fit into one of the other standard classification of investigations at either public venues or private residences. Even if we stretch the meaning of those phrases a bit, some of the things in which we involve ourselves just seem to belong more properly in this kind of a grab bag part of the case files books.

Though this part is relegated to the back of the book, don't be fooled into thinking it's less important than the others. Every case in which we participate offers something interesting and educational, and we hope you enjoy reading about some of our less typical activities here.

Essay

The Paranormal and the Law

If you're reading this book straight through, you've just encountered a horrible case in which paranormal claimants ended up behind bars for child abuse. This essay isn't going to dwell too heavily on the psychology or the horror of such cases, but it does point out that every once in a while, paranormal beliefs do happen to intersect with law and public policy. How are legislatures and courts to deal with paranormal claims? How does a state or a country even know where to begin in those situations?

True crime afficionados have surely seen plenty of cases in which the police began to suspect someone (rightly or wrongly—these cases go both ways) in part because their house was full of paranormal paraphernalia. Fortunately for those of us who've never broken a (serious) law in our lives but nevertheless collect all manner of weird paranormal oddities, courts would not accept possession of something like a Ouija board as sufficient evidence to send someone to prison. Police might jump to conclusions on faulty logic from time to time, but there are good reasons our courts have strong evidentiary standards.

Plus, we Americans are fortunate enough to have the freedom of religion enshrined in our First Amendment. Paranormal beliefs often coincide with religious matters, so government at any level is rightly reluctant to persecute those who hold even the most unusual of beliefs, as long as they don't step beyond certain well-established legal boundaries. Performing a weird ritual in your home or church is perfectly protected; performing human sacrifice is not, for example.

Beyond the obvious cases, though, there are times when the paranormal and the law are destined to intersect in ways wherein the outcome is not always immediately clear. How should a court react if a defendant claims to have been demonically possessed? What if someone sues a realtor because the house they just bought turns out to be haunted? Can legislatures ban psychics and fortune tellers from operating within their jurisdictions, or is this a violation of the Free Exercise Clause?

These can be difficult questions to answer even for legal experts—and we at Rocky Mountain Paranormal, though some of us do enjoy reading legal opinions for fun—are not. We're not lawyers, nor do we play them on television. Nothing we're about to say should be construed under any circumstances as legal advice. But we have been thinking about paranormal issues for a long time, so we've gathered some interesting ideas and cases that will help you *begin* to understand the complexities of how paranormal belief and legal doctrine sometimes interact.

Let's begin with real estate law. Are real estate sellers required to disclose if a property is haunted? What if they didn't know about it? What if they did but lied about it? Under what circumstances can buyers sue or find their way out of a real estate transaction? We actually already addressed that question in Volume 1 of this series (albeit in the context of determining whether owning a haunted property is or isn't profitable), but our answer still holds and so we will begin this essay by quoting ourselves:

> Most real estate agents will tell you that a house with a reputation for being haunted is typically slower to sell (and for a lesser price) than an equivalent property without such a reputation. The same is true for houses that have been the site of murders or other notorious crimes. Sometimes, this makes sense in a purely naturalistic way; a former meth house is likely not as valuable due to the possibility of health hazards or structural damage. But in the case of a haunted house, it seems to mostly be people's psychological unwillingness to stay in such a house that makes it lose its value.
>
> These are called "stigmatized properties," because they have some kind of reputation or history that makes them potentially undesirable for buyers. Stigmas include crimes committed at the house, debts, and, yes, allegations of alleged paranormal phenomena. In the United States, state law governs what (if anything) sellers or real estate agents must disclose regarding stigmatized properties prior to a sale.
>
> Of course, those laws vary from state to state. Most states require "physical defects" to be disclosed. These are potential defects that may affect the safety, habitability, or structural soundness of the property. But some states also require disclosure of "emotional defects," which may include criminal histories or paranormal claims.
>
> Several states have laws concerning the disclosure of deaths that may have occurred on the property. Alaska requires disclosure of any murder or suicide on the property in the year preceding the sale, for example.[1] Conversely, Arkansas statute specifically relieves sellers of the burden to disclose a murder or suicide on the property.[2] It's quite strange how widely the laws vary.
>
> Interestingly, four states specifically mention allegations of paranormal activity in their laws. These merit individual mention. According to Massachusetts law, sellers do not need to disclose any psychological stigmas, including potential paranormal activity.[3] The same is true in Minnesota: sellers don't need to disclose facts that may affect a buyer's psychological enjoyment of the property, in-

---

1   Alaska Statutes 08.88.615 c. 1-2.

2   Arkansas Code 17-10-101.

3   Massachusetts Law, Part 1, Title XV, Chapter 93, Section 114.

cluding "perceived paranormal activity."[4] In New Jersey, sellers are not required to volunteer these kind of facts, including related to paranormal activity, but must disclose them if the buyer specifically asks about them.[5] New York law is a bit of an odd man out. According to statutory law, there is no duty to disclose,[6] except as established by court precedent. According to that precedent, established in the 1991 New York Supreme Court ruling in *Stambovsky v Ackley* (commonly known as the Ghostbusters ruling), the court will rescind a sale if the seller himself or herself created or publicized the reputation for paranormal activity (such as by holding haunted tours of the property) and takes unfair advantage of the buyer's ignorance of that reputation.[7]

Since we're based in Colorado, we should also note that Colorado is one of the states that does not require disclosure of psychologically stigmatizing facts such as deaths or murders within the property.[8] While state law does not specifically address paranormal claims, the most likely interpretation given the statute concerning disclosure of deaths is that disclosure of paranormal activity or reputation is not required in the state of Colorado.[9]

So the real answer is: mostly it depends on what state you're in.

Law governing things like the sale of supposedly haunted objects is even less well-defined. It may be that if one bought an object specifically advertised as being haunted and then found it was not, one might have a tort claim against the seller. But that would likely require the buyer to prove to the court that the object was indeed not haunted. A clever seller would be able to reply that it is haunted and the buyer is just insufficiently psychic to perceive it. Arguments like these are likely the reason courts try very hard to stay clear of such ill-defined fields as the paranormal and chalk up any disputes to mere disagreements over religious or philosophical ideas.

Next, we must consider the question of fortune telling or psychics. Skeptics will argue that these are fraudulent practices which deprive people of their hard-earned money all for an empty promise of communication with the deceased or a glimpse into the future. Believers will argue that they're legitimate religious practices which must be protected by the First Amendment. And others taking some kind of middle-ground might argue that they're mere novelty entertainments not meant to be taken seriously and so are therefore protected more by the free *speech* protections of the First Amendment than the freedom of religion clauses. What are we to think and where does the law actually stand on the matter?

Once again, states and municipalities have handled this in different ways.

4   Minnesota Statute 513.56.

5   New Jersey Administrative Code 11:5-6.4(d).

6   Consolidated Laws of New York, Real Property (RPP), Chapter 50, Article 12-A, Section 443.

7   169 A.D.2d 254 (N.Y. App. Div. 1991).

8   C.R.S. 38-35.5-101.

9   Quoted from "The Profitability of a Haunted Business" in Volume 1 of this series.

Some jurisdictions have sought to ban the practice of fortune-telling outright while others have completely ignored the issue and assumed customers had the duty to educate themselves on the matter and to make up their own minds. Where governments have tried to ban psychic readings, fortune-telling, palmistry, and so forth, they have generally chosen to be (or have been legally forced to be) cautious to avoid banning any similar practices carried out in accordance with a legitimate religious belief.

Of course, that raises the question of what qualifies as a legitimate religious belief in the first place. Most religious practices seem unusual to those who belong to other religions, so you can't judge just based on what makes sense to a judge or a legislator. Even if you get past the free exercise test, you have then to face the question of whether fortune-telling counts as a fraudulent practice for the purposes of the law if it's advertised as "for entertainment purposes only." Technically, that does negate any claims as to the reality of the service, but many customers might take the psychic claims at face value and ignore the legal disclaimer, especially if it's printed small enough at the bottom of the advertising. It's a complicated question, and unlikely to be finally solved any time soon.

In Colorado, there is no statewide ban on fortune-telling (and, indeed, when you drive through certain neighborhoods you'll see plenty of shops advertising such services. Certain municipalities do have fortune-telling bans on the books, though, and some of them even enforce them. The City of Vail, Colorado, has a decades-old ordinance on the books that they started enforcing strongly in 2007 (much to the psychics' surprise) when they shuttered the doors of a lot of psychic and fortune-telling shops.[10] Their law specifically prohibits fortune-telling "for gain or lucre," meaning only commercial operations are affected; those who practice just for fun, as a hobby, or as part of religious observance are not.

That latter exception is probably their way to make this a part of commercial regulation rather than an encroachment on freedom of religion. The law stands, even decades later, so they might have found the balance between conflicting legal imperatives.

Psychics and fortune-tellers have tried to challenge similar ordinances on Constitutional grounds. Back in 1985, the California Supreme Court decided a case called *Spiritual Psychic Science Church v. City of Azusa*, in which they upheld a similar municipal ordinance as not violating free speech guarantees, but noted that First Amendment limitations to such ordinances would apply if they interfered with practices tied to religious beliefs.[11]

Other paranormal claims are not banned by law but are regulated by it. In this category are things like alternative medical practices based more in spiritual theory than in medical science. These are governed by both Federal and State law, and the regulatory landscape is far more convoluted than we have time to address here, but the general practice (with minor variations from jurisdiction to jurisdiction) is similar to the approach taken by the State of Colorado, which permits these alternative medical practices (such as Reiki or other "energy-based" practices), but requires practitioners to provide written disclaimers as to their

---

10   Vail Daily (2007). Anti-fortune telling law enforced. *Vail Daily*. <https://www.vaildaily.com/news/anti-fortune-telling-law-enforced/> (accessed August 24, 2025).

11   39 Cal.3d 501.

non-medical status and prohibits them from explicitly claiming to diagnose or treat illness.[12] In our informal experience studying these practitioners, many but not all of them fail to provide the required disclaimer (or bury it where patients are unlikely to see it) and many do claim to diagnose and treat illness.[13]

Meanwhile, Colorado law also affirmatively endorses patients' rights to either refuse medical treatment or to pursue treatments in accordance with their religious or spiritual orientation, provided those treatments are not prohibited by law.[14]

The darker side of paranormal and the law shows up in cases such as the one described in Chapter 22, wherein individuals prioritized paranormal belief over the health of their child. The law did not, obviously, take their paranormal beliefs into consideration when administering justice. But throughout the years, a few people have tried a variant on the "devil made me do it" defense against criminal charges. How has the law responded to those kinds of claims?

There have been more such cases than we have time to discuss, but we'll start off with just a few highlights to set the stage.

Probably the most famous of these thanks to the *Devil Made Me Do It* film in the *Conjuring* horror movie franchise is the defense of Arne Cheyenne Johnson, who was tried for first degree murder after fatally stabbing his landlord. Famed demonologists and either paranormal investigators or media chasers (depending on who you ask) Ed and Lorraine Warren took the case and, as usual, supplied a paranormal explanation. Johnson attempted to use demonic possession as his defense—the only time we know if in which demonic *possession* has been attempted as a defense in an American court of law—but Judge Robert Callahan disallowed the defense as "incompetent evidence" and Johnson was ultimately convicted of the lesser charge of manslaughter.[15]

More commonly, demonic influences are mentioned in criminal cases as part of an insanity defense. This is what happened in the case of David Berkowitz, better known as the "Son of Sam" serial killer, whose attorneys attempted to claim his *belief* that he was acting in accordance with diabolic instruction made him incompetent to stand trial. When multiple psychiatrists found him competent, he changed his plea to guilty.[16]

Most readers will probably be relieved to find these kinds of defenses tend not to work. Courts might entertain a *belief* in demonic influence as justification for an insanity defense, but, though there is no specific rule prohibiting demonic possession as a standalone defense, there are several evidentiary hurdles which

---

12   C.R.S. 6-1-724.

13   A more detailed inquiry into occult, spiritual, and energy-based medical practices is on our to-do list. Note also that we're not lumping all of alternative and complimentary medicine into one basket here. These practices exist on a spectrum of plausibility. For now, we advise caution and detailed study.

14   C.R.S. 15-14-504.

15   Miller, G. (2024). The Devil Made Me Do It: The Viability of Demonic Possession as a Murder Defense. *Vermont Law Review*. <https://lawreview.vermontlaw. edu/the-devil-made-me-do-it-the-viability-of-demonic-possession-as-a-murder-defense/> (accessed August 24, 2025).

16   *Ibid.*

are probably impossible to overcome.

State rules of evidence in criminal procedure vary a little bit, but most are substantially similar to Federal Rules of Evidence (FRE), which we'll explore for a moment here. First, FRE 402 prohibits the introduction of any irrelevant evidence. A court might conclude that demonic possession is irrelevant. On the other hand, if it's the central claim in a defendant's defense, it might be relevant, but FRE 403 further bars the admission of even *relevant* evidence "if its probative value is substantially outweighed by a danger of one or more of the following: unfair prejudice, confusing the issues, misleading the jury, undue delay, wasting time, or needlessly presenting cumulative evidence."[17]

Readers inclined to believe that demonic possession does take place might find these hurdles easily surmountable. Fair enough. But the real challenge is how to prove it. Claiming demonic possession may be one thing, but for a claim to be admitted in a court of law, it must actually constitute *evidence*. Short of the Devil himself showing up in a court of law and claiming responsibility for a crime, which seems highly improbable, the only way a case of demonic possession could be demonstrated to a court would be through the introduction of expert testimony. But FRE 702 requires that any expert testimony admitted must be based on reliable scientific or technical knowledge supported by sufficient facts or data, which simply isn't the case in demonic possession claims. After all, we've been searching for reliable facts and data in support of paranormal claims for more than a quarter of a century now and the best we've come up with are some intriguing but inconclusive cases.

Ironically, the best chances of a demonic possession defense being successful are cases in which it is assumed to be a false claim. Then the court could legitimately take a *delusion* about demonic influences as evidence of insanity. Psychiatric testimony along those lines can easily overcome the hurdles of FRE 402, 403, and 702. Insanity defenses are not a get out of jail free card, though. Defendants found to be insane are often confined to psychiatric institutions for at least as long as they would have been confined to prison—though arguably the conditions are somewhat more comfortable and there may be a possibility of early release if doctors find the insanity has been cured.

Examples of "the devil made me do it" being used successfully as part of an insanity defense include the 1974 trial of Michael Taylor (who attacked a friend and destroyed all the religious items in his house before being exorcised by the Catholic church before murdering his own wife) in England, where the defense successfully argued his mental health was compromised by the trauma of the exorcism itself,[18] and the infamous case of Andrea Yates who drowned her four children in 2001 but successfully argued insanity on appeal, claiming she murdered her children to save them from Satan's and her own evil influences.[19] Sort

---

17   Federal Rules of Evidence, Rule 403.
18   Miller, G. (2024). The Devil Made Me Do It: The Viability of Demonic Possession as a Murder Defense. *Vermont Law Review*. <https://lawreview.vermontlaw. edu/the-devil-made-me-do-it-the-viability-of-demonic-possession-as-a-murder-defense/> (accessed August 24, 2025).
19   Denno, D. W. (2003). Who is Andrea Yates? A Short Story About Insanity. *Duke Journal of Gender Law & Policy, 10.*

of the opposite case occurred when Deanna Laney claimed God told her to stone her children to death in 2004 as a test of faith, which she dutifully did; she was found not guilty by reason of insanity manifesting itself in her religious convictions and was ultimately released from the state hospital in 2012.[20]

Other times when the paranormal must intersect with the law include cases in which people claim the government is withholding paranormal information from the public. Most often this is in relation to UFO or extraterrestrial claims, though there have been some other variants. In Volume 2 of this series, we discussed our involvement in a case wherein individuals sought to create an Extra-terrestrial Affairs Commission for the City of Denver. And in 2023, Congress passed the "UAP Disclosure Act" requiring the government to declassify documents related to UFOs and related phenomena[21] which was lauded with much fanfare among the UFO enthusiasts but which has (at least so far) produced little more than some of the most embarrassing lines of questioning we've ever heard in Congressional hearings—and that is saying something.

Alas, a full treatment of these kinds of interactions between law and the paranormal would require an entire book of their own, so we have to cut our discussion short, just hinting at those ideas.

Ultimately, it's a good thing that the law and the paranormal usually stay in their own spheres. As we've seen, when they are forced to interact, the results are often messy and confusing. Still, we hope this gives the reader some idea of the kinds of issues lawyers and paranormal enthusiasts alike have to grapple with in those rare but unfortunate cases in which paranormal beliefs and legal doctrines meet.

Spencer, S. (2015). *Breaking Point*. United States: Diversion Books.

20   Bender, E. (2004). Forensic Experts Probe Mind Of Mother Who Killed Kids. *Psychiatric News, 39*(24).

Hughes, J. (2012). Deanna Laney released from state mental hospital. *CBS 19*. <https://www.cbs19.tv/article/news/deanna-laney-released-from-state-mental-hospital/501-266929900> (accessed August 24, 2025).

21   Lagatta, E. (2023). Did America get 'ripped off'? UFO disclosure bill derided for lack of transparency. *USA Today*. <https://www.usatoday.com/story/news/nation/2023/12/18/ufo-disclosure-bill-what-to-know/71960193007/> (accessed August 24, 2025).

23
The Red Rocks UFO

This is the most recent UFO case we've had a chance to look into. It took place in 2024, caught international media attention, and involves a combination of media analysis, on-site investigation, witness interviews, and armchair sleuthing. It all started when multiple witnesses saw a UFO flying over the hogback ridge near the Red Rocks Amphitheatre.

Figure 23.1. Photo of Red Rocks modified to show approximate location of alleged UFO (photo & modification: Robert Lewis).

While we were preparing the manuscript of our last book (Volume 2 in this series), which itself included quite a slew of UFO and extraterrestrial stories, another one came to our attention. This time, it didn't come to us through a request for help. Rather we simply noticed a bunch of news articles started popping up describing an incident in which people saw a UFO flying over the Red Rocks Amphitheatre in Morrison, Colorado.

First, we saw a local news report on the incident.[1] Then we noticed it was starting to get picked up in the national media.[2] And then, much to our surprise, it started even getting international press attention.[3]

Normally, we don't pay much attention to every UFO report. There are simply too many of them to keep track of. There are organizations which catalog as many such reports as possible, but for us, it's just not what we're interested in doing. Yes, we're interested in the stories. But if we tired to investigate every time someone saw something unusual in the sky, we'd spend our entire lives chasing the stories and finding in almost every case that no real investigation could be done because only one witness saw the phenomenon, no photos were taken, and no other evidence could be found.

When a case gets international attention, though, we often feel like we need to jump in and see if we can learn anything of interest. This case also had some other things going for it from an investigative point of view: multiple witnesses saw the same thing and we knew exactly when and where it took place.

The story told by the various news reports was fundamentally the same as the story we were told by an eyewitness later on, so we'll amalgamate all those versions into a single narrative just so you know what the claim was all about and then we'll describe how we set about trying to investigate it.

According to the reports, about a dozen employees were busy on the loading docks outside of Red Rocks Amphitheatre and across from the Red Rocks Trading Post loading musical and sound equipment into trucks on the morning of June 5, 2024, following a June 4 evening concert. At approximately 1:00 in the morning, one of the employees alerted his colleagues to something unusual. They all turned to look and saw what they described as looking like a spaceship hovering about a mile or a mile and a half north of Red Rocks and a few hundred

1   KDVR (2024). Dozens saw bizarre UFO sighting at Red Rocks. *KDVR*. <https://kdvr.com/video/dozens-saw-bizarre-ufo-sighting-at-red-rocks/9824437/> (accessed August 24, 2025).

2   Robledo, A. (2024). Red Rocks employees report seeing UFO in night sky above famed Colorado concert venue. *USA Today*. <https://www.usatoday.com/story/news/nation/2024/06/28/ufo-sighting-colorado-red-rocks-employees/74247738007/> (accessed August 24, 2025).

Shoaib, A. (2024). UFO Reported Over Colorado Concert Venue. *Newsweek*. <https://www.newsweek.com/ufo-spotted-colorado-red-rocks-1921200> (accessed August 24, 2025).

3   Rathore, B. (2024). Colorado concert venue workers claim UFO sighting: Black metallic disk seen in sky. *Hindustan Times*. <https://www.hindustantimes.com/world-news/us-news/colorado-concert-venue-workers-claim-ufo-sighting-black-metallic-disk-seen-in-sky-101719626167446.html> (accessed August 24, 2025).

feet off the ground, as near as they could tell.

Figure 23.1 contains a photo we took from the location where the employees were located with a UFO shape digitally added to show its approximate location according to their story.

Witnesses describe the object as a disc-shaped metallic object that seemed to have lights or windows arranged in three levels. They say it was completely silent. When they started paying attention to it, though, and one of them turned his flashlight in its direction, the object adjusted its orientation and then moved in an eastward direction until it vanished. It didn't just move out of sight, they said. It disappeared. The entire sighting lasted about thirty seconds, though the witnesses couldn't be sure how long the object was there before they noticed it.

One of the witnesses reported the case to the National UFO Reporting Center and assigned sighting number 181776. According to the description submitted, the witnesses said it was a dark metallic object, probably several hundred feet in diameter and moved five to ten miles per hour at a distance of approximately one mile from their vantage point.[4]

Witness reports like these are both credible and not all at the same time. Because there were multiple witnesses and none of them ever disagreed with the description, we take the report as reasonably credible. It's always possible they could have cooked up a story just to get some attention—such things have happened in the past—but that does not appear to be the case here. We have every reason to believe they're all being truthful about what they saw. On the other hand, gauging size and distance is almost impossible without specialized equipment. Because the object in question didn't pass directly in front of or behind any buildings, mountains, or trees—we asked and the witnesses confirmed this—it could very easily have been a smaller object closer than they thought or a larger object further than they thought, or they could have nailed its size and distance exactly. There's just no good way to tell.

But the whole story caught our attention, so we started to dig in. As luck would have it, one of our investigators is acquainted with a gentleman who happens to work contract jobs at Red Rocks performing similar tasks to what the witnesses were doing at the time of their encounter. We reached out to him and asked if he happened to know anything about the case. While he wasn't one of the witnesses, he was friends with one of them and agreed to put us in contact. Never let it be said that networking doesn't pay off. We always tell people we have the weirdest Rolodexes in the world and we end up needing to call upon just about everyone at some point for a paranormal investigation.

We were glad to find one of the witnesses was happy to talk to us so we began an online correspondence for a short time just to get some questions answered. First, since the sighting took place just after midnight, some of the reports were a little confused about the date, and he was able to confirm it was June 5, just after midnight, following a June 4 concert. Armed with that information, we'd be able to look for any known aerial or atmospheric disturbances in the area around the time in question.

He was also able to confirm that they'd pointed their flashlights in the ob-

4   NUFORC (2024). NUFORC UFO Sighting 181776. *National UFO Reporting Center*. <https://nuforc.org/sighting/?id=181776> (accessed August 25, 2025).

ject's direction and it didn't reflect any light back to them, suggesting it was some distance away, that it moved smoothly—not in a jerking pattern—that the weather was calm and clear, and that the object first seemed to be motionless and only several seconds later did it begin to move before vanishing. When asked about his prior beliefs and background, he said he'd always been—much like us—an interested and hopeful skeptic but had never seen anything like this before. One of the other witnesses, though, had seen a UFO in the past, and this was the witness who first pointed out that the object appeared to have three rows of windows.

Then he said a couple days hence, he was meeting some field investigators from MUFON—the Mutual UFO Network, one of the largest groups of people who try to trace and study UFO sightings—and invited us to join them. We agreed and dispatched one of our investigators to Red Rocks for the meeting, where we photographed the scene and conducted another interview.

Between us and the MUFON people, we spent probably an hour asking questions but didn't gain a whole lot of additional information. Because we'd only identified one of the dozen or so witnesses so far, we told him to pass along our contact information to his colleagues and ask if any of them wanted to share their story independently (and anonymously, if that's how they would prefer it), but none of them ever reached out to us.

While on scene, we did notice the nearby Red Rocks Trading Post building (see Volume 1, Chapter 11 for its own paranormal stories, of a more ghostly nature) had roof-mounted security cameras pointed in roughly the direction of the sighting. The witness said he didn't think they would have caught anything because they're pointed too low to the ground (so as to surveil the parking lot and road) but that he'd check with his employer's security officers to see if they captured anything. We never received any follow-up so we have to assume the cameras didn't capture anything of interest.

Our next step was to check the weather at the time in question. We went online to find historic weather reports for Morrison. The nearest report we could find was from a weather station in Broomfield, Colorado, several miles away but still close enough that we assume the patterns were similar. According to that report, at approximately 1:00 in the morning of June 5, 2024, the temperature was about sixty-three degrees Fahrenheit, humidity dropped from thirty-four percent to twenty-six percent between 12:55 and 1:15, and winds were nine to ten miles per hour from the west or north-northwest, with gusts up to sixteen miles per hour beginning at about 1:15. There were few clouds and no precipitation.[5]

We also looked through as much air traffic data as we could find publicly available, but came up empty. We couldn't find any reports of airplanes or anything in the area at that time, but too little of the information is published to say anything for certain. We also reached out to the military and Federal Aviation Administration for a comment but never heard back from any of them.[6]

---

5   Weather Underground (2024). Broomfield, CO Weather History: June 5, 2024. *Weather Underground*. <https://www.wunderground.com/history/daily/KBJC/date/2024-6-5> (accessed August 25, 2025).

6   Personally, we think governments should be legally obliged to respond to citizens' or journalists' requests for information—even if the response needs to be "no comment"—but the powers that be don't listen to our suggestions. Yes, there are

At this point, the trail seemed to be going cold. We had some ideas. For instance, the idea that a prior UFO witness was the first one to notice the object's three rows of windows and the others only spotted them after he pointed it out could mean some power of suggestion was taking place. If the visual information they could see were ambiguous, they might only have decided what they were seeing was a craft with windows because he planted the idea in their minds. That's a common psychological phenomenon, but we can't prove it. We couldn't even explore it in any further detail without being able to contact the other witnesses, and they were choosing to remain silent.

We lurked around social media posts concerning the incident for a while and read the comments sections on news articles about it, hoping other witnesses might come forward. None did, but several skeptical commenters suggested it might have been some kind of balloon, and a few of them mentioned that, though they didn't see a balloon at the specific time in question, they had seen several balloons operating in the area in the weeks prior.

We do know from past experience that some UFO sightings have turned out to be things like scientific or weather balloons. People aren't used to seeing those in the sky. Especially at a distance, they could easily be confused for something otherworldly.

So we started researching balloons that could potentially have been confused for a spacecraft. We found a potential contender: the Thunderhead Balloon System. Developed and operated by Aerostar, this is a steerable balloon system which consists of a large nearly translucent balloon which carries a gondola containing a payload, flight and navigation systems, and solar panels.[7]

Now, this ballon does not look like what the witnesses described. Not exactly. However, if we assume that on a dark night the ballon itself might not be visible, the gondola and payload does look somewhat similar to what the witnesses described. It's more rectangular than disc-shaped, true, but in low-light and at a distance, people might not be able to tell the difference. Furthermore, the solar panels could have been confused for windows if they reflected some ambient light from the surrounding area. It's also capable of hovering in place for a while or moving in a slow, gentle manner similar to what the witnesses described. Add to all that, the winds that evening came from the west, which could have blown a balloon in the easterly direction described by the witnesses.

It's not a perfect explanation, but it's a plausible one and we wanted to follow up on it some more. We reached out to Aerostar to ask if they'd been operating one of their systems at the time and location in question and we received the following reply: "I can confirm Aerostar recently operated several Thunderhead High Altitude Balloon Systems in your area. I generally can't provide details about individual customer projects, but generally speaking, Thunderhead Balloon Systems serve various purposes including extending communications across wide distances, environmental monitoring, earth observation, and scientific research."

public disclosure laws and we could have pursued those avenues, but we determined at this point that if they had any useful information for us, it would almost certainly be classified.

7   Aerostar (n.d.). Thunderhead Balloon Systems. *Aerostar*. <https://aerostar. com/products/balloons-airships/thunderhead-balloons> (accessed August 25, 2025).

That's not a confirmation, nor is it a denial. Though we can't be certain, it sounds a little bit like they're saying "we can't say it was us, but it might have been us."

Case closed? Hardly. We think we might have found the culprit, and we've certainly supplied an explanation at least as plausible as extraterrestrial visitation, but we really don't know if such a balloon is what those witnesses saw or not. Unfortunately, at this point, the case has gone cold, interest has waned, and we're unlikely to be able to learn any more about it. So we have to leave it open, but with the addendum that we think we *might* have found the solution.

Which is rather unfortunate, because we wanted to meet the aliens.

24
A UFO in the News

In November of 2012, one of the local news stations, responding to claims of UFOs which appeared to be taking off around the intersection of 56th Avenue and Clay Street in Denver, sent an investigative reporter to film the area. Sure enough, they captured some UFOs of their own. It sparked so much interest, locally at first and eventually nationwide, that we felt like we needed to analyze their footage and perform some tests of our own.

Figure 24.1. Our recreation of the News UFO (photo: RMP archives).

As indicated in the last chapter, we don't get involved in every UFO sighting we hear about. For us to get involved we need to either be personally invited, notice it's getting a lot of press attention, or think we have something important to add to the investigation. In this case, the final two of those criteria were met. It was getting a lot of press attention and we thought we might have some insights a lot of the coverage was missing.

The story began in November of 2012 when the Denver local Fox affiliate (Fox 31 then, now part of KDVR along with Colorado's CW 2) ran a story about a UFO sighting near Denver. An anonymous witness had sent them video footage they couldn't quite explain, so they dispatched an investigative reporter to the same location and managed to film some UFOs of their own.[1] In this case, it's important that we—and they—refer to a UFO in the proper sense of the initialism. That is, we literally mean an *unidentified* flying object. We're not saying it was an alien, and they stopped short of saying that, too. But they did emphasize just how unusual it was and that they couldn't figure it out.

One of the remarkable claims in this case was that this wasn't a one-and-done UFO sighting. This was a case in which people were reliably capturing video of these objects—whatever they were—between noon and 1:00 p.m. across multiple days. What's more, it wasn't a single witness reporting it. A single witness may have got everyone's attention, but even the news reporter managed to witness and record the same phenomenon.

This wasn't something people were seeing in person, though. Even on video, they noted it was nearly impossible to see what they were talking about at normal speed. Only when the video was slowed down could the UFO be noticed, which suggested the object was moving incredibly fast. The original anonymous videographer said it appeared the objects were taking off and landing near 56th Avenue and Clay Street in Denver, though the actual video footage was shot from an open field closer to 84th Avenue and Federal Street, with the camera pointed in the direction of the aforementioned intersection.

After recording their video, they didn't just stop there. To their credit, they did some investigation of their own (though we think they stopped short of calling in the "right" experts…if we do say so ourselves). They started by contacting both the Federal Aviation Administration (FAA) and the North American Aerospace Defense Command (NORAD), both of whom indicated they had no knowledge of any flight paths or aerial anomalies in the area at the time in question.[2]

They called in a former pilot and aviation expert to further study the case. He said he couldn't identify what he was seeing. All he could say was that it didn't look like an airplane, helicopter, bird, or insect. We took him at his expert word on the first two points, but weren't sure if he'd be qualified to identify insects or

---

1   KDVR (2012). Mile High mystery: UFO sightings in sky over Denver. *KDVR*. <https://kdvr.com/news/problem-solvers/mile-high-city-mystery-ufo-sightings-in-sky-over-denver/> (accessed August 25, 2025).

2   Apparently, journalists affiliated with major broadcasting networks have an easier time getting a response from the government than do your friendly neighborhood paranormal investigators. We're disappointed but unsurprised that this is the case.

birds, so we wanted further study on those points.

In a follow-up article some weeks later, the same media outlet did reach out to an entomologist in response to numerous social media posts claiming the footage probably just showed an insect (other popular suggestions were remote controlled aircraft or military drones). According to their aviation expert, it didn't look like a military drone or any sort of unmanned aircraft. And according to their insect expert, the shape of the objects in the videos just didn't look like insects, though she did admit it was a tough call.[3]

We weren't quite satisfied. While that expert certainly knows more about insects than us, we think what they really needed was at least one more expert—someone who knew all about photography. Indeed, the shapes on the video didn't look particularly bug-like, but small out of focus objects moving at high speeds might not show up as their own true shape on a video recording. We still thought insects were a potential explanation.

Though these videos were taken in November, it was a warm season and the videos were taken between noon and 1:00 in the afternoon—usually a warm part of the day. According to historic weather archives, the average high temperature throughout the month of November, 2012 was just over fifty-eight degrees Fahrenheit, and some days got as warm as seventy-three degrees.[4] Furthermore, the videos were shot in an open dirt field with numerous bodies of water nearby, which we thought made this a prime location for insects despite some witnesses and photographers claiming they hadn't seen many insects flying around during the UFO sightings.

What's more, we also thought birds might be involved in the story. In fact, in one of the videos, there are actually two objects seen, moving in opposite directions. Though they only focused on one object, the other appeared to our admittedly non-expert eyes to be a bird flying through the frame. Likely they didn't focus on that one because it was more readily identifiable, but we also thought it looked largely similar to the other footage.

So we decided to head out to the same location and do some studying of our own, operating under the working hypothesis that birds or insects might be the true culprits. We knew there were several native animal species in the area. Great blue herons, white pelicans, Canadian geese, and cormorants are all common, particularly around the bodies of water near the location. Eagles and hawks, while rarer, are also sometimes seen nearby. When they're seen flying in the distance, they often appear as dots or blurs in the video footage. On the day of our experiment, we didn't see too many of the birds.

We did, though, see insects. For the hour between noon and 1:00 in the afternoon, we spent our time doing nothing but collecting bugs. In that hour, we managed to catch several: green peach aphids (*Myzus persicae*), damsel bugs (*Nabis*

3   KDVR (2012). Insect expert: UFOs over Denver not bugs; images on video remain a mystery. *KDVR*. <https://kdvr.com/news/problem-solvers/insect-expert-ufos-over-denver-not-bugs-images-on-video-remain-a-mystery/> (accessed August 25, 2025).

4   Weather Underground (2012). Broomfield, CO Weather History: November, 2012. *Weather Underground*. <https://www.wunderground.com/history/monthly/us/co/broomfield/KBJC/date/2012-11> (accessed August 25, 2025).

americoferus), black-flies or buffalo gnats (*Simulium sp.*), and dung beetles (*Aphodius distinctus*). Additionally, we observed but were unable to capture specimens of green tree hoppers (*Atymna inornata*) and green bottle flies (*Lucilia sp.*). A few bites also proved there were mosquitoes about, though we didn't actually ever see any of the bloodsuckers. For just one hour of searching—and by non-professional entomologists, to boot—that's a lot of insects. Yeah, we'd say there are plenty of flying bugs in the area.

But we weren't finished yet. Just showing that insects are in the area does not necessarily mean they can account for the footage people recorded. To test that idea, we employed the most high-tech methodology we could think of—we glued a dead fly to a piece of glass and held it in front of our cameras.

Figure 24.2. Behind the scenes: we glued a fly to a piece of glass (photo: RMP archives).

With this cutting-edge piece of equipment, we were able to see what it would look like if an insect were in front of the camera at various distances, various speeds, and various camera settings.

Obviously, the more distant the fly was from the camera, the less of it we could see. Much further than a foot away, it shrank to just a dot and then disappeared entirely. But we found that a distance of about ten inches from the lens was the most consistent with the size of the anomalous objects in the UFO videos (at least with our own camera set-up; other cameras may yield different results).

Figure 24.3. The fly (10 in. from lens) is clearly visible when in focus (photo: RMP archives).

We had a bit of a problem with that footage, though. At that distance, it looked clearly like a fly. This was, we think, exactly what the news station's expert entomologist was looking for. If this is what footage of a fly looks like, then whatever was in the UFO footage wasn't a fly (or any other kind of an insect, for that matter).

This is where having a photography expert on hand also would have helped. We noticed that in the UFO videos, the buildings or landscape in the distance were in focus. In our fly photo, the fly was in focus and the landscape was blurred. So we adjusted the focus so the landscape became clear. The result? The fly became just a blurry spot. No longer did its image clearly show any of its insectoid anatomy. Indeed, it looked pretty much identical to the UFO footage.

Figure 24.4. The fly (10 in. from lens) when the landscape is in focus (photo: RMP archives).

By now, we were pretty sure we'd solved the case. Numerous flying insects inhabited the field where the videos were taken, and as long as the focus was set correctly, photographs of a fly looked a lot like the images in the videos.

Still, those were static images and the original claim was about video footage. We had just a little more work to do.

In fact, one of the claims some witnesses had made was that in some (not all) of the footage, you can see "thrusters" fire partway through the video. Our resident forensic video analyst immediately knew what caused those flashes of light. The moving object reflected sunlight back toward the lens which created a bright spot where the image got overexposed.

But as for the rest of the video evidence, we wanted to do one final experiment in the field with our fly stuck to our pane of glass. We wanted to move the fly past the camera and see how the image would resolve in video if the fly were in motion. As expected, this further blurs the image, removes all final vestiges of "fly shape" from the black dot on the video, and looks even more like the original UFO footage.

Does this mean the UFO video from the news report absolutely was nothing but an insect? We can't say that with certainty. For what it's worth, we've convinced ourselves that's what we saw in the video, but the quality of the footage just didn't allow for definitive proof. Even if not an insect, though, it could also have been a bird (albeit much more distant). Ockham's Razor suggests we should accept the simplest explanation, all else being equal. Here, the simplest explanation, as much as we keep hoping to find genuine proof of aliens, is that a bunch of people recorded insects flying in front of their cameras.

Figure 24.5. The "UFO" appears to move at high speed when the fly is moved (photo: RMP archives).

This case is a good example of the kind of extra effort we think is called for in paranormal claims investigation, though. We thought it was a bug right from the start, but rather than sitting in our armchairs and just proclaiming it so, we put in the time to go out to the location and test our hypothesis. Science progresses when people put in the work. It's the best way to debunk claims that turn out to be false, yes. More importantly, if any truly solid and incontrovertible evidence of the paranormal is ever going to be found, it will be by someone working out in the field and testing every possible idea.

25

The Mysterious Case of the Haunted Ouija Board

A lot of people are terrified of Ouija boards (or spirit boards). Maybe they saw *The Exorcist* and that was enough to frighten them. Religious people in certain denominations think they're the tools of the Devil. Others just think they're dangerous because you never know what kind of spirit or entity you might be talking to. In this case, we're going to tell you about one that was filmed moving by itself.

Figure 25.1. A still frame from the "Haunted Ouija Board" video. (photo: RMP archives).

The story behind our Haunted Ouija Board is a strange one. In addition to our investigative activities, we always have a to-do list packed with all manner of spooky and paranormal-related activities. Some of it is educational outreach. Some is purely entertainment. Once in a while, we find ways to combine the two. And we're always happy to just engage in a bit of good-natured shenanigans whenever the opportunity strikes.

This is the story of how one of our side projects accidentally escaped our archives and we re-discovered it (so to speak) "in the wild." It's going to be a short story, but we think you'll find it as amusing as we do and it also teaches an important lesson about media literacy.

Throughout our books, we've mentioned that hoaxes, while they do exist, are rare. That's true. We know of a few, but by and large, the paranormal claims that make it to our desk are either legitimately mysterious or they're the result of honest people who simply misunderstood something happening in their homes or businesses. Sometimes, though, something that was never even meant to be a true hoax gets taken out of context. That's exactly what happened here.

Several years ago, we were in the early stages of negotiating the production of a television series to be called *Colorado X: Case Files of the Paranormal*. The idea was to produce something similar to all the ghost hunting shows we all love to watch (and sometimes to turn into a drinking game), but with legitimate science, real history, and a skeptical twist. Our pitch, along with our partners in crime from the film production side of things, was that the show would be just as scary as the ghost hunting shows can be at their best, but also legitimately educational by the end. We thought we had the magic formula to give audiences the best of both worlds.

It didn't happen. As a side note, any television producers reading this should drop us a line. We have even more ideas now that we've been at this for a few more years.

As part of the pitch process, though, we began shooting a pilot episode we could show television executives. Several of the individual spots we shot, we figured, would end up in the pilot episode itself, and several might be used in marketing materials as stand-alone clips. A lot of it, though we were doing a professional project, involved us messing around with different ideas to see which ones might work.

Because we had consistent access to the building at the time, a lot of the footage involved staged ghostly manifestations at the Yak and Yeti restaurant (see Chapter 3). Indeed, some of that footage is still knocking around the Internet as of this writing.[1]

Much of it got left on the cutting room floor. A few other pieces are still preserved in our video archives in case we ever want to do anything with it. But one piece of video took on a life of its own.

As part of the same film project, we decided to make a video of a haunted Ouija or spirit board that moved by itself. Several of our members have trained

1   Bryan & Baxter Video Collection (2015). Rocky Mountain Paranormal is blind at the Yak and Yeti. *YouTube*. <https://www.youtube.com/watch?v=i0VeCOamIXA> (accessed August 25, 2025).

as either amateur or professional magicians so we knew we could do it well, and we could do it live to camera so it wouldn't look like mere video trickery. A plan was developed.

Because this was originally going to be part of a television pilot, we had concerns about copyright protections for images used on the commercially available Ouija boards. No problem there. We commissioned an artist to make one for ourselves. Being the *Exorcist* fans that we are, instead of "Ouija" or "Spirit Board," the title at the top of ours would bear the name of the demon Pazuzu (a real demon from Mesopotamian lore but also the demon chosen as the villain in *The Exorcist*). And of course because the video was going to be used for a program called *Colorado X*, the letter "X" on the board would be subtly stylized.

That also served a practical purpose. Our method for making the planchette move by itself was going to be the magicians' old favorite: invisible thread. Apologies to any magicians reading this, but the existence of threads and wires isn't a well-kept secret anymore anyway. We drilled a tiny hole through the letter X, disguised by the design, which allowed an off-camera operator to pull a thread and cause the planchette to move. We designed it so that the thread would break when pulled so there would be no evidence left after the fact. And of course, because of the way we set it up, the planchette would always move so that it pointed to the letter X. We weren't being subtle.

After the TV show died, we put the video online just to see if it might get any attention. We never claimed it was either real or fake—we just posted the video and let it speak for itself. We figured that would let us study how people reacted to it, and if the TV show ever got resurrected, we'd have it available as a potential promotional device. Then we largely forgot about it.

Forgot about it, that is, until an entirely different TV show happened across the video. The producers for *Fact or Faked*—a SyFy Channel program dedicated to (supposedly) skeptical paranormal investigations—found it and, without realizing it was ours, reached out through the account on which the video was hosted to see about featuring it on their show. The kicker, though? They wanted us to re-shoot the video. Without bothering to figure out whether it was real or fake in the first place, they asked us to film a more sensationalized version which would emphasize that no magnets or strings were used.

We reached out to our friends at the James Randi Educational Foundation for ethical advice, and they encouraged us to play along for a while and ultimately published an expose about our experiences with the television producers.[2]

Playing along, we re-shot the video as they requested. Apparently even this wasn't quite sensationalized enough and they asked for another shot showing even more footage. To sweeten the pot, they offered a license fee of $1,500 for the footage. While that money would have been incredibly useful to us (we are not wealthy people and ever dollar counts), we backed out of the deal at that point. Money is nice, but we wouldn't allow footage we knew to be faked to potentially be portrayed as real (we were not sure how they were going to portray it, but something smelled funny to us), nor did we want to lose control of our

2   Stollznow, K. (2010). Fact or Faked: Faked! *The James Randi Educational Foundation*. <https://archive.randi.org/site/index.php/swift-blog/1103-fact-or-faked.html> (accessed August 25, 2025).

footage. The story was never used on their show.

That's a pretty interesting result for a video we never really expected to use, but believe it or not, its saga doesn't end there!

The video resurfaced again in an online "reaction" video by a self-professed exorcist in 2021.

Bob Larson is an evangelist with programs on radio and television and the pastor of the Spiritual Freedom Church in Arizona. He's probably best known either for his books on cults and Satanism or for his work as an exorcist (though personally we'd differentiate his form of deliverance ministry from the more formal rite of exorcism performed by Catholics and some other churches). He's even performed exorcisms via the video conferencing platform Skype, though he said "tweeting an exorcism is ridiculous."[3]

A few years back, we were watching his YouTube channel—we like to stay informed, and though we find many of his practices suspect, we also consider him to be a great source of information on cults—and were quite surprised when we saw our own Pazuzu board in his video thumbnail!

As a part of his "The Real Exorcist Reacts" series, he spent about five minutes watching our video, which he hadn't seen before and whose source was unknown to him, and describing his reaction.[4] He gets quite excited when he sees that planchette move by itself, though to his credit he did pick up on how strongly we tried to emphasize that we weren't using trickery, which he took as potential evidence the video might have been faked. With apologies to the Bard of Avon, we doth protest too much. He got that part right. Even still, he admitted he didn't know whether it was real or fake, and took the video as an opportunity to warn people about the dangers of using spirit boards.

Finally, because the saga of this particular Pazuzu board never ends, it how has a permanent home in our archives. In the years since, we've had it autographed by Eileen Dietz, the actress who portrayed Pazuzu in *The Exorcist*.

Do us a favor. If you ever spot a video of our haunted board in the while, let us know. We have a feeling its story might not yet be over.

---

3   Gupta, P. (2014). Must-see morning clip: "The Daily Show's" Jessica Williams gets an exorcism via Skype. *Salon.* <https://www.salon.com/2014/10/14/must_see_ morning_clip_jessica_williams_gets_an_exorcism_via_skype/> (accessed August 25, 2025).

4   Bob Larson…The REAL Exorcist (2021). Bob Larson Reacts to Haunted Ouija Board. *YouTube.* <https://www.youtube.com/watch?v=Ujd4xiw-z7s> (accessed August 25, 2025).

26

Bryan the Gazer

Time for a confession. Despite constantly claiming hoaxes are rare, we've participated in one or two ourselves. It's important to note that we never do so maliciously and we never profit from them. Rather, we do them when we think they're necessary either to prove a point or to study how people react to certain types of media or situations. In this case, we learned of a sort of faith healer whose gimmick was simply staring at people and decided to create a "gazer" of our own to see whether people could be fooled by the performance.

Figure 26.1. Bryan in his "Bryan the Gazer" costume (photo: RMP archives).

Several years ago, while attending a metaphysical fair (events we periodically attend just to stay abreast of what's going on in the paranormal world), we wandered into an event that featured what promised to be a new kind of spiritual healing experience unlike anything we'd ever seen before. The individual claimed to be a "gazer." During the performance, he never spoke. He simply gazed out at the audience. People were deeply affected. Some claimed to have been healed. Others were brought to the point of tears. We didn't feel anything so spectacular ourselves, but we wondered what could possibly be going on.

To thoroughly test such a faith healer (though we should emphasize he carefully avoided direct claims of healing illnesses), we would need to conduct a properly controlled scientific test. Over the years we've tested a number of psychics, healers, and other types of paranormal practitioners, but we didn't have access to this individual, so that was going to be out of the question.

As such, we should start this off by saying we never tested him directly and aren't saying we've proved his claims false. Articles have been written which have called his claims into serious question, true, and done so from both the perspectives of skeptical science and mainstream religion.[1] We share the skepticism. Extreme skepticism, even. However, readers of these books will note that we don't presume to either endorse nor debunk any claims we have not personally examined.

What, then, is this case file all about? Well, lacking the ability to test this miraculous gaze directly, we started to wonder whether we could artificially engineer something similar on our own.

Hoaxes, we always tell people, are rare. But we felt like in this case one was going to be necessary. A plan began to develop that we'd create a "gazer" of our own and then see first if we could get him a featured performance at a major metaphysical event and, if so, whether attendees would describe similar experiences to those believers have described after a "real" gazer performance.

Rocky Mountain Paranormal founder (and co-author of this series) Bryan Bonner was chosen as our most likely gazer. A major metaphysical convention was coming to town shortly, and we knew that was going to be our best opportunity, but time was short. We quickly dispatched a letter to their organizers explaining that "Bryan the Gazer" had grown up in town and was going to be visiting only briefly during a break from his world tour, during which his gaze was seen by thousands of people whose lives had been changed by the experience.

Of course, we didn't want to actually pay to rent a room for this event, so our letter went on to explain that Bryan the Gazer normally charges tens of thousands of dollars for his appearances. However, because he still has a great connection to his home state and would only be in town for a sort interlude which happened to coincide with their event, he'd be more than happy to waive

---

1   Stollznow, K. (2011). Braco the Gazer: The Silent Evangelist. *Skeptical Inquirer.* <https://skepticalinquirer.org/exclusive/braco-the-gazer/> (accessed August 25, 2025).

Armstrong, P. M. (2018). Take It From An Exorcist: The New Age Performer 'Braco the Gazer' Is No Joke. *EWTN.* <https://ewtn.co.uk/article-take-it-from-an-exorcist-the-new-age-performer-braco-the-gazer-is-no-joke/> (accessed August 25, 2025).

his normal fee and gaze at their attendees for free if only they could provide a banquet hall suitable for the occasion.

They fell for it. Never underestimate the power of social engineering.

With the event booked, we needed to transform Bryan into "Bryan the Gazer." We sourced some white clothing (strangely enough, Bryan didn't have anything white in his wardrobe at the time). So off to the thrift shop we went and found something that looked vaguely "spiritual" in his size. The plan was coming together.

One problem we were a little bit concerned about is that Bryan has been one of the public faces of Rocky Mountain Paranormal for a long time and is sometimes recognized in paranormal circles. Both for that reason and to make him seem more "spiritual" in appearance, we recruited a special effects makeup artist of our acquaintance to transform him into the Gazer. The makeup guy at first didn't seem to quite understand what we were going for and the initial makeup looked more like a horror movie character (see Figure 26.1, the only surviving photograph of the event, before the makeup was corrected). But we quickly softened the touch on the makeup and Bryan looked to all the world like some kind of spiritual leader. We also had him walk with a cane, to further mask his gait in case anyone might recognize him and because we thought that somehow just seemed more fitting for this kind of character we were creating.

On the day, we agreed Bryan would not speak at all—not even backstage. This was consistent with the original gazer's performance but even more importantly for us, it was to help Bryan keep a straight face and not burst into laughter in the middle of it all. Another Rocky Mountain Paranormal member accompanied him and did all the speaking.

When he took the stage and saw an audience of hundreds of faces looking at him expectantly, the urge to laugh was momentarily almost insurmountable. But as he gazed at them and saw genuine reverence on their faces, the urge to laugh faded rapidly. These people genuinely believed in Bryan the Gazer and the mystical healing powers of his gaze!

We learned a lot from that experience about the psychology of people who do these kinds of things professionally. On the one hand, we could be tempted to think it pathetic that people would believe so deeply in Bryan the Gazer on the basis of so little information. But on the other hand, we can also understand how being on the receiving end of that kind of unquestioning reverence can become addictive. Were we not doing this for skeptical purposes, we can even see how that kind of response might convince someone he really does have some kind of mystical power.[2]

Bryan stared at the audience for ten to fifteen minutes or so and then went backstage, leaving some of our other people to wrap things up. We provided people with an email address we'd created for just this purpose so they could send their thoughts, comments, and reports of any effects from Bryan's gaze.

In the days and weeks following the event, we were inundated with hundreds of letters. People claimed their feet stopped hurting. Their relationships had been repaired. Bad backs were no longer bad. Financial situations had turned around.

---

2   For the record, lest there be any doubt, Bryan has no mystical power. He's as psychic as a teapot.

People genuinely believed the good that had happened was the result of Bryan's gaze.

Many of them thanked us for making the event free (we certainly would never profit off of a hoax like this or deprive people of their hard-earned money just for the privilege of being unwitting subjects in our experiment) and inquired as to how they could make donations to Bryan's organization. Of course, we never took the money, but we were blown away by the number of generous offers we received.

Initially, we planned to write back to everyone who responded with a complete confession and short one-page explanation of what we were doing and why. It seemed only fair to let them in on the joke. We changed our minds, though. The people who wrote to us had become so invested in the story and so believed that they'd been healed, we were afraid a confession could mentally destroy some of them. We reached out to a psychological ethicist friend of ours to see if he thought we were thinking correctly about the issue and he said he agreed.[3]

Does any of this prove anything about the reality of any other psychic performers or faith healers? No. All we did was to show how easy it can be to convince people that you're the real thing, and to learn from personal experience how powerful the emotions can be during such a performance.

Perhaps there are real psychics out there. We've never found the proof we're looking for, but we keep searching. Before you devote your life to such a person or give them a lot of money, though, we urge everyone to remember Bryan the Gazer and double- or triple-check any claims with open but skeptical minds.

---

3   When we reached out to the same individual for ethical advice on another case in which we were afraid we'd gone too far (which story will be in a future volume of this series), his response was (paraphrasing here): "But was it funny?" This time, though, he understood the gravity of the situation and gave us a serious answer.

27

The Evergreen Cemetery Affair

Usually when Rocky Mountain Paranormal shows up at a cemetery, it's to investigate claims of ghosts. Or maybe even a vampire (see Volume 1, Chapter 8). But we're also interested in making sure that paranormal groups present themselves well and don't damage property—especially historic property. This story takes place at the historic Evergreen Cemetery in Colorado Springs. But rather than a normal paranormal investigation, it involved our attempt to save a property from abuse by another paranormal group.

Figure 27.1. Evergreen Cemetery (photo: RMP archives).

Our case files attest to our luck over the years. We have been fortunate enough to visit some of the most remarkable places in Colorado and beyond. Often, we've been invited after hours and into back rooms the public doesn't normally get to see. Much of this good fortune can be attributed to our reputation, both as solid investigators and as the kinds of people who are at least as interested in preserving the properties we visit as we are in investigating them.

Alas, not all paranormal groups think the same way, and this case involves our brief entanglement with one that we think behaved abominably. Normally, we don't throw other teams under the bus even when we disagree with them, but we thought this went far enough beyond the pale that we had to get involved and that it's worth including as the final chapter in this book as plea for others to do things properly.

While preparing this manuscript, we went back and forth regarding whether we should name and shame the other players or whether we should preserve their anonymity. A good case could be made for the former, but we ultimately decided instead of bringing them any extra publicity (though the group is now defunct, some of the individuals may still be around) that we would either change or omit names to protect the guilty.

The group in question formed a short time before our involvement began. They claimed to be a scientifically minded team, much like ourselves, though instead of carefully recruiting only the best members based on their skills and backgrounds (as we do), they openly recruited on social media and even told people they didn't mind whether they were joined by believers or skeptics. That suggested to us that they were more "ghost hunters" and less true "paranormal investigators." The lines can become blurry at times, and there's nothing wrong with having a skeptic on a team of mostly believers or vice versa, but it just didn't sound like a serious forensic organization. It sounded more like an excuse to have fun exploring some haunted places.

Fair enough. That is fun, and we don't have anything against it as long as you don't claim it to be genuine science and as long as you treat the property you visit with all due respect. Unfortunately, as we'll see, neither was the case here.

They also claimed to be the official paranormal team for the Cities of Colorado Springs and Manitou Springs. We reached out to the cities' public information offices. They'd never heard of these people. Nor do cities tend to have official paranormal teams. Despite being around for so long and so well respected (if we do say so ourselves) by believers and skeptics alike, we've never received such an honor.[1]

Their reputation came further into question when they posted on social media that they'd been offered the opportunity to teach a class at one of Colorado's major universities. When called on it, they followed up to explain that, essentially, they'd been given access to the campus to host one of their lectures and a ghost hunt. Not a formal class and certainly not accredited by the university.[2]

---

1   Any mayors or city councilmen reading this: feel free to drop us a line.

2   Again we add, in case any university administrators might be reading: we have ideas for *legitimate* college classes. Feel free to reach out if you'd like us to teach the most popular course in your department.

This didn't sound like a particularly trustworthy group. And then they started offering public "haunted tours" of Evergreen Cemetery. Their event would last three hours and promised to provide attendees tales from the cemetery's history as well as an opportunity to explore and see if they could find any ghosts, with ticket sales to benefit the Evergreen Cemetery Benevolent Society.

Evergreen Cemetery is one of the oldest historic cemeteries in the state. Normally at this point we'd provide a full history, but we'll save that until we have the opportunity to do a proper paranormal investigation there. For now, what you need to know is that it's a beautiful cemetery with a lot of old (and modern) graves and several old buildings, and that it's been owned and operated by the City (through the Evergreen Cemetery Benevolent Society) since our state's founding in 1876. It's the final resting place of many noteworthy characters from Colorado history as well as the friends and family of people who still live in the area.

Though we weren't particularly impressed by what we'd read about them as a group, we became interested in these events when they came to our attention. We figured looking into this might be a good way to check up on what some other paranormal groups were getting up to. Call it professional curiosity.

We also had some ethical concerns. Investigations with small groups of people are one thing, but we were a little worried about a large group of people wandering around an active cemetery, telling spooky stories, and trying to communicate with the dead. It can be done respectfully. We don't object in principle. But our skeptical antennae were starting to perk up a bit and we had concerns that this might not be the kind of respectful operation we'd hope for.

Once it all came to our attention, we started by looking at some of the evidence they'd collected on prior investigations at the same cemetery, which they offered as part of their publicity for their tours. It included an EVP session in which one of the ghosts from the Cemetery supposedly threatened an investigator with "certain death." After listening to it numerous times, we couldn't hear the message they claimed to have heard.

Another video clip, offering sort of a "best of" montage from their Cemetery tours and investigations included video of attendees sitting on the gravestones while they conducted a séance in which they talked to their flashlights and asked the spirits to turn the light off. From an investigative perspective, we don't find that technique very useful. Often, it's done by just barely screwing the battery compartment on the light shut so that the slightest movement will cause the connection to form or break, turning the light on or off. Not exactly scientific.

Worse, they were sitting and climbing *on the tombstones*. We didn't see any of them actually cause any damage in that clip, but they easily could have. Plus, it's just disrespectful to the dead and to their living relatives who might still be in the area.

One of their key pieces of evidence to show the Cemetery was haunted was a video taken in the crypt by an old elevator in the main chapel in which they saw a black shadow move momentarily across the frame. They never seemed to consider that the only source of light in the building was their own infrared camera, and that the shadow was cast by one of their own "investigators" moving through the infrared beam.

In other words, we found their evidence shoddy and unimpressive.

A bit more digging revealed, from one of their forum posts, that they'd been given permission to access the entire Cemetery after hours, which they'd been doing for nearly a year at the time we got involved.

Going back in time just a bit, this all first came to our attention courtesy of a concerned citizen who'd seen some of the group's activities in the Cemetery and didn't approve. She reached out to us to see what we thought of potential ethics concerns surrounding their operations. We took an interest and started looking into the evidence we described, but said she should probably talk directly to the Benevolent Society because they were the ones in charge.

After they got her letter, they seemed to panic a little bit and asked her to come in to share her concerns and discuss what was happening. But she didn't really want to get that involved, so she said she'd rather let Rocky Mountain Paranormal handle it because she knew we sometimes acted as ethics watchdogs in the paranormal world. We were copied on some correspondence between her and the Benevolent Society and then reached out to the latter with an introduction and a quick statement of our concerns.

Specifically, we told them we were concerned that the haunted tours were more like parties, and this seemed inappropriate given the sanctity of a burial ground (especially one whose occupants still have bereaved living in the area). Further, we thought their ghostly evidence was of poor enough quality that their perceived professional association with the Cemetery could damage the Cemetery's reputation. Finally, we were concerned they were not showing due respect to the property itself and could cause some damage.

In response, the Benevolent Society shared with us some of their own concerns, which apparently had been growing ever since they allowed the paranormal group access to their property. Their concerns were many. Vandalism had happened after hours in the past and they weren't sure who was responsible but noted that it coincided with some of the haunted events. They gave us several examples. A statue of an angel had its hand broken off. Grave markers were tipped over. Crypts had been broken into. Tour guides were giving inaccurate historical information and spreading urban legends, including tales of witches. Attendees at the event had been disrespectful to the property and to the deceased. Guides had on at least some occasions encouraged others to break into the Cemetery after it was closed for the night (their events were sanctioned but the Cemetery normally closes to the public overnight). Finally, though the group was claiming to share proceeds from their events with the Benevolent Society, we were told the Benevolent Society received some but nowhere near all of the money.

Other than the last point, we know these kinds of issues are common for a lot of cemeteries. Vandalism and break-ins do happen. Unfortunately some people just don't know how to show respect. We also know that, unless carefully run, paranormal events can sometimes unwittingly encourage or escalate this kind of behavior. So we decided, with the Benevolent Society's blessing, that we needed to conduct a more thorough investigation and figure out if things really were as bad as they'd begun to suspect.

How to do it? Easy. We disguised ourselves as another newly formed paranormal group (you wouldn't believe how many domain names we own with variations on "Colorado" and "Paranormal" in the name so we can quickly assume a false identity in these kinds of cases—no, we won't tell you which ones are ours other than rockymountainparanormal.com), initiated contact, and asked if we could pay in cash instead of through their normal payment processor. Two reasons. First, we wanted to see if a bunch of people showed up and paid in cash, whether they would report that to the Benevolent Society and pay what they owed (spoiler alert: they didn't). Second, all of our banking information was tied to our real names and we were going under cover.

They fell for it.

Before people could go on their tour, though, they required a release of liability. Pretty standard operating procedure. We use them on some of our own events. No problem. We were a little surprised, though, that their release included a limitation of liability specifically for "spiritual injuries." We almost wish we suffered such a spiritual injury so we could see how a judge would react to their document. One of the Benevolent Society's concerns about doing this sting on their own was that they might be identified when they signed the release and then their cover would be blown. We solved this problem by signing the release forms illegibly and trusting they would never ask to see our identification, which they did not.

While we're on the topic of their legal structure, we checked out their website, on which they claimed to be registered with the State of Colorado as a non-profit. That's true, but essentially meaningless. All they did was check the "non profit" box when they filed their business paperwork. They were not a 501(c)3 or any other kind of tax-exempt group. They have subsequently fallen delinquent in their reporting with the Secretary of State, but were active at the time of our interactions with them. They also claimed that 100% of all proceeds would go to a non-profit organization and not to any member of their organization. That sure sounds like they were going to give all the money to the Benevolent Society. But they also claimed some of the proceeds would go to a paranormal research-related non-profit, so it sounded to us like they were "donating" at least some of the money to themselves. Given what the Benevolent Society told us, we wanted to follow the money as best we could and see what we could learn on this topic as well. Not to get too far ahead of ourselves, but we were never able to get into any forensic accounting, but the Benevolent Society confirmed they never received all of the money they were promised.

They claimed to be licensed and insured. However, shortly after our sting operation, their filings with the Secretary of State fell delinquent for the first time, which likely would have voided any insurance policy. They cured the delinquency shortly after we gave a public lecture outlining our findings but have since fallen delinquent again (and remained so for a long time, leading us to think they're completely defunct, though paranormal groups do sometimes resurrect themselves).

Finally on the legal and administrative side of things, we also discovered that another group split off from them which claimed to manufacture and sell paranormal investigation equipment. That business was only ever registered as a trade name under which an individual was operating, and it expired a year later.

That's a decent segue to talk about their approach to science. You might expect the manufacturers of paranormal equipment use some truly revolutionary stuff. No, they use the standard cheap EMF meters, ghost boxes, and flashlights everyone else uses and which we've discussed our thoughts about in prior volumes so we won't belabor the point here except to say that none of what they used has ever been reliably shown to detect ghosts. But they were all cheap—especially considering how many tickets they were selling—so they were comfortable letting their attendees handle their tools on the tour.

When the tour began, they started by once again reminding the attendees that they weren't keeping any of the money for themselves. Then they handed the floor over to their technical specialist, who tried to explain how all their various gizmos worked, and to instruct attendees on how they'd be able to use them later. To say their "expert" didn't understand what an electromagnetic field even was would be an understatement. We gave them another chance and asked about the EMF meters directly. We were told how to turn the sound on and off on one of them and then told that one goes up to five thousand while the other goes up to twenty thousand. No units were specified. When we asked about different sensitivities between the two, we were just told that they pick up on different things. They don't. One, we were told, answers "yes or no" questions better than the other. Complete nonsense.

Another thing we noticed during their introductory lecture was that they went to great lengths to try to set the spooky mood by burning candles all throughout the building. That was another point in our report that stuck in the Benevolent Society's craw because the building was over a century old; they didn't allow open flames in there, much less in front of a bunch of strangers who were getting excited about running around and scaring themselves.

Once everyone was "instructed" about how to use the equipment, the tour began downstairs in the crypt.

We were told about a mysterious vortex down there, as well as doors that open and close by themselves. All we found was that the EMF meters successfully detected the electrical box. They told us plenty of stories about how the bodies would be stored down there when the ground was too frozen to bury people.

Then we went out to the Cemetery itself. Their "historian" was reading his historic stories from a book and notecards about the location's history. Perhaps we could forgive this as just someone trying to remember his script, but given everything else we witnessed, we honestly doubt he really knew his history very well.

Worse, throughout the tour, we documented quite a bit of disrespectful behavior. People climbed and sat on crypts and headstones to use their meters. Others banged and rattled on crypt doors while calling out to the spirits. One walked up to a crypt and asked, "what's that smell?"

For their "spirit box" sessions, in which a radio scanner pans through all the stations and ghost hunters listen for meaningful messages from the ghosts in the noise, they suggested that because FCC regulations ban use of certain words on the radio, a good way to confirm a real ghostly interaction would be to convince the spirits to cuss on the radio. We're not offended by the language, but it does seem somewhat disrespectful considering the spirits they might be trying to talk with might have been deeply religious people—or, perhaps more importantly, their living family members might be. After the tour, people posted results on their forum confirming that the ghosts did as requested (though of course the audio files were not provided for us to examine them).

At one point we overheard a conversation in which someone was looking for a restroom and was informed there wasn't one nearby. Some people expressed their dismay until one person said, "guys can just go around; just don't go on the graves."

Many people stood on graves and even headstones. Someone else poked and prodded at the headstone on a child's grave and then monkeyed with the grave decorations the child's family had left.

We'll take a brief side note here and say that (thank God) we didn't see anyone actually steal any of the grave decorations on this tour. However, we have seen cases in which *other* paranormal groups have gone around to cemeteries to steal objects people have left on their loved ones' graves and then sold those online as "haunted" items! It's enough to make even us start to lose our sense of humor about the paranormal sometimes.

Then we heard about the urban legends the Benevolent Society was concerned about. When the tour visited the crypt of one of the local wealthy and influential families—a very religious family, no less, named the Bacon family—they explained the door was chained shut because in the past the crypt had been vandalized, a skull belonging to Mary Bacon had been stolen, and that it was rumored she'd been involved in witchcraft. They did admit the stories were speculation, but shared the rumor that local witches wanted her skull (and specifically her skull) for some kind of ritual. Further, they said the witches liked to sneak into the Jewish section of the Cemetery and bury their own people under those graves because the "energies" would be more appropriate.

There's no evidence of any of it. All of the witchcraft stories are pure fiction, and we think they confused this Mary Bacon with one who was executed in the 1645 witch trial at Bury St. Edmunds in England. Grave robbing is certainly not unheard of, unfortunately, but we don't know of any case matching the description they provided.

After the tour was over, we typed up a quick report of what we'd seen and hand-delivered it to the Benevolent Society and said it needed to get to the manager. When we received no reply, we contacted the manager directly to share our findings and our concerns.

Almost immediately, he wrote back to tell us he'd written to the leader of the other paranormal group in question to ask about our findings but that in the meantime and until he was satisfied that our (and now his) concerns had been addressed, they had been banned from any after hours activities in the Cemetery, effective immediately.

We learned that the other group responded by sending letters to everyone they ever did any business with—or indeed anyone they ever had any correspondence with, it would seem—asking for reference letters. We received such a request ourselves. We did not provide a reference. Instead, we sent the mayor and city council a nice commendation letter congratulating the manager for addressing the problem so quickly and effectively.

A bit later, we followed up on the group's website and found that they had started making their activities secret. They would advertise haunted tours but wouldn't specify what location they'd go to until people already signed up and paid for a ticket. The Cemetery manager said if they were caught after hours, they'd be cited for trespassing, so we assume they weren't foolish enough to return to Evergreen Cemetery, but had concerns their behaviors might continue at other locations. We were, however, unable to follow up on those events.

Some people might think this is unimportant. We disagree on several levels. Most importantly, we think a society which fails to honor its dead and its ancestors is a pretty sick society in the first place. Paranormal investigations and ghost stories can be a great way to honor our dead. But when it's done disrespectfully, when it's done just as a party, it can damage the physical property and bring dishonor to the very place where we're meant to remember our lost loved ones.

We've seen people posting openly on the Internet that they intended to go to the Cemetery after hours to look for ghosts. No, they weren't involved with this group we briefly interacted with, but the persistence of the ghost stories and the view that the Cemetery is just a playground can contribute to a culture in which people want to break in to the property when it's closed. We like to explore after hours, too, but only respectfully and only with permission from the property owner.

Afterword

Well, that was quite another adventure, was it not? We've explored ghost, demons, aliens, and more throughout these pages. And we feel like we've still only just barely scratched the surface. Now we're three volumes into this series and we still have more stories to tell. Beyond that, we have an ever-growing to-do list of all the locations and cases and experiments we'd like to add to our case files but just haven't gotten to yet. If we were to just list all of the supposedly haunted locations in Colorado (never mind the rest of the world), the list alone would probably be as long as this book. And that doesn't even count the aliens and UFOs, the cryptids, the psychics, and all the other things we'd like to work on. We've certainly got our work cut out for us!

Speaking of having our work cut out for us, we have to admit, this book took a lot of effort. Somehow it seemed even harder to write than the first two for some reason. Not really sure why. Maybe it's because all of our other activities have been picking up some steam and it's been harder to find the time to write. Or maybe it's because this volume, as fun as some of the cases were, also involved some that deeply disturbed us and were genuinely difficult to write about. You've read it by now; you know which ones we're talking about.

Still, this is why we keep doing what we do. Nothing worth doing is ever easy, and we continue to believe that our brand of paranormal investigation is not only fun for us, but important for the world. That's not just us patting ourselves on the back. We'd say the same of any of the other researchers who take a similar approach. We think it's important to teach people about history and science, and the ghost stories are a great way to do it. Further, because so many people do believe in the paranormal, it's important that people should be willing to investigate those claims, as we always say, with open but skeptical minds. There will always be paranormal beliefs, but approaching them with both respect and an insistence upon quality evidence remains the best way to separate the wheat from the chaff.

All of which is to say: we're not going anywhere.

On the investigative front, we have new investigations in our case files all the time. Many end up being abandoned email correspondences, but some get followed to fruition in on-site work. We're still out there, fighting the good fight, and there's a lot more to come. The same goes when paranormal claims reach the media. Our ears are always to the ground and we're keen to jump in whenever it seems necessary. In fact, just in the week or so before we finished this manuscript, we saw a new wave of news articles that provided further confirmation for a longstanding theory we've had about the origin story of a famous cryptid. That tale, though, will have to wait for a future volume.

In terms of those future volumes, we still have plenty more stories to tell. We haven't yet told you about the Devil's Tree, the ghost named George, or that time paranormal investigation accidentally got us involved in a casino heist. Those will have to wait for Volume 4—and perhaps even future volumes beyond that. In the meantime, we've also been working hard on our much-anticipated "how to" book for would-be paranormal investigators. Our goal is to make that the definitive resource for anyone interested in doing anything like what we do, and we're making good progress.

On the educational side, we're still giving our lectures, now supplemented by readings and book signings. Wherever there's a paranormal story to tell, we're happy to be there. Children in particular, though the same is true of adults, love a good ghost story, and we love using those as an excuse to talk about history and science. Not that we don't like the ghost stories for their own sake, but why not do both, right?

But for now, as we close the book metaphorically on this chapter in our saga and as you prepare to close this book physically, we have to just thank you for reading and following along with all of our crazy adventures. We'd do these things even if no one listened to us, but that you are interested enough to buy our books (or borrow them from your local library) and give us your attention really does mean the world to us. You therefore have our deepest thanks and eternal gratitude.

Until next time, then, we bid you farewell and hope you enjoy some good ghost stories!

If you've enjoyed this book, please don't forget to leave us a review online. Reviews of even just a few words are immeasurably helpful to independent authors and publishers. And the ego boost will help get us through the process of writing the next one.

Acknowledgements

Though writing a book is often seen as a solitary endeavor, it necessarily involves the help and support of numerous people. That's doubly the case when the book describes real-world case files spanning more than a quarter of a century. As such, the number of people to whom we owe our sincere thanks is beyond our ability to count, and any attempt at a complete listing would be doomed to failure. Nevertheless, we'd like to begin by offering our sincere thanks to all of those who've allowed us to conduct our investigations in their homes or businesses over the years. Without them, this book clearly would never have happened. Similarly, we'd like to thank everyone who has joined us on investigations over the years. Both groups of people would require another full-length book to list, so we thank them all collectively. We would like to specifically highlight the ongoing support and hard work of Carol Olivacz, Jack Hanley, and Kathy Josey, who've been tirelessly working with us for years, as well as all the former members of Rocky Mountain Paranormal.

Special thanks are due to the entire crew at the Colorado Festival of Horror. Their support has been immeasurable, not only as we prepared this book but as we've delivered our lectures and more over the years. And readers are encouraged to attend their annual festival and show them some love.

Thanks also to Kealan Patrick Burke of Elderlemon Design for the truly excellent interior artwork that helps bring some of these stories to life.

Finally, thanks to everyone who bought and read a copy of this book and/or its predecessors. Without you, none of this would mean anything.

About the Authors

Robert (Bob) Lewis is a Colorado-based author, editor, paranormal investigator, scholar, magician, and more. He holds degrees from the University of Colorado Denver in Biology, English, Mathematics, and Psychology, and a Master of Education from the University at Buffalo in Science and the Public. A dedicated polymath, he likes to tell people that his hobby is to collect new hobbies. He's (obviously) a member of the Rocky Mountain Paranormal Research Society. Additionally, he's a co-host of the *Do You Like Scary Movies* horror podcast, host of the *Phobophile* YouTube channel, and is always looking for more projects to whittle away at what little time he has left for sleep.

In addition to this series, his publications include *In the Woods: A Fiction Foundry Anthology* and *Arithmophobia: An Anthology of Mathematical Horror*, both of which are available from Polymath Press.

He can be found online at www.robertlewisauthor.com.

Bryan Bonner is a founding member of the Rocky Mountain Paranormal Research Society. For nearly three decades, Bryan has dedicated himself to the examination of a wide range of paranormal phenomena, including ghosts, poltergeists, psychics, UFOs, conspiracy theories, urban legends, and much more. Setting himself apart from others in the field, Bryan has always maintained a grounded approach, refraining from running around cemeteries at night and needlessly scaring others with imaginative tales. Instead, he meticulously tests bizarre beliefs and practices, conducts experiments and on-site investigations, and even recreates unusual events, all with the aim of uncovering the truth. Bryan's work has garnered the respect of believers and skeptics alike, while simultaneously instilling fear in fraudsters and charlatans. You can read more about Bryan and his work at www.rockymountainparanormal.com.

Also available from Polymath Press

Case Files of the Rocky Mountain Paranormal Research Society Volume 1 by Robert Lewis & Bryan Bonner

Don't miss the first volume of the *Case Files of the Rocky Mountain Paranormal Research Society* series!

Included in this volume:

- How Rocky Mountain Paranormal involved multiple departments of the United States Federal government in a paranormal investigation
- The haunting of an active nuclear military base
- A paranormal investigation inside a jail
- A paranormal investigation deep inside a cave
- A location that is still home to thousands of unidentified human bodies—and how Rocky Mountain Paranormal worked to tell their stories
- Colorado's own vampire legend
- The time one of our investigators got slapped by an unseen entity
- The inspiration behind a classic horror movie
- And many more!

Case Files of the Rocky Mountain Paranormal Research Society Volume 2 by Robert Lewis & Bryan Bonner

And also don't forget the exciting second volume of the *Case Files of the Rocky Mountain Paranormal Research Society* series!

Included in this volume:

- Denver's most haunted mansion
- A murder that led to ghost stories
- A paranormal investigation lasting more than 15 years
- An infestation of invisible jellyfish monsters
- An alien sighting that led to political action
- Ghosts and tommyknockers in an old mine
- The haunting of Denver's first fire station
- An investigation in an active casino
- And many more!

In the Woods: A Fiction Foundry Anthology
Edited by Robert Lewis

Strange things can happen in the woods.
Sometimes they're frightening.
Sometimes they're funny.
Sometimes they're just plain weird.

The authors of the Fiction Foundry writers' critique group have taken it upon themselves to explore all the strange things that happen in the often majestic and yet often harsh woodlands.

Fiction Foundry, established 2012, is a group of writers dedicated to helping prepare one another's work for professional publication. In this anthology, the group's authors show off their eclectic visions of life among the trees.

Featuring contributions by John H. Howard, Sangita Kalarickal, Josh Snider, Carolyn Kay, Robert Lewis, Charli Cowan, Henry Snider, Shiloh Silveira, Kari J. Wolfe, Christophe Maso, and Hollie Snider, this anthology brings us out of urban life and shows a world of forest spirits, haints, mental illness, parasitic spiders, werewolves, out of control plants, evil forces, reincarnation, humans with animal ears, witches, and Lovecraftian horrors.

And all of them can be found...*In the Woods.*

Arithmophobia: An Anthology of Mathematical Horror
Edited by Robert Lewis

"Arithmophobia," *n.*: The fear of numbers or mathematics.

Whether you love mathematics or find it terrifying, this anthology of original tales of terror is sure to send a chill down your spine. With an unlucky thirteen brand new horror stories and a bonus poem in case any readers suffer from triskaidekaphobia, these pages combine the talents of some of the genre's most experienced award-winning practitioners of terror and some of the literary world's most promising new voices.

Featuring contributions by Elizabeth Massie, Miguel Fliguer, Mike Slater, Patrick Freivald, Liz Kaufman, Damon Nomad, Sarah Lazarz, Martin Zeigler, Josh Snider, Rivka Crowbourne, Joe Stout, Brian Knight, Wil Forbis, David Lee Summers, and Maxwell I. Gold.

These stories tell us of strange and horrifying new geometries, crazed and violent mathematicians, sentient and malevolent numbers, and even some new mathematical twists on some classic monsters. You needn't be a mathematician to experience these new forms of mathematical terror, though students of the discipline might recognize some familiar names and ideas lurking in the shadows.

So pull up a chair, dust off your abacus and slide rule, and prepare to experience…

Arithmophobia.

Get your copies of these and other Polymath Press
titles online at
www.polymathpress.com
or wherever fine books are sold!

www.ingramcontent.com/pod-product-compliance
Lightning Source LLC
Chambersburg PA
CBHW021608120626
46545CB00001B/130